This tenth edition of the NATIONAL ELECTRICAL CODE BLUEPRINT READING has been revised to reflect many of the important changes which appear in the 1987 *National Electrical Code*. The changes are presented as they pertain to Single-Family Dwellings, Multi-Family Dwellings, Commercial Locations, Industrial Locations, and Specialized and Hazardous Locations. Besides appropriate changes and additions in text and illustrations, the APPENDIX has been expanded to provide additional supplementary material to aid the student.

NATIONAL ELECTRICAL CODE BLUEPRINT READING is designed to enable the student to learn electrical blueprint reading, and at the same time receive exposure to and become familiar with applicable sections of the 1987 *National Electrical Code*. Complete references to the latest *Code* are presented throughout the text. Trade Competency Tests appear at the end of each Unit to help students check their understanding of the text material and the *Code*. There are also Final Examinations included in the back of the book.

The author has carefully prepared this tenth edition of NATIONAL ELECTRICAL CODE BLUEPRINT READING so that it will continue to serve as a standard for training electrical inspectors, electricians, apprentices, technicians, drafters, designers, contractors, and all persons interested in acquiring a knowledge of electrical installations which comply with the 1987 *National Electrical Code*.

AMERICAN TECHNICAL PUBLISHERS, INC.

Material presented here deals with that body of regulations known as the *National Electrical Code*. This set of rules, abbreviated NEC throughout the text, has been revised and published at more or less regular intervals of about three years ever since the original NEC of 1897. The task of reviewing and revising the NEC is necessary because of changes in technology, methods of installation, and new electrical materials put into use from time to time.

The increased demand for electricity and the rapid technical changes in the materials and procedures used for electrical installation have caused the National Electrical Code Committee to incorporate into the 1987 NEC many new minimum provisions considered necessary to safeguard persons and buildings using electricity for light, heat, power, radio, signalling, and other purposes.

The 1987 NEC is divided into the introduction and nine chapters. The introduction states the purpose, scope and other general interest items. Each chapter supplies numerous paragraphs devoted to its basic subject matter. The specific provisions are stated in a semi-legal style. Chapters 1 through 8 deal with rules. Chapter 9 provides some tables and examples.

Although this method is perhaps best suited for presenting the specific rules, it does not readily lend itself to the business of learning or teaching. Therefore, instead of following the NEC order, this text groups the requirements around five specific fields of application: Single-Family Dwellings, Multi-Family Dwellings, Commercial Locations, Industrial Locations, and Specialized and Hazardous Locations. Basically, the method used here is pictorial, with the appropriate Code Sections linked to these visual examples. The subject of Electrical Blueprint Reading is discussed throughout the text, as part of the instructional procedure, and electrical plans are inserted at key points.

Examinations are provided at the end of each Unit. They are designed to be an effective learning tool and provide the student with the opportunity to carefully study various Sections of the NEC. They should be equally helpful to self-study students and students in the classroom situation.

The 1987 edition of the NEC contains many important changes from the 1984 NEC. (The new Code indicates major changes and new rules with a vertical mark in the margin where the changes occur.) Naturally, it is not possible to cite all of these changes here. However, as in previous editions, this tenth edition of *National Electrical Code Blueprint Reading* provides a select list of key and interesting changes in the NEC so that students and teachers alike can become familiar with new material at a glance.

1. The 1987 NEC follows the format of previous codes in that a horizontal line in the margin indicates a code change. A new marking has been established in that a bullet in the margin indicates that material has been deleted.
2. 90-2(a)(1), 90-2(b)(1): Revised to change floating dwelling units to floating buildings.
3. 90-3: Revised that subscript letter "X" indicates the text is from an NFPA document.
4. Art. 100: Definitions. Scope revised that only definitions and terms relative to the NEC are included. Other terms in IEEI Dictionary or standard dictionaries.
5. Art. 100: Definition of a feeder revised to indicate that feeder can originate in a separately derived system.
6. 110-11: Revised to prohibit lubricants that have a deleterious effect on conductors.
7. 110-16(a): Now requires work space about elect. equip. be adequate to permit a 90° opening of doors or hinged covers.
8. 200-6(d): (New) Grounding conductors of different systems to be identified (This now includes feeders).
9. 210-4: New (FPN) to indicate continuity of grounding conductor on multiwire circuits be as per 300-13(b).
10. 210-7(d) Ex.: Revised to permit downstream receptacles where GFCI replaces existing rec. on ungrounded ckt.
11. 210-8(a): Revised to require kitchen counter top receptacles within 6′ of sink to be on GFCI.
12. 210-8(a): All receptacles in dwelling unit boathouses to have GFCI protection.
13. 210-8(a): In dwelling units at least one identified GFCI basement receptacle is required.
14. 210-8(a): Grade level access to outdoor dwelling unit recept. is not 6′-6″ above grade and accessible (not thru unit).
15. 210-8(a)(6): All 125 volt 15- or 20-amp receptacles in dwelling unit boathouses require GFCI.
16. 210-19(a), Ex: New (FPN) refer to 310-10 for ampacity and conductor size when conductor is installed in thermal insulation.
17. 210-24: New EX permits branch ckts. to feed more than one dwlg.-unit if for alarm, signal, communications, etc.
18. NEC 210-52(a): New (FPN) warning that baseboard heaters may include instructions not permitting baseboard heaters to be installed below dwelling unit receptacles. (This would also prohibit installing receptacles above existing baseboard heaters.)
19. 210-60(b) Ex.: Rec. in hotels and motels to comply with 210-52 but may be located convenient to permanent furniture location.
20. 210-63: Requires 125-volt rec. within 75′ of rooftop air-conditioning, heating and ventilating equipment.
21. 220-22: Data processing and similar equip. supplied by a 3-phase, 4-wire circuit does not permit a reduced neutral.
22. 225-22: New ex. permits flexible metal conduit on exterior surface of bldg's as per 350-2.
23. 230-40: Ex. 1 now defines a multiple occupancy as 2 or more dwellings and commercial bldgs. with multiple occupancies.
24. 230-43: New (13) permits 6′ of flexible metal conduit between raceways or between raceways and service equipment if bonding jumper installed around flexible metal conduit as per 250-79(a), (b), (c), and (e). Also see 350-2.
25. 230-44: New (3) revised to consider a transformer vault (as per 450 Part. C) outside the building.
26. 230-7 (old 230-47): Ex. 2 changed time sw. conductors permitted in service raceway to load management conductors.
27. 230-51(a): Supports for SE cable changed from 4½ ft. to 30 inches.
28. 230-71(a): New exception excludes disconnect for control circuit of ground-fault protection system as one of the 6 service disconnecting means.
29. 230-72(c): New ex. In multiple occupancy bldgs. occupants not required to have access to their service equip. provided management limits accessibility to authorized personnel only. (also see 240-24(b) Ex.)
30. 230-82, Ex. 5: Change time sw. to load management devices as permitted to be connected on supply side of service.
31. 230-202(b)(3): Revised to require Schedule 80 PVC or Schedule 40 PVC in 2″ of concrete for services over 600 volts.
32. 240-24(a), New Ex. 3: Overcurrent devices to be readily accessible except for service as per 230-92.
33. 250-5(d): New (FPN) that on-site generator not a sep. derived system if solid connection of neutral to service neut.
34. 250-51: Revised to indicate that the earth shall not be used as sole equipment grounding conductor. [Also see 250-91(c)].
35. 250-72(c): Revised to clarify that on supply side of service threadless connectors also require bonding locknuts or bushings for grounding continuity.
36. 250-91(b): Revised to require fittings be listed for grounding for greenfield flex metal tubing and sealtight.
37. 250-113: Exothermic (Cadweld, One-Shot) connections now permitted for grounding and bonding connections.

national electrical code

blueprint reading

10th edition

based on the
1987 national
electrical code

Kenneth L. Gebert

AMERICAN TECHNICAL PUBLISHERS, INC.
HOMEWOOD, ILLINOIS 60430

Copies of the 1987 National Electrical Code® may be ordered directly from its publisher:

National Fire Protection Association
Batterymarch Park
Quincy, Massachusetts 02269

Printed in the United States of America

10 11 12 13 14 15 - 86 - 9 8 7 6 5 4 3 2 1

Library of Congress Cataloging-in-Publication Data

Gebert, Kenneth L.
  National electrical code blueprint reading.

  Includes index.
  1. Electric wiring—Insurance requirements.
2. Electric wiring—Charts, diagrams, etc. I. Title.
TK3275.G4 1986   621.319′24   86-22143
ISBN 0-8269-1552-3

38. 250-115: Exothermic (Cadweld, One-Shot) connections now permitted for connecting grounding conductor to grounding electrode.
39. 300-2: New (b) to refer to temperature limitations of 310-10 where conductors are embedded in thermal insulation.
40. 300-3(b)[Old 300-1(c)]: Now requires all conductors of same ckt. (including neutral and equip. ground conductors) to be in same raceway except as per 250-57(b), 250-79(e), 300-5(i), 318-7(d), 300-20(a), 339-3(a)(2).
41. 300-4(b): Requires protection where NM cable or electrical nonmetallic tubing passes thru metal framing members. A $^1/_{16}$" steel plate, sleeve, or clip required where nails or screws likely.
42. 300-5(a), Ex. 3: Areas subject to vehicular traffic where cables, conduit etc. are to be 24" underground now defined as streets, highways, alleys, driveways, and parking lots.
43. 300-5(d): Buried conductors and cables emerging from the ground now require below grade physical protection not greater than the minimum cover required by Table 300-5 but never greater than 18".
44. 300-8: New section prohibiting wiring in the same raceway or cable tray with piping for steam, water, air, gas, etc. unless identified for such application.
45. 300-11(b): New Ex. 3 permits Class 2 ckt. cables to be secured to raceways for power to heating and air-conditioning.
46. 300-22(c): Now permits metal cable trays with solid bottom and cover in "other spaces used for environmental air."
47. 300-22(c): Now permits a dry-type transformer in environmental air handling areas if ventilated as per 450-2, Ex. 2.
48. 305-4(b)(c): Requires hard usage or extra hard usage cords for feeders or branch circuits on construction sites.
49. 305-4(e): Requires temporary multiwire branch circuits have a simultaneous disconnect. (Handle ties OK)
50. 305-4(f): Changed to require all lamps for temporary wiring to have a suitable fixture or guard.
51. 310-4: New Ex. 3 permits conductors smaller than 1/0 in parallel for 360 hertz or higher if as per Ex. 2.
52. Table 310-13: Delete Types RUH, RUW, and T because they are obsolete.
53. 310-14: Aluminum conductor material in sizes 8, 10, and 12 to be of an AA-8000 series aluminum alloy material. (AKA ACM).
54. Table 310-16: Heading revised to include Types AC, NM, NMC, and SE based on ambient of 30°C. Delete Types RUH, RUW, T.
55. 318-2(b)(1): New Ex. to permit welding cables in cable trays as per 630 Part E.
56. 324-4: Concealed knob and tube not permitted in hollow spaces of walk, ceilings, etc. where insulation is used.
57. 328-2: Type FCC revised that a bottom shield between floor and cable need not be an integral part of the cable.
58. 331-3: Revised to permit electrical nonmetallic tubing in buildings over 3 floors in height.
59. 333-5: Type AC cable in insulated walls, ceilings, and attics to have conductor rating of 90°C with ampacity as per 60°C.
60. 333-7: New Ex. 3 permits not over 6' of Type AC cable without support for lighting or equip. in accessible ceilings.
61. 339-3(a)(4): When UF cable is used as NM cable Art. 336 to apply which requires 90°C conductor and 60°C ampacity.
62. 350-3: New Ex. 4 permits $^3/_8$" flexible metal conduit for manufactured wiring systems as per 604-6(a).
63. 350-5: New Ex. 2 requires a grounding conductor when flexible metal conduit is used for flexibility.
64. 351-23(a): New (3) which permits liquidtight flexible metal conduit in outdoor locations if listed and marked.
65. 351-27: Revised to permit 6' of equipment grounding conductor on outside of liquidtight flex nonmetallic conduit.
66. 354-15: Revised to permit electrical nonmetallic tubing (ENT) as a wiring method for underfloor raceways.
67. 356-11: Revised to permit electrical nonmetallic tubing (ENT) as a wiring method for cellular metal floor raceways.
68. 362-2: Revised to permit wireways to be in concealed spaces for sound recording equipment as per 640-4, Ex. C.
69. 362-10: Revised to permit ENT as a wiring method for wireways where grounding is as per 250-112 and 250-118.
70. 364-8: Busways now permit electrical nonmetallic tubing as a wiring method if grounded as per 250-113 and 250-118.
71. Article 366: Electrical Floor Assemblies deleted because it is not being used and is not being manufacturered.
72. 370-3: Revised to permit electrical nonmetallic tubing to be used with a nonmetallic box.
73. 370-6(a)(1): Deduct one additional conductor where grounding conductor run to insulating rec. as per 250-74, Ex. 4.
74. 370-7(b): Prohibits splices, taps or devices in capped elbows (LB'S) or service-entrance elbows. (SLB). Also revised to require an acceptable size to provide free space for all conductors in the fitting.
75. 370-13(b): Prohibits support of boxes solely by wires supporting suspended ceilings.
76. 370-13(c): Permits support of boxes to nonstructural framing members if secured to framing member by bolts, screws, or rivets.
77. 370-17(c): Revised to prohibit outlet boxes as sole support for ceiling (paddle) fans.
78. 373-1: Scope of Cabinets and Cutout Boxes revised to include meter sockets.
79. 373-6(b)(1): New Ex. 2 and (FPN) permits less bending space than required by Table 373-6(a) for lay-in type terminals in meter sockets.
80. 373-6(c): Revised to permit both insulating bushing or PVC connector with rounded surface to protect No. 4 and larger where entering a raceway in a cabinet, pull box, junction box, gutter etc. This change appears throughout the 1987 NEC.
81. 380-2(a): New Ex. added to indicate that switch loops in metal raceways do not require a grounded conductor.
82. 384-4: Revised to clarify that dedicated space for switchboards etc. does not end at suspend ceiling.
83. 384-4: New (FPN) added to indicate that it was not intentent of CMP 9 to require a dedicated room.
84. 384-4: New FPN to clarify that this Section not intended to prohibit sprinkler protection for elect. installation.
85. 384-3(f): Ex. permits phase arrangement of metering equipment throughout single or multisection switchboard.
86. 384-27: Revised to permit a connection between the equipment grounding terminal bar and the neutral bar at the first disconnect means for a separately derived system.
87. Table 400-4 and Note 5 revised to accept Types G and W cable as "Extra hard usage."
88. 400-5: Revised to add info. on ampacity of neutral for cords on 3ϕ circuits similar to NOTE 10 to Tables 310-16 through 310-19.
89. 410-66(a): Revised to require all recessed fixtures to have $^1/_2$" clearance from combustibles.
90. 410-73(f): HID fixtures installed indoors or are operated by a remote ballast shall have thermal protection.
91. 410-101(a): Revised to prohibit general purpose receptacles on lighting tracks.
92. 410-104: Lighting track permitted to be mounted on suspended ceilings or suspended from the ceiling.
93. 422-8(d)(3)(New): Portable high-pressure spray washing machines and supply cords to have GFCI protection.
94. 422-17(a)(b): Wall-mounted ovens and counter-mounted cooking units permitted to be cord-and plug-connected for ease in servicing but disconnect means also required as per 422-20.
95. 422-18: New Section requires ceiling fans to be supported independent of outlet box. (Also see 370-17(c))
96. 422-22-(a): Revised to permit cord-and-plug connection as appliance disconnect if accessible.
97. 422-22(d)(4): Revised to require polarized or grounding attachment plug where appliance has single-pole switching.
98. 422-27(e): Revised to read: A single nonmotor-operated appliance overcurrent rating shall not exceed the overcurrent rating marked on the appliance or, if not marked and appliance rated over 13.3 amps the overcurrent device not to exceed 150% of single appliance rating. Appliance rating on 20-amp ckt. not to exceed 13.3 amps.

99. NEC 424-99: New Section added to cover heating panels or heating panel sets installed under floor covering.
100. 430-9(b)(New): Copper conductors are now required for motor controllers and terminals of control circuit devices unless otherwise identified.
101. NEC 430-9(c): Revised to require control circuit devices with screw-type terminals with 14 or smaller copper conductors to be torqued to a minimum of 7 pound-inches unless identified for a different torque value.
102. 430-102: Revised by moving 430-86 of 1984 NEC to Section 430-102 of the 1987 NEC. This locates requirements for the disconnecting means for both motors and controllers in the same section. No code changes were made.
103. NEC 430-109, Ex. 5: Revised to require a hp rated plug and receptacle for cord-and plug-connected motors where plug is used as a disconnect means.
104. 440-4(c)(FPN): for hermetic equipment revised to add that the branch-circuit selection current will always be equal to or greater than the marked rated-load current.
105. 440-14: Revised to permit the disconnect for air-conditioning or refrigeration equip. on or within the equipment.
106. 450-3(a): Overcurrent protection for transformers over 600 volts has been completely revised but transformers 600 volts or less remains the same as the 1984 NEC.
107. 500-2: Intrinsically safe circuits to be physically separated from circuits that are not intrinsically safe.
108. 501-4(b), 502-4(b), 503-3(a): PLTC cable permitted in Class I, Div. 2, Class II, Div. 2 and Class III, Div. 1 areas.
109. 501-16(a), 502-16(a), 503-16(a): In Class I, II, & III locations bonding jumpers with proper fittings or other approved means to apply to all intervening raceways, fittings, boxes, enclosures, etc. between the hazardous area and the point of grounding for service equipment.
110. 511-10: New Section requires GFCI protection in commercial repair garages for all 125 volt, 1 phase, 15- and 20-amp receptacles where electrical automotive diagnostic equip., electrical hand tools and portable lighting devices are used.
111. 517: Health Care Facilities: Superscript letter "X" indicates material is from another NFPA document.
112. 517-104(b)(1): Minimum alarm level on line isolation monitor for fault hazard current increased from 0.7 to 3.7 mA. Total hazard current trip level increased from 1.7 mA to 5.0 mA.
113. 517-105(a)(1): The voltage level defining low voltage equipment has been increased from 8 volts to 10 volts.
114. 520-44(b): Theater stage equipment border lights revised to permit Types G and W flexible cable.
115. 545-5: New (FPN) added to see 310-10 for temperature limitation of conductors in a Manufacturer Building.
116. 547-8: New (b) to recognize use of an equipontential plane to limit the difference of voltage in a dairy barn.
117. 550-5(i)(2) and 550-10(i): Revised to permit rigid nonmetallic conduit on the underside of a mobile home.
118. 550-8(a)(3): Revised to permit both 15 and 20-amp receptacles in a mobile home.
119. 550-8(c): Revised to permit a duplex grounding type receptacle for fixed appliances in a mobile home.
120. 550-23: New (e) to require mobile home service equipment mounted no less than 2 ft. or over 6½ ft. above ground.
121. 551-22(b): Enclosed and gasketed type, listed fixtures over a bathtub or shower stall in RV's require GFCI.
122. 553: NEW Article replaces 555 part B of 1984 NEC and covers wiring, services, feeders, and grounding for floating buildings. A floating bldg. floats on water, is moored in a permanent location, has a premises wiring system served by permanent wiring from a supply system not located on the premises.
123. 600-8(g): Revised to permit wiring connections on bottom of sign enclosures exposed to weather.
124. 604-6(a)(2); Manufacturered wiring systems revised to permit not over 6 ft. of greenfield with No. 12 copper ground.
125. 605-5(b): "Office Furnishings" revised to require cord-and plug-connection to lighting accessories not over 9 ft. long, not smaller than No. 18, be of extra hard usage type and contain an equipment grounding conductor.
126. 620-21: Wiring methods for elevators revised to permit rigid nonmetallic conduit. (See Ex. 1 thru 6)
127. 630: Electric Welders has a new Part E covering welding cables for use in dedicated cable trays.
128. 680-1: Hydromassage bathtubs, permanently installed or storable are now covered by Art. 680-Swimming Pools.
129. 680-6(a)(3): FPN Revised to include a doorway with hinged or sliding door or window opening as a permanent barrier when locating receptacles within 10 ft. from the inside walls of a pool.
130. 680-41(b)(1), Ex. 2: Revised to permit lighting fixtures below 7½ ft. over an indoor spa or hot tub if protected by GFCI and (a) recessed with lens and nonmetallic trim suitable for wet location or, (b) surface fixture with globe and nonmetallic body suitable for wet location.
131. 680: New Part G "Hydromassage Bathtubs" added and new 680-70 requires GFCI protection. 680-71 does not require other electrical equipment in same room to be an GFCI.
132. 690: Solar Photovoltaic Systems has a new Part H added for the installation of storage batteries.
133. 700-1: Emergency Systems revised by new FPN to reference NFPA 110 "Standard for Emergency and Standby Power Systems."
134. 700-5(b): Revised for the capacity of generators for selective load pickup and load shedding has been revised to permit the alternate power source to be used for peak load shaving.
135. 700-9(a): New section now requires all boxes and enclosures for emergency circuits to be marked so they will be readily identified as part of an emergency circuit.
136. 700-12(f): New exception permits unit equip. in an open area supplied by 3 normal lighting ckts. to be fed from a separate ckt. originating from same panelboard if provided with a lock-on feature.
137. 705: Interconnected Electric Power Production Sources. New Article covers installation of one or more electric power production sources operating in parallel with a primary source (s) of electricity.
138. 725-1: New FPN indicates that Class 1, 2, and 3 circuits are characterized by usage and electrical power limitations which are different from normal light and power circuits and therefore, alternate requirements to Chapters 1 thru 4 are given with regard to minimum wire size, derating factors, overcurrent protection, insulation requirements, and wiring methods and materials.
139. 760-30(f): New section to permit listed nonconductive and conductive optical fiber cables for fire-protective signaling circuits.
140. 760-31: New section requires cable marking for listed power-limited fire-protective signaling cables.
141. 800-3(b)(1): Communication wiring in bldgs. to be listed as Type CM, CMR, CMP, or CMX depending on where installed.

NOTE: In general, within the National Electrical Code, the term *watts* has been superseded by the term *volt-amperes* for the computation of loads. However, references to nameplate ratings still reflect the term *watts*. The terms *watts* and *volt-amperes* are used interchangably in this text.

Table of Contents

Unit   Page

1 • SINGLE-FAMILY DWELLINGS 1

Floor plans with calculations based on N.E.C.—Clearance for Service Drops—Service Disconnects—Service Heads and Conductors—Grounding and Bonding—Ground Fault Circuit Interrupters (GFCI) for Personnel Protection—Outlet Boxes—Overcurrent Protection—Ground Fault Protection of Equipment—Concealed Knob and Tube Wiring—Nonmetallic Sheathed Cable—Metal-Clad and Armored Cable—Attic and Roof Spaces—Fishing Cable Into Masonry Walls—Rigid Nonmetallic Conduit—Lighting Fixtures, Appliances and Receptacles.

  TRADE COMPETENCY TEST NO. 1A 35
  TRADE COMPETENCY TEST NO. 1B 37
  TRADE COMPETENCY TEST NO. 1C 41

2 • MULTI-FAMILY DWELLINGS 43

Allowable Current Carrying Capacities of Conductors—Use of N.E.C. Tables—Load and Service Calculations—Services and Feeders—Grounding Multiphase Systems—Material and Wiring Rules—Nonmetallic Extensions—Branch Circuits—Heating Panels—Heating Cables—Fixtures and Appliances—Voltage Drop Formulas.

  TRADE COMPETENCY TEST NO. 2A 61
  TRADE COMPETENCY TEST NO. 2B 63
  TRADE COMPETENCY TEST NO. 2C 67

3 • COMMERCIAL LOCATIONS (Wiring Plans for a Store Building) 69

Excerpts from N.E.C. Specifications—Main Service Requirements—Conductor Specifications—Wiring Materials—General Equipment—Telephone and Speaker Systems—Miscellaneous Equipment—Notes on Architectural Blueprints—Load and Service Calculations—Floor Plans—Service Conduit Size—480 Volt Systems—Aluminum Conduit—Schedules and Drawings.

  TRADE COMPETENCY TEST NO. 3A 95
  TRADE COMPETENCY TEST NO. 3B 97
  TRADE COMPETENCY TEST NO. 3C 101

4 • INDUSTRIAL LOCATIONS (Power Installations) 103

Basic N.E.C. Rules for Motor Circuits—Elevator Motors—Crane Motors—Machine Tools—Transformers—X-Ray Units—Resistance Welders—Transformer Arc Welders—Motor-Generator Arc Welders—Feeder and Service Calculations—Power Installation for Restaurant.

  TRADE COMPETENCY TEST NO. 4A 135
  TRADE COMPETENCY TEST NO. 4B 137
  TRADE COMPETENCY TEST NO. 4C 139

5 • SPECIALIZED and HAZARDOUS LOCATIONS 141

Classes of Hazardous Locations—Methods for Reducing Hazards—Especially Listed Occupancies —Class Divisions—Garages—Aircraft Hangars—Gasoline Storage—Finishing Processes—Hospitals —Theatres—Radio and TV Studios—Capacitor Formulas—Transformer Formulas.

TRADE COMPETENCY TEST NO. 5A 161
TRADE COMPETENCY TEST NO. 5B 163
TRADE COMPETENCY TEST NO. 5C 167

• FINAL EXAMINATIONS 169

FINAL EXAMINATION — TEST NO. 6A 169
FINAL EXAMINATION — TEST NO. 6B 171
FINAL EXAMINATION — TEST NO. 6C 175

• APPENDIX 177

Enclosures, Electrical Equipment 177

Electrical and Fire Alarm Symbols and Abbreviations Commonly Used on Blueprints 179

Incandescent and Fluorescent Lamp Construction, Configurations for Plugs and Receptacles 182

General Formulas 186

Fault Current Form 187

Index 188

Single-Family Dwellings

The first drawing shows a single-family residence, the dwelling area being confined to one floor. Lighting outlets, plug receptacles, and switches are marked. It is not customary to furnish a complete wiring plan for such an installation, the common practice being to indicate switching arrangements by means of broken lines connecting controlled outlets with their switch or switches.

Starting at the entry, two three-way switches, one at the front door, the other at the far side of the living-room arch control the entry light and one in the front hall. Two three-way switches are connected to a pair of lighting outlets in the central hall. Bedroom, laundry, and bath lights operate by means of single-pole switches. Two three-way switches are used with a four-way in the kitchen, and two three-ways are employed in the dining room. In the living room, one single-pole switch is connected with two bracket lights at the fireplace, while another single-pole switch controls two plug receptacles.

Paragraphs dealing with applicable NEC Sections are placed around the figure. Observe that certain plug receptacle outlets are marked with one or more of the small letters, from *a* to *f* inclusive, and a certain point on the wall of Bedroom *2* is marked with a small *e*. These indications are explained in the paragraphs. Calculations necessary to determine feeder and service requirements under the general rules of Articles 220 and 230 are worked out immediately below the figure.

The second illustration, somewhat more involved than the first, represents a dwelling which occupies a floor and a half. Switching arrangements are indicated, as before, by means of broken lines. The optional method of load determination, permitted by NEC 220-30, is followed here. All necessary calculations are shown on the drawing.

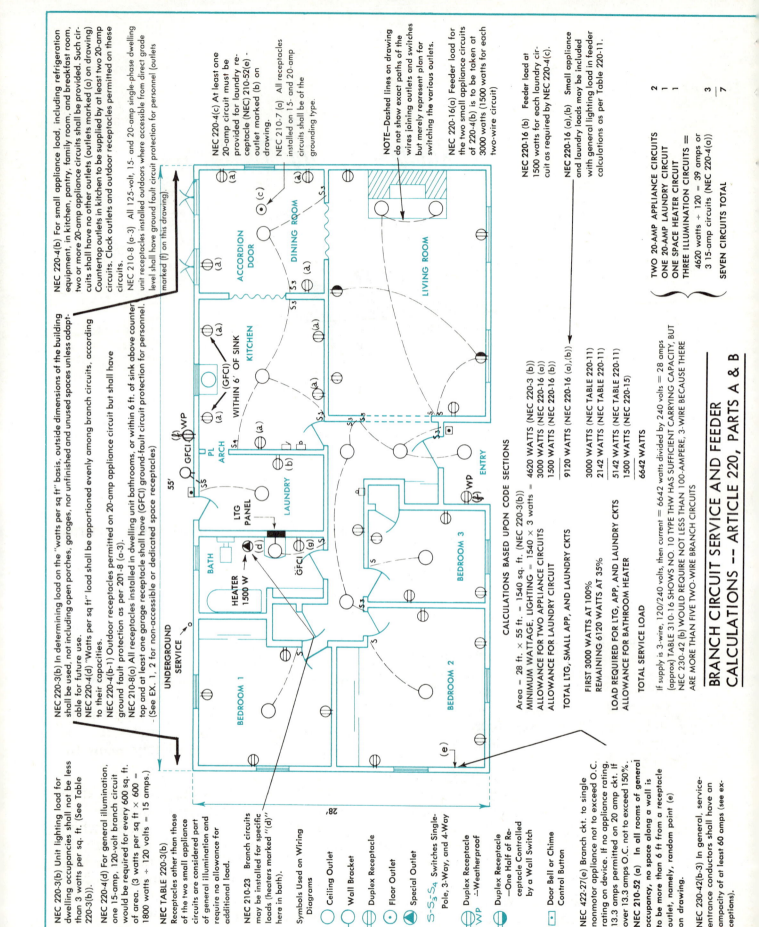

NEC 220-3(b) In determining load on the "watts per sq ft" basis, outside dimensions of the building shall be used, not including open porches, garages, nor unfinished and unused spaces unless adaptable for future use.

NEC 220-4(d) "Watts per sq ft" load shall be apportioned evenly among branch circuits, according to their capacities.

NEC 220-4(b-1) Outdoor receptacles installed on 20-amp appliance circuit but shall have ground fault protection as per 201-8 (a-3).

NEC 210-8(a) All receptacles installed in dwelling unit bathrooms, or within 6 ft of sink above dwelling top and at least one garage receptacle shall have (GFCI) ground-fault circuit protection for personnel - (See Ex. 1, 2 for non-accessible or dedicated space receptacles.)

NEC 220-4(b) For small appliance load, including refrigeration equipment, in kitchen, pantry, family room, and breakfast room, two or more 20-amp appliance circuits shall be provided. Such circuits shall have no other outlets (outlets marked "a" on drawing). Countertop outlets in kitchen to be supplied by at least two 20-amp circuits. Clock outlets and outdoor receptacles permitted on these circuits.

NEC 220-4(c) At least one 20-amp circuit must be provided for laundry receptacle (NEC) 210-52(e) - outlet marked (b) on drawing.

NEC 210-7 (a) All 125-volt, 15- and 20-amp single-phase dwelling circuits installed on 15- and 20-amp circuits shall be of the grounding type.

NOTE—Dashed lines on drawing do not show exact paths of the wires joining outlets and switches but merely represent plan for switching the various outlets.

NEC 220-16(a) Feeder load for the two small appliance circuits of 220-4(b) is to be taken at 3000 watts (1500 watts for each two-wire circuit).

NEC 220-16 (b) Feeder load at 1500 watts for each laundry circuit as required by NEC 220-4(c).

NEC 220-16 (a),(b) Small appliance and laundry loads may be included with general lighting load in feeder calculations as per Table 220-11.

NEC 220-3(b) Unit lighting load for dwelling occupancies shall not be less than 3 watts per sq. ft. (See Table 220-3(d)).

NEC 220-4(d) For general illumination, one 15-amp, 120-volt branch circuit would be required for every 600 sq. ft. of area. (3 watts per sq ft × 600 = 1800 watts ÷ 120 volts = 15 amps.)

NEC TABLE 220-3(b) Receptacles other than those of the two small appliance circuits are considered part of general illumination and require no allowance for additional load.

NEC 210-23 Branch circuits may be installed for specific loads (heaters marked "(d)" here in bath).

Symbols Used on Wiring Diagrams

- ◉ Ceiling Outlet
- ⊢ Wall Bracket
- ⊟ Duplex Receptacle
- ⊙ Floor Outlet
- ◀ Special Outlet
- S-S_3-S_4 Switches Single-Pole, 3-Way, and 4-Way
- Duplex Receptacle WP -Weatherproof
- ◖ Duplex Receptacle -One Half of Receptacle Controlled by a Wall Switch
- ▫ Door Bell or Chime Control Button

NEC 422-27(e) Branch ckt. to single nonmotor appliance not to exceed O.C. rating on device. If no appliance rating, 13.3 amps permitted on 20 amp ckt. If over 13.3 amps O.C. not to exceed 150%.

NEC 210-52 (a) In all rooms of general occupancy, no space along a wall is to be more than 6 ft from a receptacle outlet, namely, random point (e) on drawing.

NEC 230-42(b-3) In general, service-entrance conductors shall have an ampacity of at least 60 amps (see exceptions).

TWO 20-AMP APPLIANCE CIRCUITS   2
ONE 20-AMP LAUNDRY CIRCUIT   1
ONE SPACE HEATER CIRCUIT   1
THREE ILLUMINATION CIRCUITS =
4620 watts ÷ 120 = 39 amps or
3 15-amp circuits (NEC 220-4(a))   3
—
SEVEN CIRCUITS TOTAL   7

CALCULATIONS BASED UPON CODE SECTIONS

Area = 28 ft. × 55 ft. = 1540 sq. ft. (NEC 220-3(b))

MINIMUM WATTAGE. LIGHTING = 1540 sq. ft. × 3 watts =	4620 WATTS	(NEC 220-3 (b))
ALLOWANCE FOR TWO APPLIANCE CIRCUITS	3000 WATTS	(NEC 220-16 (a))
ALLOWANCE FOR LAUNDRY CIRCUIT	1500 WATTS	(NEC 220-16 (a))
TOTAL LTG, SMALL APP, AND LAUNDRY CKTS	9120 WATTS	(NEC 220-16 (a),(b))
FIRST 3000 WATTS AT 100%	3000 WATTS	(NEC TABLE 220-11)
REMAINING 6120 WATTS AT 35%	2142 WATTS	(NEC TABLE 220-11)
LOAD REQUIRED FOR LTG, APP, AND LAUNDRY CKTS	5142 WATTS	(NEC TABLE 220-11)
ALLOWANCE FOR BATHROOM HEATER	1500 WATTS	(NEC 220-15)
TOTAL SERVICE LOAD	6642 WATTS	

If supply is 3-wire, 120/240 volts, then current = 6642 watts divided by 240 volts = 28 amps (approx) TABLE 310-16 SHOWS NO. 10 TYPE THW HAS SUFFICIENT CARRYING CAPACITY, BUT NEC 230-42 (b) WOULD REQUIRE NOT LESS THAN 100-AMPERE, 3-WIRE BECAUSE THERE ARE MORE THAN FIVE TWO-WIRE BRANCH CIRCUITS

BRANCH CIRCUIT SERVICE AND FEEDER CALCULATIONS -- ARTICLE 220, PARTS A & B

Floor plan room labels: ACCORDION DOOR · DINING ROOM · LIVING ROOM · KITCHEN · WITHIN 6' OF SINK · PL ARCH · LAUNDRY · LTG PANEL · BATH · HEATER 1500 W · ENTRY · BEDROOM 1 · BEDROOM 2 · BEDROOM 3 · UNDERGROUND SERVICE · 28' · 55'

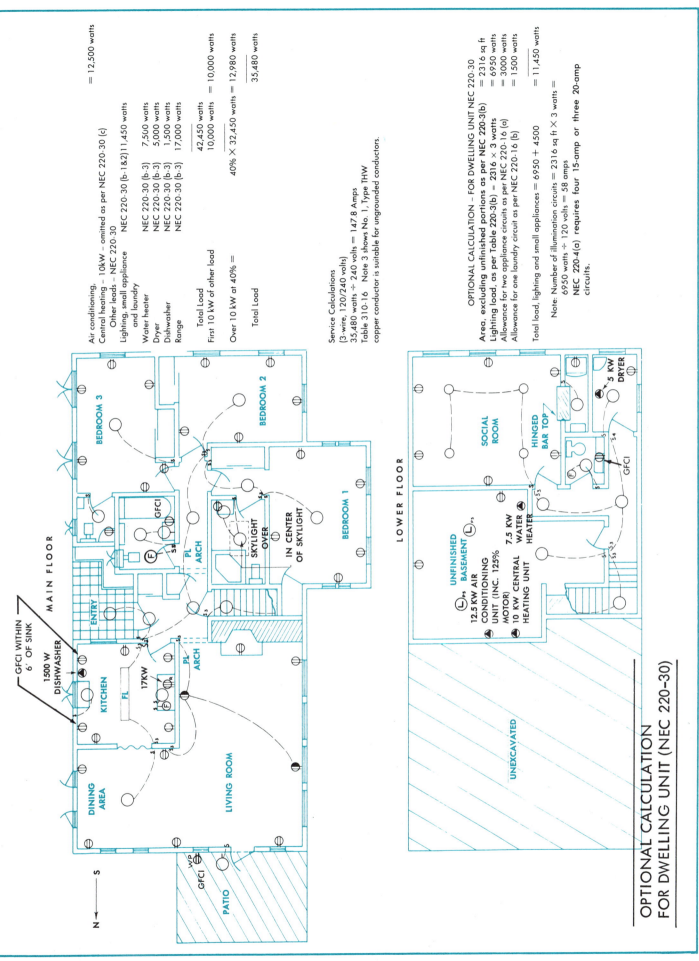

Air conditioning,
Central heating – 10 kW – omitted as per NEC 220-30 (c)   = 12,500 watts
Other leads – NEC 220-30
Lighting, small appliance   NEC 220-30 (b-1&2)11,450 watts
and laundry
Water heater   NEC 220-30 (b-3)   7,500 watts
Dryer   NEC 220-30 (b-3)   5,000 watts
Dishwasher   NEC 220-30 (b-3)   1,500 watts
Range   NEC 220-30 (b-3)   17,000 watts

        Total Load   42,450 watts
        First 10 kW of other load =   10,000 watts = 10,000 watts
        Over 10 kW at 40% =   40% × 32,450 watts =   12,980 watts

        Total Load   35,480 watts

Service Calculations
(3-wire, 120/240 volts)
35,480 watts ÷ 240 volts = 147.8 Amps
Table 310-16 Note 3 shows No. 1, Type THW
copper conductor is suitable for ungrounded conductors.

OPTIONAL CALCULATION – FOR DWELLING UNIT NEC 220-30
Area, excluding unfinished portions as per NEC 220-3(b)   = 2316 sq ft
Lighting load, as per Table 220-3(b) = 2316 × 3 watts   = 6950 watts
Allowance for two appliance circuits as per NEC 220-16 (a)   = 3000 watts
Allowance for one laundry circuit as per NEC 220-16 (b)   = 1500 watts

Total load, lighting and small appliances = 6950 + 4500   = 11,450 watts

Note: Number of illumination circuits = 2316 sq ft × 3 watts =
6950 watts ÷ 120 volts = 58 amps
NEC 220-4(a) requires four 15-amp or three 20-amp
circuits.

OPTIONAL CALCULATION
FOR DWELLING UNIT (NEC 220-30)

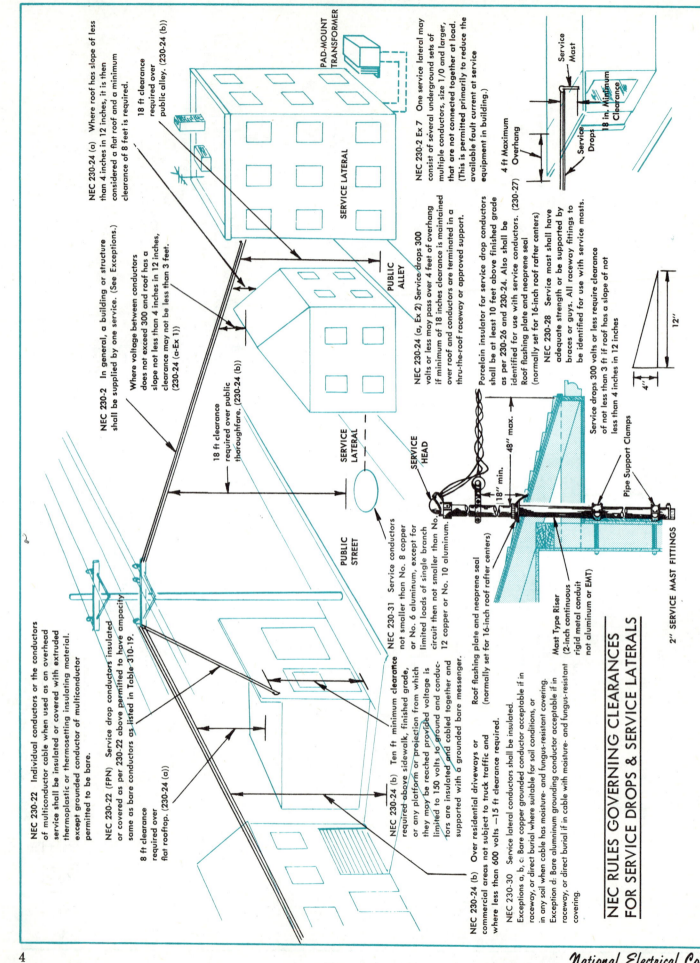

NEC 230-24 (a) Where roof has slope of less than 4 inches in 12 inches, it is then considered a flat roof and a minimum clearance of 8 feet is required.

18 ft clearance required over public alley. (230-24 (b))

PAD-MOUNT TRANSFORMER

SERVICE LATERAL

NEC 230-2 Ex 7 One service lateral may consist of several underground sets of multiple conductors, size 1/0 and larger, that are not connected together at load. (This is permitted primarily to reduce the available fault current at service equipment in building.)

Service Mast

Service Mast

18 in. Minimum Clearance

Service Drops

4 ft Maximum Overhang

NEC 230-22 Individual conductors or the conductors of multiconductor cable when used as an overhead service shall be insulated or covered with extruded thermoplastic or thermosetting insulating material, except grounded conductor of multiconductor permitted to be bare.

NEC 230-2 In general, a building or structure shall be supplied by one service. (See Exceptions.)

Where voltage between conductors does not exceed 300 and roof has a slope not less than 4 inches in 12 inches, clearance may not be less than 3 feet. (230-24 (a-Ex 1))

NEC 230-24 (a, Ex 2) Service drops 300 volts or less may pass over 4 feet of overhang if minimum of 18 inches clearance is maintained over roof and conductors are terminated in a thru-the-roof raceway or approved support.

Porcelain insulator for service drop conductors shall be at least 10 feet above finished grade as per 230-26 and 230-24. Also shall be identified for use with service conductors. (230-27)

Roof flashing plate and neoprene seal (normally set for 16-inch roof rafter centers)

NEC 230-28 Service mast shall have adequate strength or be supported by braces or guys. All raceway fittings to be identified for use with service masts.

NEC 230-22 (FPN) Service drop conductors insulated or covered as per 230-22 above permitted to have ampacity same as bare conductors as listed in Table 310-19.

8 ft clearance required over flat rooftop. (230-24 (a))

18 ft clearance required over public thoroughfare. (230-24 (b))

SERVICE LATERAL

SERVICE HEAD

PUBLIC STREET

NEC 230-31 Service conductors not smaller than No. 8 copper or No. 6 aluminum, except for limited loads of single branch circuit then not smaller than No. 12 copper or No. 10 aluminum.

48" max.

18" min.

12"

4"

Service drops 300 volts or less require clearance of not less than 3 ft if roof has a slope of not less than 4 inches in 12 inches

Pipe Support Clamps

NEC 230-24 (b) Over residential driveways or commercial areas not subject to truck traffic and where less than 600 volts —15 ft clearance required.

NEC 230-24 (b) Ten ft minimum clearance required above sidewalk, finished grade, or any platform or projection from which they may be reached provided voltage is limited to 150 volts to ground and conductors are insulated and cabled together and supported with a grounded bare messenger.

Roof flashing plate and neoprene seal (normally set for 16-inch roof rafter centers)

Mast Type Riser (2-inch continuous rigid metal conduit not aluminum or EMT)

NEC 230-30 Service lateral conductors shall be insulated.
Exceptions a, b, c: Bare copper grounded conductor acceptable if in raceway, or direct burial where suitable for soil conditions, or in any soil when cable has moisture- and fungus-resistant covering.
Exception d: Bare aluminum grounding conductor acceptable if in raceway, or direct burial if in cable with moisture- and fungus-resistant covering.

2" SERVICE MAST FITTINGS

NEC RULES GOVERNING CLEARANCES FOR SERVICE DROPS & SERVICE LATERALS

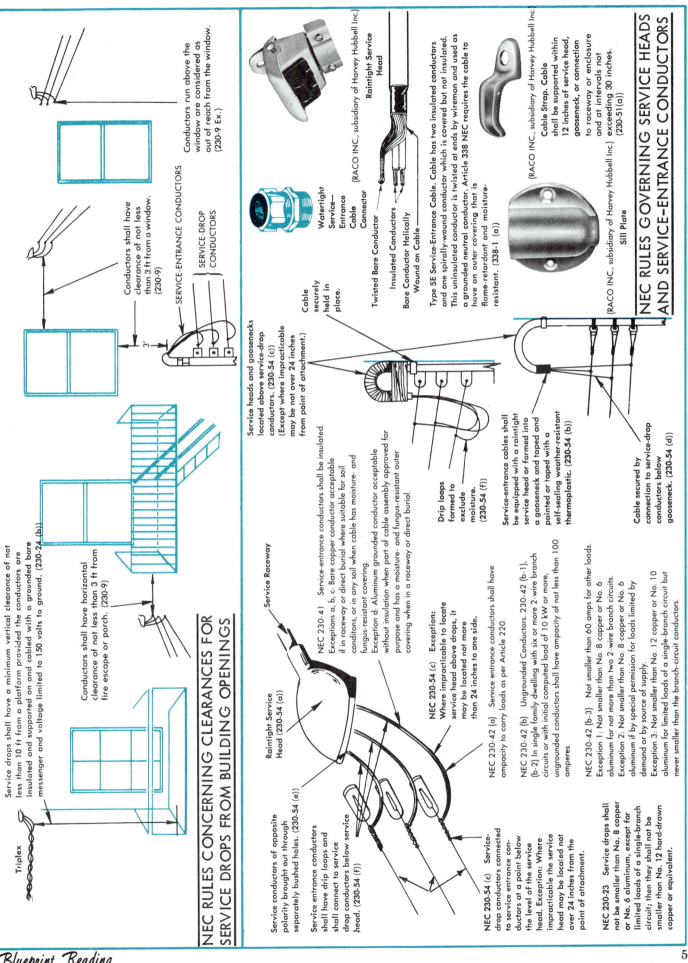

Conductors run above the window are considered as out of reach from the window. (230-9 Ex.)

Conductors shall have clearance of not less than 3 ft from a window. (230-9)

SERVICE-ENTRANCE CONDUCTORS

SERVICE-DROP CONDUCTORS

3'

Triplex

Service drops shall have a minimum vertical clearance of not less than 10 ft from a platform provided the conductors are insulated and supported on and cabled with a grounded bare messenger and voltage limited to 150 volts to ground. (230-24 (b))

Conductors shall have horizontal clearance of not less than 3 ft from fire escape or porch. (230-9)

NEC RULES CONCERNING CLEARANCES FOR SERVICE DROPS FROM BUILDING OPENINGS

Service conductors of opposite polarity brought out through separately bushed holes. (230-54 (e))

Service entrance conductors shall have drip loops and shall connect to service drop conductors below service head. (230-54 (f))

Service Raceway

Raintight Service Head (230-54 (a))

NEC 230-54 (c)   Service-drop conductors connected to service entrance conductors at a point below the level of the service head. Exception: Where impracticable the service head may be located not over 24 inches from the point of attachment.

NEC 230-23   Service drops shall not be smaller than No. 8 copper or No. 6 aluminum, except for limited loads of a single-branch circuit; then they shall not be smaller than No. 12 copper or equivalent.

Service heads and goosenecks located above service-drop conductors. (230-54 (c)) (Except where impracticable may be not over 24 inches from point of attachment.)

NEC 230-41   Service-entrance conductors shall be insulated. Exceptions a, b, c: Bare copper conductor acceptable if in raceway or direct burial where suitable for soil conditons, or in any soil when cable has moisture- and fungus-resistant covering. Exception d: Aluminum grounded conductor acceptable without insulation when part of cable assembly approved for purpose and has a moisture- and fungus-resistant outer covering when in a raceway or direct burial.

NEC 230-54 (c)   Exception: Where impracticable to locate service head above drops, it may be located not more than 24 inches to one side.

NEC 230-42 (a)   Service entrance conductors shall have ampacity to carry loads as per Article 220.

NEC 230-42 (b)   Ungrounded Conductors. 230-42 (b-1), (b-2) In single family dwelling with six or more 2-wire branch circuits or with initial computed load of 10 kW or more, ungrounded conductors shall have ampacity of not less than 100 amperes.

NEC 230-42 (b-3)   Not smaller than 60 amps for other loads. Exception 1: Not smaller than No. 8 copper or No. 6 aluminum for not more than two 2-wire branch circuits. Exception 2: Not smaller than No. 8 copper or No. 6 aluminum if by special permission for loads limited by demand or by source of supply. Exception 3: Not smaller than No. 12 copper or No. 10 aluminum for limited loads of a single-branch circuit but never smaller than the branch-circuit conductors.

(RACO INC., subsidiary of Harvey Hubbell Inc.)
Raintight Service Head

Watertight Service—Entrance Cable Connector

Twisted Bare Conductor
Insulated Conductors
Bare Conductor Helically Wound on Cable

Type SE Service-Entrance Cable. Cable has two insulated conductors and one spirally-wound conductor which is covered but not insulated. This uninsulated conductor is twisted at ends by wireman and used as a grounded neutral conductor. Article 338 NEC requires the cable to have an outer covering that is flame-retardant and moisture-resistant. (338-1 (a))

(RACO INC., subsidiary of Harvey Hubbell Inc.)
Cable Strap. Cable shall be supported within 12 inches of service head, gooseneck, or connection to raceway or enclosure and at intervals not exceeding 30 inches. (230-51(a))

(RACO INC., subsidiary of Harvey Hubbell Inc.)
Sill Plate

NEC RULES GOVERNING SERVICE HEADS AND SERVICE-ENTRANCE CONDUCTORS

Cable securely held in place.

Drip loops formed to exclude moisture. (230-54 (f))

Service-entrance cables shall be equipped with a raintight service head or formed into a gooseneck and taped and painted or taped with a self-sealing weather-resistant thermoplastic. (230-54 (b))

Cable secured by connection to service-drop conductors below gooseneck. (230-54 (d))

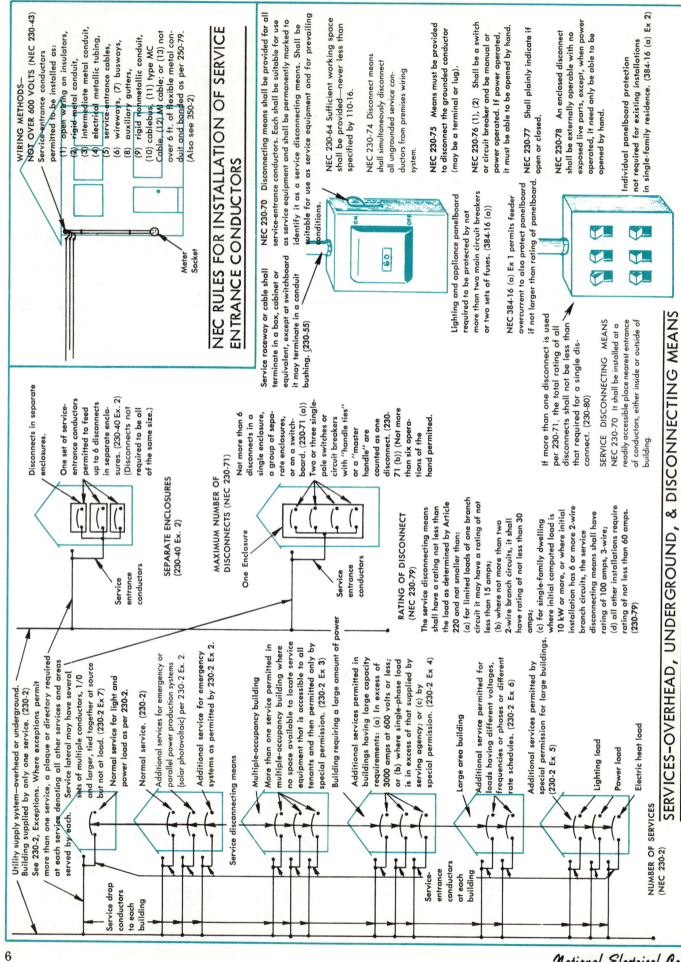

WIRING METHODS— NOT OVER 600 VOLTS (NEC 230-43) Service-entrance conductors permitted to be installed as:

(1) open wiring on insulators,
(2) rigid metal conduit,
(3) intermediate metal conduit,
(4) electrical metallic tubing,
(5) service-entrance cables,
(6) wireways, (7) busways,
(8) auxiliary gutters,
(9) rigid nonmetallic conduit,
(10) cablebus, (11) type MC
Cable. (12) MI cable; or (13) not
over 6 ft. of flexible metal conduit and bonded as per 250-79.
(Also see 350-2)

Meter Socket

Service raceway or cable shall terminate in a box, cabinet or equivalent, except at switchboard it may terminate in a conduit bushing. (230-55)

NEC 230-70 Disconnecting means shall be provided for all service-entrance conductors. Each shall be suitable for use as service equipment and shall be permanently marked to identify it as a service disconnecting means. Shall be suitable for use as service equipment and for prevailing conditions.

NEC 230-64 Sufficient working space shall be provided—never less than specified by 110-16.

NEC 230-74 Disconnect means shall simultaneously disconnect all ungrounded service conductors from premises wiring system.

NEC 230-75 Means must be provided to disconnect the grounded conductor (may be a terminal or lug).

NEC 230-76 (1), (2) Shall be a switch or circuit breaker and be manual or power operated. If power operated, it must be able to be opened by hand.

NEC 230-77 Shall plainly indicate if open or closed.

NEC 230-78 An enclosed disconnect shall be externally operable with no exposed live parts, except, when power operated, it need only be able to be opened by hand.

Lighting and appliance panelboard required to be protected by not more than two main circuit breakers or two sets of fuses. (384-16 (a))

NEC 384-16 (a) Ex 1 permits feeder overcurrent to also protect panelboard if not larger than rating of panelboard.

Individual panelboard protection not required for existing installations in single-family residence. (384-16 (a) Ex 2)

Disconnects in separate enclosures.

One set of service-entrance conductors permitted to feed up to 6 disconnects in separate enclosures. (230-40 Ex. 2) (Disconnects not required to be all of the same size.)

SEPARATE ENCLOSURES (230-40 Ex. 2)

Not more than 6 disconnects in a single enclosure, a group of separate enclosures, or on a switchboard. (230-71 (a)) Two or three single-pole switches or circuit breakers with "handle ties" or a "master handle" are counted as one disconnect. (230-71 (b)) (Not more than six operations of the hand permitted.

MAXIMUM NUMBER OF DISCONNECTS (NEC 230-71)

One Enclosure

Service entrance conductors

RATING OF DISCONNECT (NEC 230-79)

The service disconnecting means shall have a rating not less than the load as determined by Article 220 and not smaller than:
(a) for limited loads of one branch circuit it may have a rating of not less than 15 amps;
(b) where not more than two 2-wire branch circuits, it shall have rating of not less than 30 amps;
(c) for single-family dwelling where initial computed load is 10 kW or more, or where initial installation has 6 or more 2-wire branch circuits, the service disconnecting means shall have rating of 100 amps, 3-wire;
(d) all other installations require rating of not less than 60 amps. (230-79)

If more than one disconnect is used per 230-71, the total rating of all disconnects shall not be less than that required for a single disconnect. (230-80)

SERVICE DISCONNECTING MEANS NEC 230-70. It shall be installed at a readily accessible place nearest entrance of conductors, either inside or outside of building.

Utility supply system—overhead or underground. Building supplied by only one service. (230-2) See 230-2, Exceptions. Where exceptions permit more than one service, a plaque or directory required at each service denoting all other services and areas served by each.

Service lateral may have several sets of multiple conductors, 1/0 and larger, tied together at source but not at load. (230-2 Ex 7)

Normal service for light and power load as per 230-2.

Normal service. (230-2)

Additional services for emergency or parallel power production systems (solar photovoltaic) per 230-2 Ex. 2.

Additional service for emergency systems as permitted by 230-2 Ex 2.

Service disconnecting means

Multiple-occupancy building

More than one service permitted in multiple-occupancy building where no space available to locate service equipment that is accessible to all tenants and then permitted only by special permission. (230-2 Ex 3)

Building requiring a large amount of power

Additional services permitted in buildings having large capacity requirements: (a) In excess of 3000 amps at 600 volts or less; or (b) where single-phase load is in excess of that supplied by serving agency; or (c) by special permission. (230-2 Ex 4)

Large area building

Additional service permitted for loads having different voltages, frequencies or phases or different rate schedules. (230-2 Ex 6)

Additional services permitted by special permission for large buildings. (230-2 Ex 5)

Lighting load

Power load

Electric heat load

NUMBER OF SERVICES (NEC 230-2)

Service drop conductors to each building

Service-entrance conductors at each building

SERVICES—OVERHEAD, UNDERGROUND, & DISCONNECTING MEANS

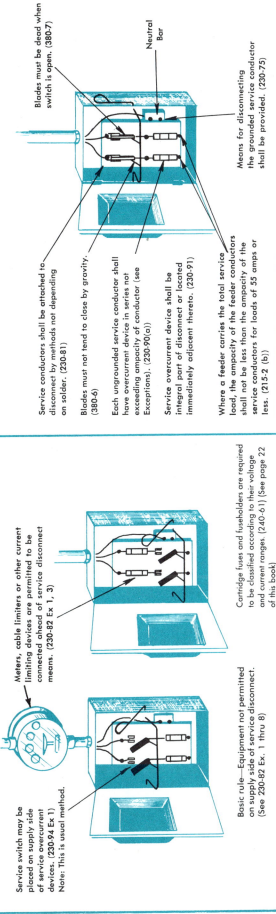

Blades must be dead when switch is open. (380-7)

Neutral Bar

Means for disconnecting the grounded service conductor shall be provided. (230-75)

Service conductors shall be attached to disconnect by methods not depending on solder. (230-81)

Blades must not tend to close by gravity. (380-6)

Each ungrounded service conductor shall have overcurrent device in series not exceeding ampacity of conductor (see Exceptions). (230-90(a))

Service overcurrent device shall be integral part of disconnect or located immediately adjacent thereto. (230-91)

Where a feeder carries the total service load, the ampacity of the feeder conductors shall not be less than the ampacity of the service conductors for loads of 55 amps or less. (215-2 (b))

NEC RULES COVERING SERVICE DETAILS

Service switch may be placed on supply side of service overcurrent devices. (230-94 Ex 1) Note: This is usual method.

Meters, cable limiters or other current limiting devices are permitted to be connected ahead of service disconnect means. (230-82 Ex 1, 3)

Cartridge fuses and fuseholders are required to be classified according to their voltage and current ranges. (240-61) (See page 22 of this book)

Basic rule—Equipment not permitted on supply side of service disconnect. (See 230-82 Ex. 1 thru 8)

NEC RULES GOVERNING SEQUENCE OF SERVICE SWITCH AND FUSES

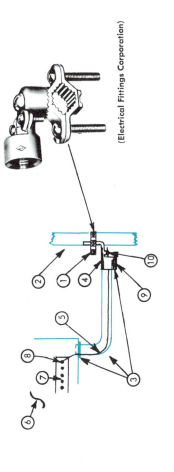

(Electrical Fittings Corporation)

NEC RULES CONCERNING SERVICE GROUNDING

1. Ground clamp suitable for the purpose shall be used to connect grounding electrode conductor to grounding electrode. (250-115)

2. Water piping grounding electrode (250-81)

3. Metal grounding electrode conductor enclosures shall be electrically continuous from cabinet to grounding electrode. (250-92(b))

4. If metal enclosure is not electrically continuous it must be bonded at each end to grounding conductor. (250-92(b))

5. Grounding electrode conductor

6. Service-entrance equipment

7. Neutral connection bar

8. Grounding electrode conductor to also ground all metal service-entrance equipment, service raceways and system grounded conductor (neutral). (250-23 (a), 250-53 (a), 250-32)

9. Grounding electrode conductor bonded to metal enclosure so as to be electrically continuous. (250-92(b))

10. Connection should be on street side of water meter or bonding required around valves, meter, unions, etc. (250-112)

(1) Service Raceways & Enclosures: Metal service equip. and enclosures to be grounded. (250-32)

(2) THREADED BOSS OR HUB—acceptable means of bonding service conduit to meter socket (250-72 (b))

(3) METER SOCKET

(4) Grounded circuit conductor can be used to ground meter enclosures, service raceways, etc., on supply side of service disconnecting means. (250-61 (a))

(5) NEUTRAL CONDUCTOR (250-25 (2))

(6) Service equipment, raceways, cable armor, cable sheaths, etc., and any service conductor that is required to be grounded by 250-5 shall be grounded as per Article 250. (230-63)

(7) For a single-phase 3-wire system the neutral conductor shall be grounded. (250-25 (2))

(8) Grounded system conductor (neutral) shall be connected to grounding electrode conductor. (250-53 (a))
Means must be provided in the service equipment to disconnect the service grounded conductor (neutral) from the premises wiring. In this case the terminal or bus to which all grounded conductors are attached by means of pressure connectors is permitted for this purpose. (230-75)

(9) To grounding electrode (see page 51 of this book).

(10) Grounding conductor connected to grounding fitting by exothermic welding, or by listed lugs, pressure connectors, clamps, or other listed means. Solder connections not permitted (250-115)

(11) When in conduit ground wire not excluded as a conductor by NEC and must be included the same as any other conductor when determining conduit size by using Tables in Chapter 9. (See Note 2 to Tables in Chapter 9.)

(12) Aluminum or copper-clad aluminum conductors not permitted where subject to corrosive conditions or in contact with masonry or earth and when outside shall not be within 18 inches of earth. (250-92 (a))

(13) GROUNDING ELECTRODE CONDUCTOR – used to connect the equipment grounding conductors, the service-equipment enclosures and the grounded service conductor (when present) to the grounding electrode [250-53 (a) (Definitions)].

(14) Size of grounding electrode conductor determined by the size of the largest service-entrance conductor or equivalent for parallel conductors as per Table 250-94. (Need not be larger than No. 6 copper or its equivalent in ampacity when connected to made electrodes as permitted in 250-83.)

(15) Grounding electrode conductor shall be copper, aluminum, copper-clad aluminum, or other corrosion-resistant material. Conductor shall be solid or stranded, insulated, covered, or bare and shall be continuous without splice or joint except splices in busbars permitted. (250-91 (a) Ex 1)
Taps are permitted to grounding electrode conductor where more than one service-entrance enclosure is installed if such tap extends inside of such enclosure. (250-91 (a) Ex 2)

(16) No. 4 or larger grounding electrode conductor needs no protection if it is not exposed to severe physical damage. A No. 6 needs no protection if it follows surface of building and is not exposed to physical damage. No. 8 shall be in rigid metal conduit, IMC, rigid nonmetallic conduit, EMT, or cable armor. (250-92 (a))

(17) Path to ground from circuits, equipment and enclosures shall: (a) be permanent and continuous, (b) have capacity to safely conduct available fault currents, (c) have low impedance so as to limit voltage to ground and to facilitate the circuit overcurrent devices. (250-51 (1), (2), (3))

(18) When underground metal water pipe used as the grounding electrode, a supplemental electrode is required. (250-81 (a))

(19) Secondary AC systems to be grounded shall have the grounding electrode conductor connected at each service. Grounding electrode conductor connected to AC system grounded conductor (neutral) on supply side of disconnecting means on premises, preferably within service enclosure. This connection not to be made on load side of disconnecting means. (250-23 (a)—see Exceptions)

(20) Equipment grounding conductor connections at service equipment shall be made: (a) for grounded systems, this connection made by bonding the equipment grounding conductor, grounded service conductor and the grounding electrode conductor, (b) for ungrounded systems, this connection made by bonding the equipment grounding conductor to the grounding electrode conductor. (250-50 (a) (b)) See Exceptions and Definitions.)

(21) MAIN BONDING JUMPER—is the connection between grounded circuit conductor and equipment grounding conductor at service. (Definitions)
Main bonding jumper shall be used to connect equipment grounding conductor and service equipment enclosure to system-grounded conductor (neutral) within Service Enclosure. (250-53(b)). Shall be a wire, bus, screw or similar suitable conductor and without splice. (250-79(a))

(22) EQUIPMENT BONDING JUMPER—Main and equipment bonding jumper shall be: (a) copper or other corrosion-resistant material; (b) attached as per 250-113 and 250-115. (250-79 (a), (b))
Equipment bonding jumper on supply side of service and main bonding jumper. Sized as per Table 250-94 but not less than 12½ percent of largest phase conductor. (250-79 (c))
Where service-entrance conductors are parallel in separate raceways as permitted in 310-4 the size of the bonding jumper for each raceway shall be based on size of conductors in each raceway or cable. (250-79 (c)) (See also page 10 of this book).

(23) SERVICE EQUIPMENT ENCLOSURE—contains service disconnecting means.

(24) GROUNDING BUSHING – Must be listed for purpose. (250-72) (See page 14 of this book.)

(25) All metal service equipment enclosures containing service-entrance conductors, meter fittings, boxes, or the like, interposed in service raceway or armor and any conduit or armor that forms part of the grounding conductor to the service raceway shall be effectively bonded together, except as permitted by 250-55 for underground service cables. (250-71 (a), (1), (2), (3))

Grounding continuity of service equipment shall be assured by: (a) bonding equipment to neutral as per 250-113; (b) threaded bosses, hubs or couplings; (c) threadless couplings for rigid conduit or EMT; (d) bonding jumpers installed as per Article 250. Bonding jumpers required around concentric or eccentric knockouts; (e) other listed devices such as bonding-type locknuts and bushings. (Standard locknuts and bushings not accepted.) Joints shall be made up wrenchtight where rigid conduit is involved. (250-72)

(26) NEUTRAL SERVICE ENTRANCE CONDUCTOR—In this drawing only the identified neutral conductor is shown, the two "ungrounded" or hot conductors are not shown.

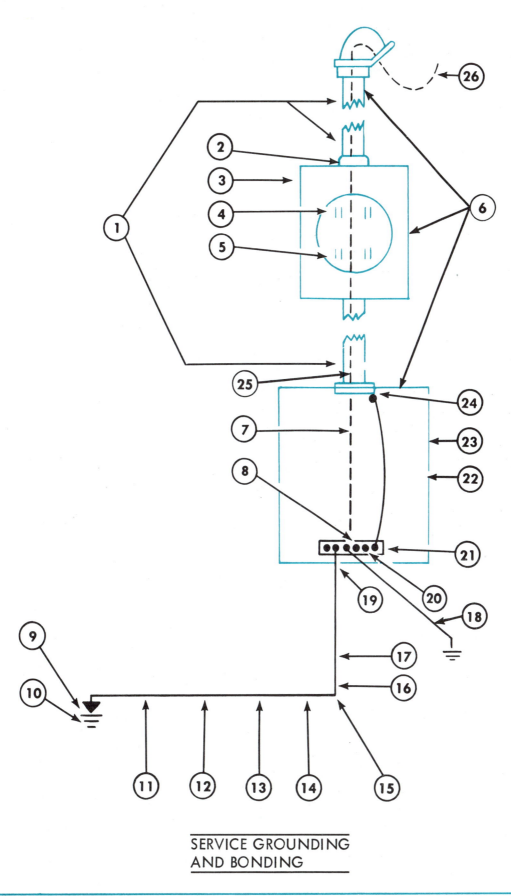

SERVICE GROUNDING
AND BONDING

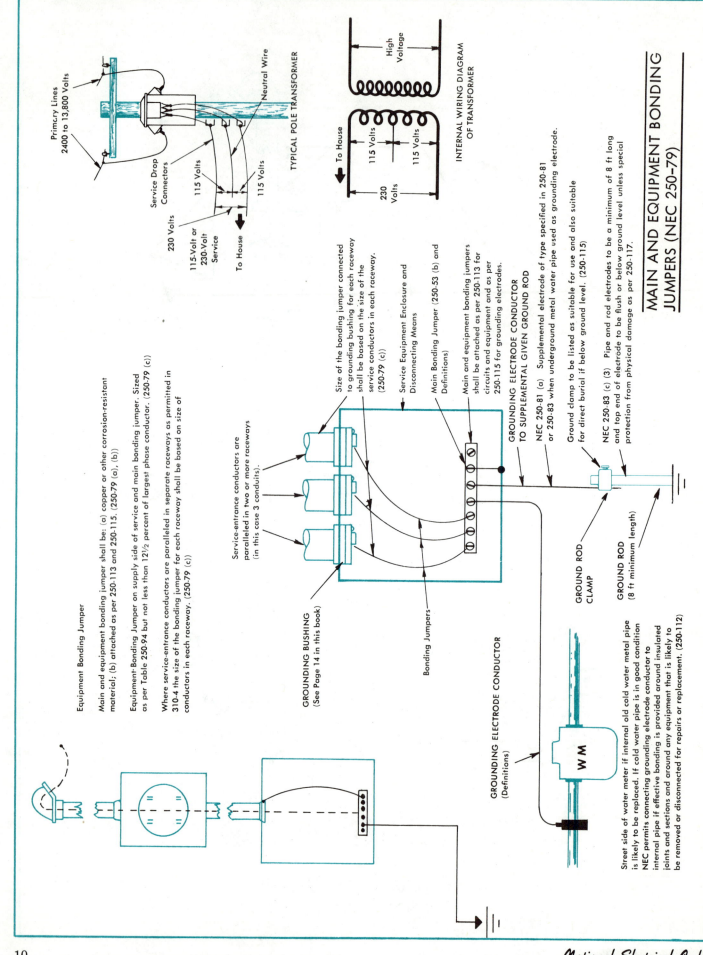

Primary Lines
2400 to 13,800 Volts

Service Drop
Connectors

Neutral Wire

TYPICAL POLE TRANSFORMER

115 Volts

115 Volts

230 Volts

115-Volt or
230-Volt
Service

To House

High
Voltage

To House

115 Volts

115 Volts

230
Volts

INTERNAL WIRING DIAGRAM
OF TRANSFORMER

MAIN AND EQUIPMENT BONDING JUMPERS (NEC 250-79)

Size of the bonding jumper connected to grounding bushing for each raceway shall be based on the size of the service conductors in each raceway. (250-79 (c))

Service Equipment Enclosure and Disconnecting Means

Main Bonding Jumper (250-53 (b) and Definitions)

Main and equipment bonding jumpers shall be attached as per 250-113 for circuits and equipment and as per 250-115 for grounding electrodes.

GROUNDING ELECTRODE CONDUCTOR TO SUPPLEMENTAL GIVEN GROUND ROD

NEC 250-81 (a) Supplemental electrode of type specified in 250-81 or 250-83 when underground metal water pipe used as grounding electrode.

Ground clamp to be listed as suitable for use and also suitable for direct burial if below ground level. (250-115)

NEC 250-83 (c) (3) Pipe and rod electrodes to be a minimum of 8 ft long and top end of electrode to be flush or below ground level unless special protection from physical damage as per 250-117.

Equipment Bonding Jumper

Main and equipment bonding jumper shall be: (a) copper or other corrosion-resistant material; (b) attached as per 250-113 and 250-115. (250-79 (a), (b))

Equipment Bonding Jumper on supply side of service and main bonding jumper. Sized as per Table 250-94 but not less than 12½ percent of largest phase conductor. (250-79 (c))

Where service-entrance conductors are paralleled in separate raceways as permitted in 310-4 the size of the bonding jumper for each raceway shall be based on size of conductors in each raceway. (250-79 (c))

Service-entrance conductors are paralleled in two or more raceways (in this case 3 conduits).

GROUNDING BUSHING
(See Page 14 in this book)

Bonding Jumpers

GROUND ROD CLAMP

GROUND ROD
(8 ft minimum length)

GROUNDING ELECTRODE CONDUCTOR
(Definitions)

W M

Street side of water meter if internal old cold water metal pipe is likely to be replaced. If cold water pipe is in good condition NEC permits connecting grounding electrode conductor to internal pipe if effective bonding is provided around insulated joints and sections and around any equipment that is likely to be removed or disconnected for repairs or replacement. (250-112)

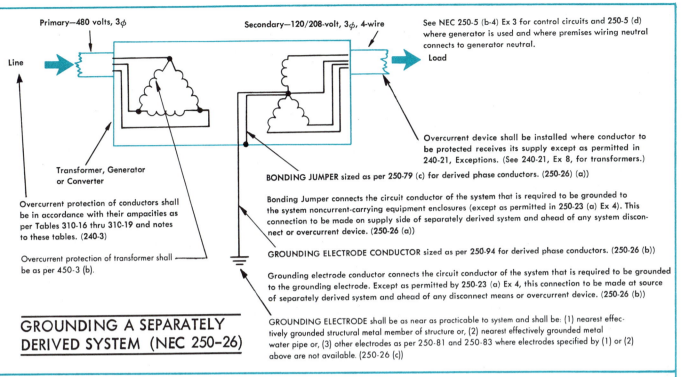

Primary—480 volts, 3φ

Line

Secondary—120/208-volt, 3φ, 4-wire

Load

See NEC 250-5 (b-4) Ex 3 for control circuits and 250-5 (d) where generator is used and where premises wiring neutral connects to generator neutral.

Transformer, Generator or Converter

Overcurrent protection of conductors shall be in accordance with their ampacities as per Tables 310-16 thru 310-19 and notes to these tables. (240-3)

Overcurrent protection of transformer shall be as per 450-3 (b).

Overcurrent device shall be installed where conductor to be protected receives its supply except as permitted in 240-21, Exceptions. (See 240-21, Ex 8, for transformers.)

BONDING JUMPER sized as per 250-79 (c) for derived phase conductors. (250-26) (a)

Bonding Jumper connects the circuit conductor of the system that is required to be grounded to the system noncurrent-carrying equipment enclosures (except as permitted in 250-23 (a) Ex 4). This connection to be made on supply side of separately derived system and ahead of any system disconnect or overcurrent device. (250-26 (a))

GROUNDING ELECTRODE CONDUCTOR sized as per 250-94 for derived phase conductors. (250-26 (b))

Grounding electrode conductor connects the circuit conductor of the system that is required to be grounded to the grounding electrode. Except as permitted by 250-23 (a) Ex 4, this connection to be made at source of separately derived system and ahead of any disconnect means or overcurrent device. (250-26 (b))

GROUNDING ELECTRODE shall be as near as practicable to system and shall be: (1) nearest effectively grounded structural metal member of structure or, (2) nearest effectively grounded metal water pipe or, (3) other electrodes as per 250-81 and 250-83 where electrodes specified by (1) or (2) above are not available. (250-26 (c))

GROUNDING A SEPARATELY DERIVED SYSTEM (NEC 250-26)

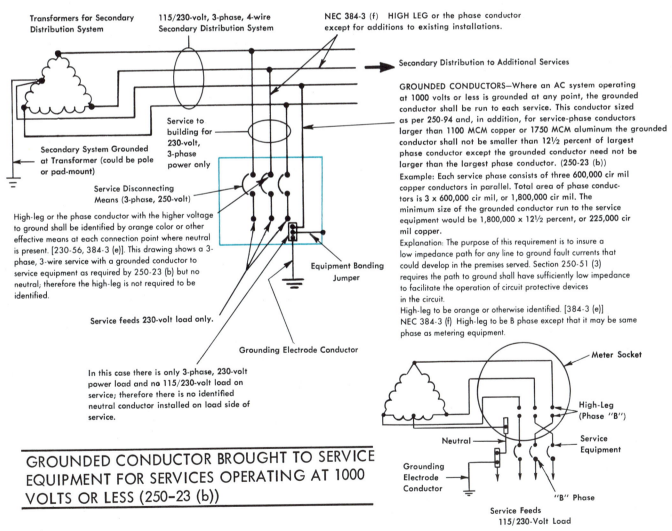

Transformers for Secondary Distribution System

115/230-volt, 3-phase, 4-wire Secondary Distribution System

NEC 384-3 (f)   HIGH LEG or the phase conductor except for additions to existing installations.

Secondary System Grounded at Transformer (could be pole or pad-mount)

Service to building for 230-volt, 3-phase power only

Service Disconnecting Means (3-phase, 250-volt)

High-leg or the phase conductor with the higher voltage to ground shall be identified by orange color or other effective means at each connection point where neutral is present. [230-56, 384-3 (e)]. This drawing shows a 3-phase, 3-wire service with a grounded conductor to service equipment as required by 250-23 (b) but no neutral; therefore the high-leg is not required to be identified.

Service feeds 230-volt load only.

Equipment Bonding Jumper

Grounding Electrode Conductor

In this case there is only 3-phase, 230-volt power load and no 115/230-volt load on service; therefore there is no identified neutral conductor installed on load side of service.

Secondary Distribution to Additional Services

GROUNDED CONDUCTORS—Where an AC system operating at 1000 volts or less is grounded at any point, the grounded conductor shall be run to each service. This conductor sized as per 250-94 and, in addition, for service-phase conductors larger than 1100 MCM copper or 1750 MCM aluminum the grounded conductor shall not be smaller than 12½ percent of largest phase conductor except the grounded conductor need not be larger than the largest phase conductor. (250-23 (b))

Example: Each service consists of three 600,000 cir mil copper conductors in parallel. Total area of phase conductors is 3 x 600,000 cir mil, or 1,800,000 cir mil. The minimum size of the grounded conductor run to the service equipment would be 1,800,000 x 12½ percent, or 225,000 cir mil copper.

Explanation: The purpose of this requirement is to insure a low impedance path for any line to ground fault currents that could develop in the premises served. Section 250-51 (3) requires the path to ground shall have sufficiently low impedance to facilitate the operation of circuit protective devices in the circuit.

High-leg to be orange or otherwise identified. [384-3 (e)]

NEC 384-3 (f) High-leg to be B phase except that it may be same phase as metering equipment.

Meter Socket

High-Leg (Phase "B")

Service Equipment

Neutral

"B" Phase

Grounding Electrode Conductor

Service Feeds 115/230-Volt Load

GROUNDED CONDUCTOR BROUGHT TO SERVICE EQUIPMENT FOR SERVICES OPERATING AT 1000 VOLTS OR LESS (250-23 (b))

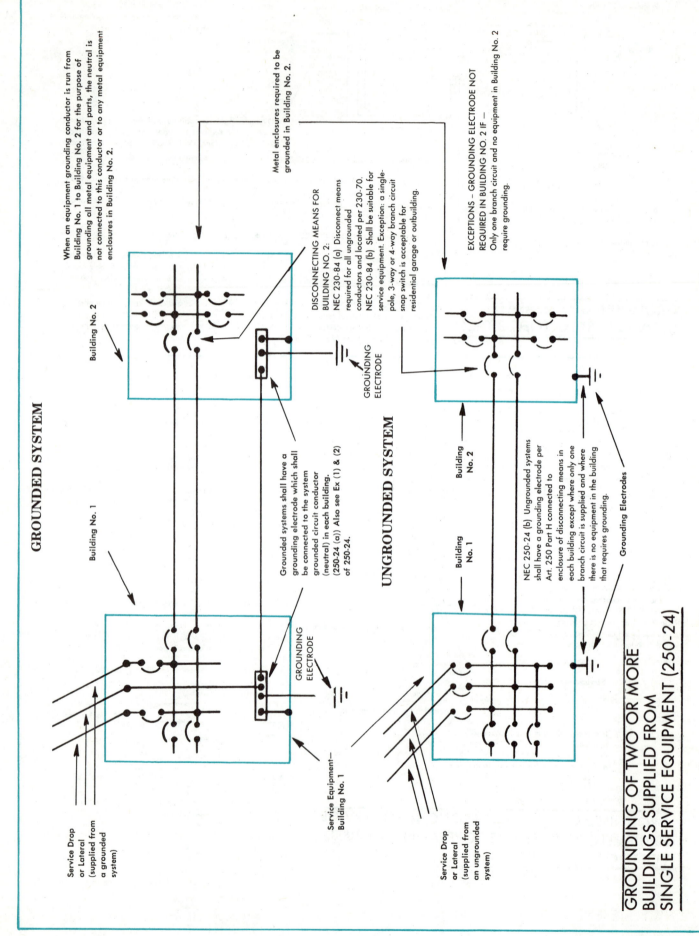

GROUNDED SYSTEM

When an equipment grounding conductor is run from Building No. 1 to Building No. 2 for the purpose of grounding all metal equipment and parts, the neutral is not connected to this conductor or to any metal equipment enclosures in Building No. 2.

Metal enclosures required to be grounded in Building No. 2.

DISCONNECTING MEANS FOR BUILDING NO. 2: NEC 230-84 (a) Disconnect means required for all ungrounded conductors and located per 230-70. NEC 230-84 (b) Shall be suitable for service equipment. Exception: a single-pole, 3-way or 4-way branch circuit snap switch is acceptable for residential garage or outbuilding.

EXCEPTIONS – GROUNDING ELECTRODE NOT REQUIRED IN BUILDING NO. 2 IF – Only one branch circuit and no equipment in Building No. 2 require grounding.

Building No. 2

Building No. 1

Grounded systems shall have a grounding electrode which shall be connected to the system grounded circuit conductor (neutral) in each building. (250-24 (a)) Also see Ex (1) & (2) of 250-24.

GROUNDING ELECTRODE

GROUNDING ELECTRODE

Service Equipment— Building No. 1

Service Drop or Lateral (supplied from a grounded system)

UNGROUNDED SYSTEM

Building No. 2

Building No. 1

NEC 250-24 (b) Ungrounded systems shall have a grounding electrode per Art. 250 Part H connected to enclosure of disconnecting means in each building except where only one branch circuit is supplied and where there is no equipment in the building that requires grounding.

Grounding Electrodes

Service Drop or Lateral (supplied from an ungrounded system)

GROUNDING OF TWO OR MORE BUILDINGS SUPPLIED FROM SINGLE SERVICE EQUIPMENT (250-24)

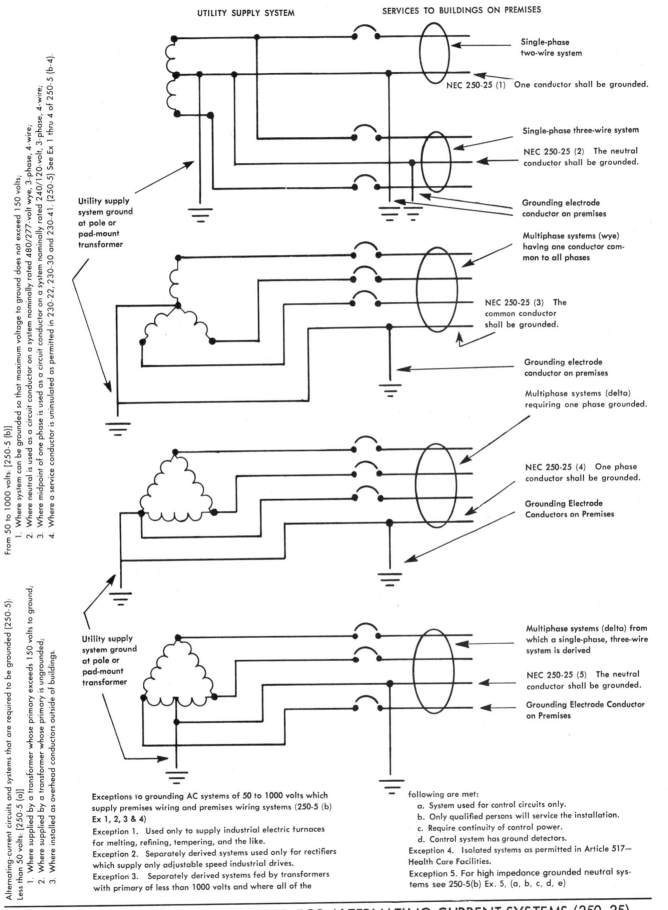

UTILITY SUPPLY SYSTEM   SERVICES TO BUILDINGS ON PREMISES

Single-phase two-wire system

NEC 250-25 (1) One conductor shall be grounded.

Single-phase three-wire system

NEC 250-25 (2) The neutral conductor shall be grounded.

Grounding electrode conductor on premises

Multiphase systems (wye) having one conductor common to all phases

NEC 250-25 (3) The common conductor shall be grounded.

Grounding electrode conductor on premises

Multiphase systems (delta) requiring one phase grounded.

NEC 250-25 (4) One phase conductor shall be grounded.

Grounding Electrode Conductors on Premises

Multiphase systems (delta) from which a single-phase, three-wire system is derived

NEC 250-25 (5) The neutral conductor shall be grounded.

Grounding Electrode Conductor on Premises

Utility supply system ground at pole or pad-mount transformer

Utility supply system ground at pole or pad-mount transformer

From 50 to 1000 volts: [250-5 (b)]
1. Where system can be grounded so that maximum voltage to ground does not exceed 150 volts;
2. Where neutral is used as a circuit conductor on a system nominally rated 480/277-volt wye, 3-phase, 4-wire;
3. Where midpoint of one phase is used as a circuit conductor on a system nominally rated 240/120-volt, 3-phase, 4-wire;
4. Where a service conductor is uninsulated as permitted in 230-22, 230-30 and 230-41. (250-5) See Ex 1 thru 4 of 250-5 (b-4).

Alternating-current circuits and systems that are required to be grounded (250-5).
Less than 50 volts: [250-5 (a)]
1. Where supplied by a transformer whose primary exceeds 150 volts to ground;
2. Where supplied by a transformer whose primary is ungrounded;
3. Where installed as overhead conductors outside of buildings.

Exceptions to grounding AC systems of 50 to 1000 volts which supply premises wiring and premises wiring systems (250-5 (b) Ex 1, 2, 3 & 4)
Exception 1. Used only to supply industrial electric furnaces for melting, refining, tempering, and the like.
Exception 2. Separately derived systems used only for rectifiers which supply only adjustable speed industrial drives.
Exception 3. Separately derived systems fed by transformers with primary of less than 1000 volts and where all of the

following are met:
  a. System used for control circuits only.
  b. Only qualified persons will service the installation.
  c. Require continuity of control power.
  d. Control system has ground detectors.
Exception 4. Isolated systems as permitted in Article 517— Health Care Facilities.
Exception 5. For high impedance grounded neutral systems see 250-5(b) Ex. 5, (a, b, c, d, e)

THE CONDUCTOR TO BE GROUNDED FOR ALTERNATING CURRENT SYSTEMS (250-25)

Blueprint Reading

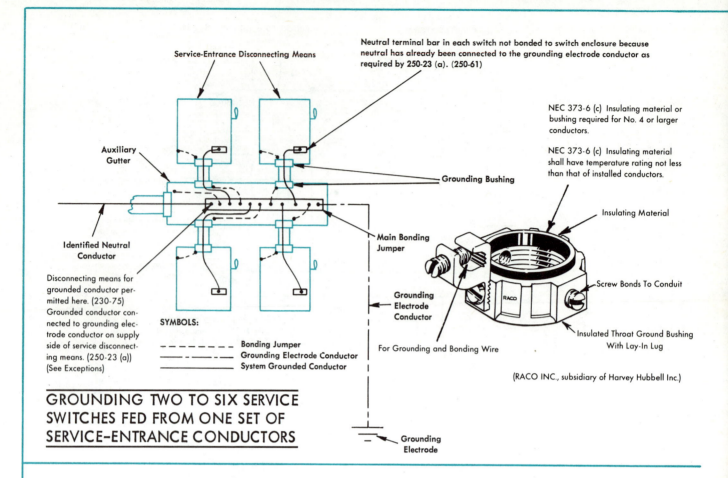

Service-Entrance Disconnecting Means

Neutral terminal bar in each switch not bonded to switch enclosure because neutral has already been connected to the grounding electrode conductor as required by 250-23 (a). (250-61)

NEC 373-6 (c) Insulating material or bushing required for No. 4 or larger conductors.

NEC 373-6 (c) Insulating material shall have temperature rating not less than that of installed conductors.

Auxiliary Gutter

Grounding Bushing

Identified Neutral Conductor

Insulating Material

Main Bonding Jumper

Disconnecting means for grounded conductor permitted here. (230-75) Grounded conductor connected to grounding electrode conductor on supply side of service disconnecting means. (250-23 (a)) (See Exceptions)

SYMBOLS:

– – – – – – – Bonding Jumper
— · — · — · Grounding Electrode Conductor
———— System Grounded Conductor

Grounding Electrode Conductor

Screw Bonds To Conduit

For Grounding and Bonding Wire

Insulated Throat Ground Bushing With Lay-In Lug

(RACO INC., subsidiary of Harvey Hubbell Inc.)

Grounding Electrode

GROUNDING TWO TO SIX SERVICE SWITCHES FED FROM ONE SET OF SERVICE–ENTRANCE CONDUCTORS

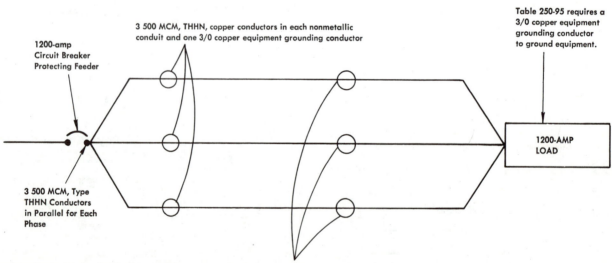

1200-amp Circuit Breaker Protecting Feeder

3 500 MCM, THHN, copper conductors in each nonmetallic conduit and one 3/0 copper equipment grounding conductor

Table 250-95 requires a 3/0 copper equipment grounding conductor to ground equipment.

1200-AMP LOAD

3 500 MCM, Type THHN Conductors in Parallel for Each Phase

EACH NONMETALLIC CONDUIT MUST CONTAIN A 3/0 COPPER EQUIPMENT GROUNDING CONDUCTOR.

NEC 250-95 requires that where conductors are run in parallel as permitted by 310-4 the equipment grounding conductor shall also be run in parallel. Each parallel equipment grounding conductor shall be sized in relation to the size of overcurrent device protecting the circuit conductor. Example: In the above diagram each of the 3-inch, nonmetallic conduits contain three 500,000 cir mil, THHN copper conductors. The three conductors are connected in parallel and are protected by a 1200-amp circuit breaker. Table 250-95 requires a 3/0 copper equipment grounding conductor and 250-95 requires that these three 3/0 conductors must be installed in each 3-inch conduit and connected in parallel. If all nine 500 MCM conductors were installed in one raceway, one 3/0 copper equipment grounding conductor would satisfy the requirements of 250-95.

EQUIPMENT GROUNDING CONDUCTORS ON LOAD SIDE OF SERVICE (NEC 250-95)

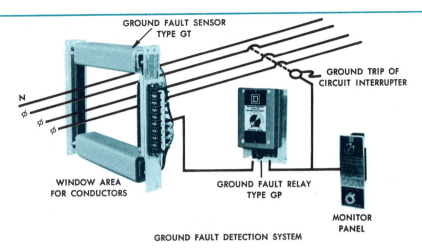

GROUND FAULT SENSOR
TYPE GT

GROUND TRIP OF
CIRCUIT INTERRUPTER

WINDOW AREA
FOR CONDUCTORS

GROUND FAULT RELAY
TYPE GP

MONITOR
PANEL

GROUND FAULT DETECTION SYSTEM

APPLICATION — CIRCUIT BREAKER

The ground current powered ground-fault detection system is designed for use on power systems which include a grounded conductor (neutral or grounded phase). When circuit conditions are normal, currents from all the phase conductors and neutral (if used) add up to zero, and the sensor produces no signal. If a ground fault occurs, the currents add up to equal the ground-fault current, and the sensor provides a signal proportional to the ground-fault current. This signal powers the ground-fault relay which activates the ground trip mechanism on the circuit breaker.

APPLICATION – SWITCH AND FUSE

A typical ground-fault protective system as applied to a manually operated shunt trip BOLT-LOC bolted pressure contact switch is shown below.

Ground-fault protection of equipment required for grounded wye services of more than 150 volts to ground but not more than 600 volts phase-to-phase for each service disconnect rated 1000 amperes or more. (230-95)

For switch and fuse combination the fuses shall be capable of interrupting any current higher than the interrupting capacity of the switch during a time when the ground-fault protective system will not cause the switch to open. (230-95 (b))

Operation of ground-fault protection shall cause service disconnect to open all ungrounded wires of faulted circuit. Ground-fault protection shall have maximum setting of 1200 amps and a maximum time delay of one second for ground-fault currents of 3000 amps or more. (230-95 (a))

Overcurrent devices shall be selected and coordinated so as to clear a fault without extensive damage to electrical components of the circuit. (110-10)

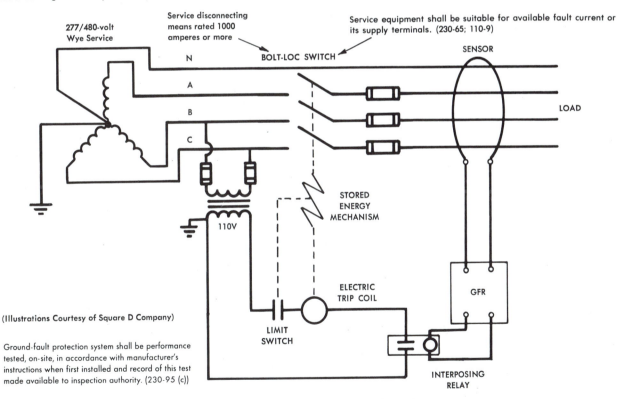

277/480-volt
Wye Service

Service disconnecting means rated 1000 amperes or more

Service equipment shall be suitable for available fault current or its supply terminals. (230-65; 110-9)

BOLT-LOC SWITCH

SENSOR

N
A
B
C

LOAD

STORED
ENERGY
MECHANISM

110V

ELECTRIC
TRIP COIL

GFR

LIMIT
SWITCH

INTERPOSING
RELAY

(Illustrations Courtesy of Square D Company)

Ground-fault protection system shall be performance tested, on-site, in accordance with manufacturer's instructions when first installed and record of this test made available to inspection authority. (230-95 (c))

DESCRIPTION OF OPERATION

Under normal operation there is no output from the ground-fault sensor and therefore the GFR contact is open. If a ground-fault current of sufficient magnitude occurs the GFR contact will close to energize the interposing relay which in turn will energize the electric trip coil of the BOLT-LOC switch to open the circuit.

GROUND FAULT PROTECTION OF EQUIPMENT (NEC 230-95)

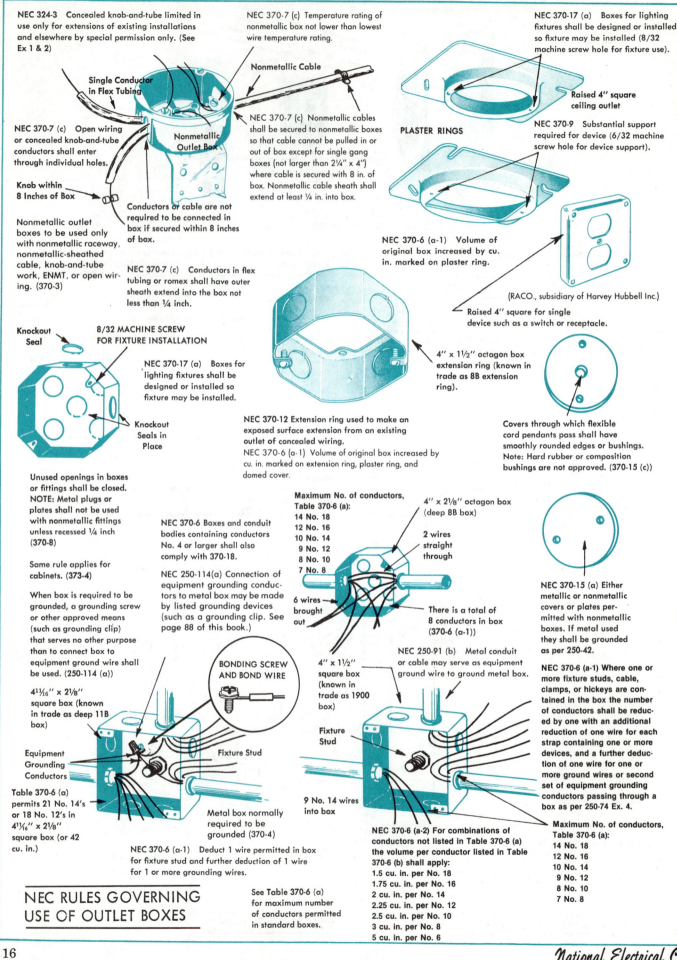

NEC 324-3 Concealed knob-and-tube limited in use only for extensions of existing installations and elsewhere by special permission only. (See Ex 1 & 2)

Single Conductor in Flex Tubing

Nonmetallic Cable

NEC 370-7 (c) Temperature rating of nonmetallic box not lower than lowest wire temperature rating.

NEC 370-7 (c) Open wiring or concealed knob-and-tube conductors shall enter through individual holes.

Nonmetallic Outlet Box

Knob within 8 Inches of Box

Nonmetallic outlet boxes to be used only with nonmetallic raceway, nonmetallic-sheathed cable, knob-and-tube work, ENMT, or open wiring. (370-3)

Conductors or cable are not required to be connected in box if secured within 8 inches of box.

NEC 370-7 (c) Nonmetallic cables shall be secured to nonmetallic boxes so that cable cannot be pulled in or out of box except for single gang boxes (not larger than 2¼" x 4") where cable is secured with 8 in. of box. Nonmetallic cable sheath shall extend at least ¼ in. into box.

NEC 370-7 (c) Conductors in flex tubing or romex shall have outer sheath extend into the box not less than ¼ inch.

PLASTER RINGS

NEC 370-17 (a) Boxes for lighting fixtures shall be designed or installed so fixture may be installed (8/32 machine screw hole for fixture use).

Raised 4" square ceiling outlet

NEC 370-9 Substantial support required for device (6/32 machine screw hole for device support).

NEC 370-6 (a-1) Volume of original box increased by cu. in. marked on plaster ring.

(RACO., subsidiary of Harvey Hubbell Inc.)

Raised 4" square for single device such as a switch or receptacle.

Knockout Seal

8/32 MACHINE SCREW FOR FIXTURE INSTALLATION

NEC 370-17 (a) Boxes for lighting fixtures shall be designed or installed so fixture may be installed.

Knockout Seals in Place

4" x 1½" octagon box extension ring (known in trade as 8B extension ring).

NEC 370-12 Extension ring used to make an exposed surface extension from an existing outlet of concealed wiring.
NEC 370-6 (a-1) Volume of original box increased by cu. in. marked on extension ring, plaster ring, and domed cover.

Covers through which flexible cord pendants pass shall have smoothly rounded edges or bushings. Note: Hard rubber or composition bushings are not approved. (370-15 (c))

Unused openings in boxes or fittings shall be closed. NOTE: Metal plugs or plates shall not be used with nonmetallic fittings unless recessed ¼ inch (370-8)

Same rule applies for cabinets. (373-4)

When box is required to be grounded, a grounding screw or other approved means (such as grounding clip) that serves no other purpose than to connect box to equipment ground wire shall be used. (250-114 (a))

NEC 370-6 Boxes and conduit bodies containing conductors No. 4 or larger shall also comply with 370-18.

NEC 250-114(a) Connection of equipment grounding conductors to metal box may be made by listed grounding devices (such as a grounding clip. See page 88 of this book.)

Maximum No. of conductors, Table 370-6 (a):
14 No. 18
12 No. 16
10 No. 14
9 No. 12
8 No. 10
7 No. 8

4" x 2⅛" octagon box (deep 8B box)

2 wires straight through

6 wires brought out

There is a total of 8 conductors in box (370-6 (a-1))

NEC 370-15 (a) Either metallic or nonmetallic covers or plates permitted with nonmetallic boxes. If metal used they shall be grounded as per 250-42.

NEC 370-6 (a-1) Where one or more fixture studs, cable, clamps, or hickeys are contained in the box the number of conductors shall be reduced by one with an additional reduction of one wire for each strap containing one or more devices, and a further deduction of one wire for one or more ground wires or second set of equipment grounding conductors passing through a box as per 250-74 Ex. 4.

BONDING SCREW AND BOND WIRE

4¹¹⁄₁₆" x 2⅛" square box (known in trade as deep 11B box)

Equipment Grounding Conductors

Table 370-6 (a) permits 21 No. 14's or 18 No. 12's in 4¹¹⁄₁₆" x 2⅛" square box (or 42 cu. in.)

Fixture Stud

NEC 250-91 (b) Metal conduit or cable may serve as equipment ground wire to ground metal box.

4" x 1½" square box (known in trade as 1900 box)

Fixture Stud

9 No. 14 wires into box

Metal box normally required to be grounded (370-4)

NEC 370-6 (a-1) Deduct 1 wire permitted in box for fixture stud and further deduction of 1 wire for 1 or more grounding wires.

NEC 370-6 (a-2) For combinations of conductors not listed in Table 370-6 (a) the volume per conductor listed in Table 370-6 (b) shall apply:
1.5 cu. in. per No. 18
1.75 cu. in. per No. 16
2 cu. in. per No. 14
2.25 cu. in. per No. 12
2.5 cu. in. per No. 10
3 cu. in. per No. 8
5 cu. in. per No. 6

Maximum No. of conductors, Table 370-6 (a):
14 No. 18
12 No. 16
10 No. 14
9 No. 12
8 No. 10
7 No. 8

NEC RULES GOVERNING USE OF OUTLET BOXES

See Table 370-6 (a) for maximum number of conductors permitted in standard boxes.

Round Box

3½ in. ceiling pan, ¾ in. deep with cable clamps – normally 5 cu. in. capacity

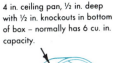

4 in. ceiling pan, ½ in. deep with ½ in. knockouts in bottom of box – normally has 6 cu. in. capacity.

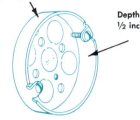

Depth ½ inch

Shallow Box (in trade known as Ceiling Pan)

Wood brace shall have a nominal 1 inch by 2 inch thickness. (370-13(b)(2))

Boxes shall be supported from a structural member either directly or by using on approved metallic or wooden brace. 370-13(b)

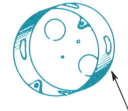

Conduit may not be fastened to side of round boxes. (370-2)

No box shall be less than ½-inch deep. Where box is intended to contain flush device it shall be at least 1⁵⁄₁₆ inch deep. (370-14) (Intent of this rule is to permit ½-inch deep box for use with a fixture with a canopy which gives added depth.

NEC 370 INSTALLATION RULES FOR OUTLET, SWITCH, AND JUNCTION BOXES

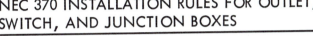

Wooden plugs driven into plaster shall not be used to secure box. (110-13 (a))

Plastered Wall or Ceiling

Boxes shall be securely and rigidly fastened to the surface upon which they are mounted. [370-13 (a)]

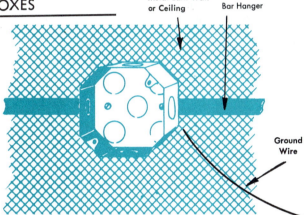

Metal Lath Wall or Ceiling

Bar Hanger

Ground Wire

When metal boxes are used with knob-and-tube or nonmetallic-sheathed cable and are mounted on or in contact with metal or metal lath they shall be grounded. Art. 250 (370-4)

Metal boxes located within 8 feet vertically or 5 feet horizontally of ground or grounded metal objects and subject to contact by persons, shall be grounded. (250-42 (a))

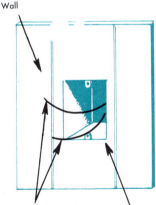

Wooden Wall

At least 6 inches of free conductor shall be left at each outlet and switch point, except for conductors pulled straight through box. (300-14)

Box in wall of combustible material must be flush with surface or project therefrom. (370-10), (373-3)

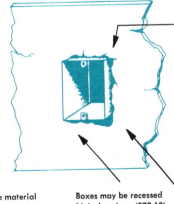

Plaster Wall

Boxes may be recessed ¼ inch or less. (370-10) Same rule applies to cabinets. (373-3)

A broken plaster surface shall be repaired. (370-11)

Wall or Ceiling

Blank Cover or Access Space

Junction boxes shall be accessible without removing any part of building. (370-19)

Boxes installed in areas above ceiling lift-out panels are considered accessible. (See Definitions of Accessible, Concealed and Exposed, as applied to wiring methods.)

Single-pole Breaker

NEC 240-81 Shall clearly indicate when open (off) or closed (on). On switchboards or panelboards where mounted vertically, the up position of handle shall be "on" position.

(Photos courtesy Square D Company)

Two-pole Breaker

Three-pole Breaker

NEC 380-8 (a) Circuit breakers used as switches shall be readily accessible and operating handle not over 6½ feet from floor or platform. (See Ex 1, 2, 3.)

NEC 380-3 Circuit breakers shall be externally operable and enclosed.

NEC 380-2 (b) Circuit breakers shall not open grounded conductor except: (1) where all conductors of circuit are disconnected simultaneously; (2) where so arranged that grounded conductors cannot be disconnected until all ungrounded conductors of circuit have been disconnected.

NEC 110-9, 230-65 Shall have interrupting capacity sufficient for available fault current.

NEC 110-10 Shall be selected and coordinated to clear without damage to electrical components of circuit.

NEC 240-82 Shall be designed so that any alteration of the trip point or time for operation will require dismantling or breaking of seal.

NEC 240-83 MARKING – (a) Shall be marked with rating and be visible after installation. Shall be visible after removal of trim or cover. (b) If 100 amps or less and 600 volts or less, rating shall be marked into handles or escutcheon areas. (c) If other than 5000 amps rating shall be on breaker. Exception: rating not required if used for supplementary protection. (d) If used to switch 120-volt or 277-volt fluorescent fixtures shall be approved for such switching duty and marked "SWD"

NEC 240-6 Standard ratings for inverse time circuit breakers shall be:
15, 20, 25, 30, 35, 40, 45, 50, 60, 70, 80, 90, 100, 110, 125, 150, 175, 200, 225, 250, 300, 350, 400, 450, 500, 600, 700, 800, 1000, 1200, 1600, 2000, 2500, 3000, 4000, 5000, 6000 amps.

NEC 240-8 Shall not be connected in parallel, except if factory-assembled in parallel and approved as a unit.

NEC 110-9, 230-65. When checking interrupting rating of a circuit breaker, also check the voltage rating because generally the I.C. is lower as the voltage rating is higher.
i.e. at 240 volts—65,000 A.I.C.
at 480 volts—25,000 A.I.C.
at 600 volts—18,000 A.I.C.
If the foregoing is not evaluated, this can result in a dangerous misapplication of the breaker insofar as interrupting rating is concerned.

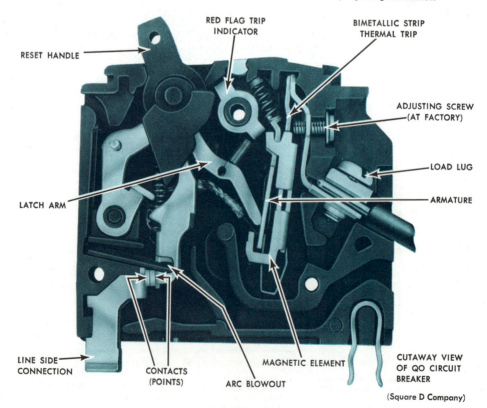

RED FLAG TRIP INDICATOR

BIMETALLIC STRIP THERMAL TRIP

RESET HANDLE

ADJUSTING SCREW (AT FACTORY)

LOAD LUG

ARMATURE

LATCH ARM

LINE SIDE CONNECTION

CONTACTS (POINTS)

ARC BLOWOUT

MAGNETIC ELEMENT

CUTAWAY VIEW OF QO CIRCUIT BREAKER

(Square D Company)

CIRCUIT BREAKERS

NEC ARTICLE 100—DEFINITION OF CIRCUIT BREAKER:
A device designed to open and close a circuit by nonautomatic means and to open the circuit automatically on a predetermined overcurrent without injury to itself when properly applied within its rating. (See Article 100 for definition of adjustable, instantaneous trip, inverse time, nonadjustable and setting.)

1-Pole
Qwik-Gard

2-pole
Qwik-Gard

(Square D Company)

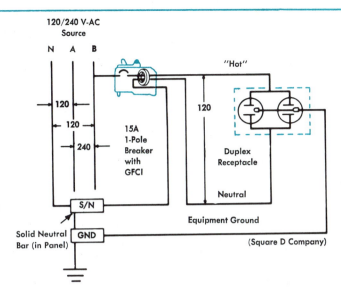

WIRING OF 1-POLE QWIK-GARD

(Square D Company)

Some of the areas where GFCI (Ground-Fault Circuit-Interrupters) are used in the NEC for the Protection of Personnel are listed below:

NEC 210-8(a) Dwelling Unit receptacles
  (a)(1) Bathrooms
  (a)(2) Garages
  (a)(3) Outdoors
  (a)(4) Basement
  (a)(5) Within 6 ft. of sink
  (2)(6) Boathouses

NEC 210-8(b) Bathrooms of hotels and motels

NEC 215-9 Feeders

NEC 305-6 Construction Sites

NEC 422-8(d)(3) Portable high-pressure spray washing machines.

NEC 426-31 EX. De-icing and Snow Melting Equip.

NEC 427-22 Pipelines and Vessels

NEC 517-90(c) Patient Care Areas

NEC 511-10 Commercial Repair Garages

NEC 550 Mobile Homes
  550-8(b)(e) Outdoors and bathrooms
  550-23(c) Outside receptacles

NEC 551 Recreational Vehicles
  551-8(e) Power supply
  551-9(c) Single phase receptacles inside unit
  551-22(b) Shower fixtures
  551-42 Site receptacles

NEC 555-3 Marinas and Boatyards

NEC 600-11 Outdoor Portable Signs

NEC 680 Swimming Pools

NEC 680-5 Transformers
NEC 680-6(a) Receptacles, (b) Receptacles, (c) Devices.
NEC 680-20 Underwater Lighting Fixtures
NEC 680-21 Junction Boxes or Enclosures
NEC 680-26(b) Electric Pool Covers
NEC 680-30 Storable Pool Pumps
NEC 680-40 Spas and Hot Tubs—Outdoors
NEC 680-41 Spas and Hot Tubs—Indoors
NEC 680-51(a) Fountains
NEC 680-62(a) Therapeutic Tubs in Health Care
NEC 680-70 Hydromassage Bathtubs

DEFINITION OF GROUND-FAULT CIRCUIT-INTERRUPTER (GFCI)
(See Definitions in NEC).
A device whose function is to de-energize a circuit within established time when current to ground exceeds a predetermined value which is less than that required to operate protective device of supply circuit.

GROUND FAULT CIRCUIT INTERRUPTERS (GFCI) FOR PERSONAL PROTECTION

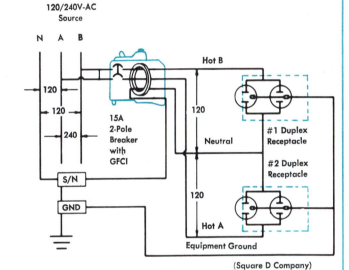

(Pass & Seymour)

All U.L. listed GFCI equipment manufactured after Jan. 1, 1976, is required to have a 5 ma (±1) fault current to ground-trip value.

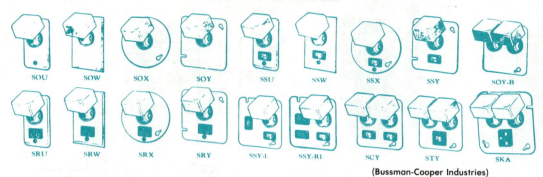

For Motors of 125 Volts or Less

SOU SOW SOX SOY SSU SSW SSX SSY SOY-B

SRU SRW SRX SRY SSY-L SSY-R.L SCY STY SKA

(Bussman-Cooper Industries)

Covers fit standard outlet or switch boxes. Units are UL listed and are made with Edison-base fuseholder. Units without switch for motor ¾ hp or smaller. Units with switch for AC motors ½ hp or smaller. Do not use on DC.

Shall be no exposed live parts after fuses or fuses and adapters are installed. (240-50 (d))

Edison-base fuseholders installed only where made to accept Type S fuses by use of adapters. (240-52)

Overcurrent devices required to be readily accessible. (240-24 (a))

FUSETRON BOX COVER UNITS

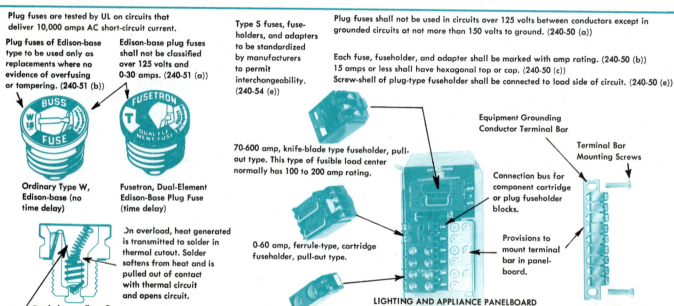

Plug fuses are tested by UL on circuits that deliver 10,000 amps AC short-circuit current.

Plug fuses of Edison-base type to be used only as replacements where no evidence of overfusing or tampering. (240-51 (b))

Edison-base plug fuses shall not be classified over 125 volts and 0-30 amps. (240-51 (a))

Ordinary Type W, Edison-base (no time delay)

Fusetron, Dual-Element Edison-Base Plug Fuse (time delay)

Dual-element Type S Fuse—Cut-Away View

On overload, heat generated is transmitted to solder in thermal cutout. Solder softens from heat and is pulled out of contact with thermal circuit and opens circuit.

On short circuit the fuse link blows and opens circuit.

Type S fuses shall be plug type and: (a) classified not over 125 volts and 0-15, 16-20, and 21-30 amps; (b) shall not be interchangeable with lower amp classification and designed to be used only with Type S fuseholder or Type S adapter inserted. (240-53 (a), (b))

Type S fuseholders and adapters designed to make tampering or shunting difficult. (240-54 (d))

Type S fuseholders and adapters to be designed only for Type S fuses. (240-54 (b))

Available in ratings of 3/10 to 30 amps.

Type S adapters to fit Edison-base fuseholders. (240-54 (a))

When once inserted Type S adapters to be nonremovable. (240-54 (c))

Dual-Element Type S Fuse

Adapter

(Illustrations, courtesy of Bussmann Mfg. Div., McGraw-Edison Co.)

Type S fuses, fuse-holders, and adapters to be standardized by manufacturers to permit interchangeability. (240-54 (e))

70-600 amp, knife-blade type fuseholder, pull-out type. This type of fusible load center normally has 100 to 200 amp rating.

0-60 amp, ferrule-type, cartridge fuseholder, pull-out type.

0-30 amp plug-type, single pole, twin plug fuseholder.

COMPONENTS FOR SERVICE PANELBOARD

Plug fuses shall not be used in circuits over 125 volts between conductors except in grounded circuits at not more than 150 volts to ground. (240-50 (a))

Each fuse, fuseholder, and adapter shall be marked with amp rating. (240-50 (b))
15 amps or less shall have hexagonal top or cap. (240-50 (c))
Screw-shell of plug-type fuseholder shall be connected to load side of circuit. (240-50 (e))

Equipment Grounding Conductor Terminal Bar

Terminal Bar Mounting Screws

Connection bus for component cartridge or plug fuseholder blocks.

Provisions to mount terminal bar in panel-board.

LIGHTING AND APPLIANCE PANELBOARD
(Enclosure and trim not shown)

Equipment Grounding Terminal Bar

See page 50 in this text for circuit breaker lighting and appliance panelboard.

(Square D Company)

Overcurrent devices shall be enclosed in cutout boxes or cabinets. (240-30)

Require ¼-inch air space between enclosure and damp or wet wall. (240-32)

Enclosures shall be mounted in vertical position except where impracticable and in compliance with 240-81. (240-33)

Classed as a lighting and appliance panelboard if 10% of its circuits are 30 amps or less and have provisions for neutral connections. (384-14)

General rule: Each lighting and appliance panelboard shall be protected at not more than the rating of the panelboard. (384-16 (a))

Not more than 42 overcurrent devices. A 2-pole device counts 2; a 3-pole device counts 3. (384-15)

PLUG FUSES, FUSES, FUSEHOLDERS, AND ADAPTERS (PART E, ARTICLE 240 NEC)

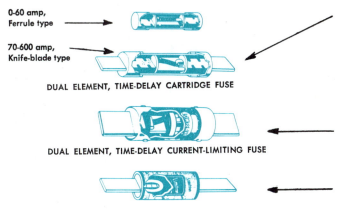

0-60 amp,
Ferrule type

70-600 amp,
Knife-blade type

DUAL ELEMENT, TIME-DELAY CARTRIDGE FUSE

DUAL ELEMENT, TIME-DELAY CURRENT-LIMITING FUSE

The dual element, time-delay fuse is designed to provide a time delay in the harmless low overload range and also to be fast-opening when exposed to dangerous short circuits. This fuse contains two fusible elements in series and the magnitude of a potentially damaging current flow determines which one of these elements opens. The thermal cutout element will open at currents of up to about 500% of fuse rating (overload range), whereas the short-circuit element will open at 500% or more of fuse rating.

Dual-element current-limiting fuses function similar to regular dual-element, time-delay fuses except that they are faster operating in short-circuit range, thus are more current-limiting. The short-circuit element is normally made of silver, with a quartz sand arc quenching filler.

Single-element current-limiting fuses function similar to regular dual-element, time-delay low-overload range. Has lowest let-thru values in comparision with other fuses. Used to protect circuit components having inadequate interrupting, bracing, or withstand rating, such as low I.C. breakers.

CURRENT-LIMITING FUSE

NEC 240-11 Current-limiting fuse when interrupting currents in its range, will reduce current in faulted circuit to magnitude substantially less than obtainable in same circuit if replaced with solid wire if same impedance.

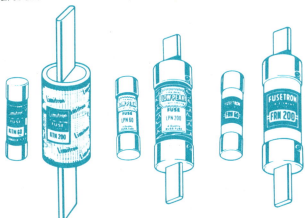

LIMITRON FAST-ACTING
FUSE, CLASS K1

LOW-LEAK DUAL ELE-
MENT FUSE, CLASS K5

FUSETRON DUAL ELE-
MENT FUSE, CLASS K5

Class K fuses have ratings of 0-600 amps in both 250 and 600 volts. Have dimensions same as Class H fuses. Grouped into three categories: K1, K5, and K9 and may have UL listing of 50,000, 100,000 or 200,000 amps symmetrical rms amps. UL has assigned a maximum level of peak-let-thru current (Ip) and amp-squared seconds (I²t) for each K rating. K rating determines degree of current limiting. K9 has least current-limiting ability and K1 the greatest. May also be listed by UL as time delay; then are required to have minimum time delay of 10 seconds at 500% of rated current.

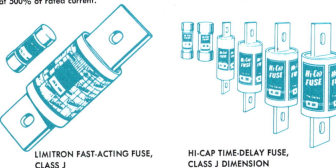

LIMITRON FAST-ACTING FUSE,
CLASS J

HI-CAP TIME-DELAY FUSE,
CLASS J DIMENSION

Class J fuses are 600-volt AC, 0-600 amps, current limiting and listed by UL with interrupting rating of 200,000 symmetrical rms amps. Fuseholders are smaller than Class H, thus are noninterchangeable.

OVERCURRENT PROTECTION-- CARTRIDGE FUSES

NEC 240-1 (Fine print note) Overcurrent protection for conductors and equipment is provided to open the circuit if the current reaches a value that will cause an excessive or dangerous temperature in conductors or conductor insulation. See 110-9, 110-10 for interrupting capacity and protection against fault currents.

NEC 110-9 Fuses shall have sufficient interrupting capacity for the voltage employed and the current that must be interrupted. (Also see 230-65.)

NEC 230-65 Service equipment shall have interrupting capacity suitable for available fault current.

NEC 110-10 Shall be selected and coordinated to clear fault without extensive damage to electrical components of circuit.

Standard ratings for fuses shall be: 15, 20, 25, 30, 35, 40, 45, 50, 60, 70, 80, 90, 100, 110, 125, 150, 175, 200, 225, 250, 300, 350, 400, 500, 600, 700, 800, 1000, 1200, 1600, 2000, 2500, 3000, 4000, 5000, 6000 amps. (240-6 Ex   Also 1, 3, 6, 10 and 601 amp sizes)

NEC 240-8   Shall not be connected in parallel. Exception: permitted if factory assembled.

NEC 240-20 Shall be connected in series with each ungrounded conductor.

NEC 240-21   Overcurrent device connected where conductor to be protected receives supply. (See Exceptions.)

NEC 240-22 Grounded conductor shall not be fused except for motors as per 430-36, 430-37.

NEC 250-24 Fuses shall be readily accessible except: (1) busways (364-12); (2) supplemental (240-10); (3) services (230-92)

NEC 240-30   Shall be enclosed in cabinets or cutout boxes. (See Exceptions 1-3).

NEC 240-32   In damp or wet locations shall be in enclosures identified for such use and ¼" air space between enclosure and mounting surface.

NEC 240-33 Enclosures shall be mounted in vertical position except where impracticable and in compliance with 240-81.

NEC 240-40 Disconnect required on supply side of cartridge fuses where accessible to other than qualified persons except: (1) service as per 230-82; (2) group operation of motors as per 430-112 and fixed space heating as per 424-22.

(Illustrations courtesy of Bussman Mfg. Division, McGraw-Edison Co.)

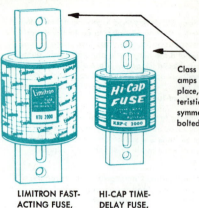

LIMITRON FAST-ACTING FUSE, CLASS L

HI-CAP TIME-DELAY FUSE, CLASS L

Class L listed by UL in sizes 601-6000 amps at 600 volts AC. They bolt into place, have a current-limiting characteristic and interrupting rating of 200,000 symmetrical rms amps and are used in bolted pressure contact switches.

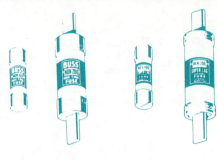

One-time Fuse

Renewable Fuse

CLASS H DIMENSION FUSES

Class H fuses are probably oldest and best known. In the past referred to as "NEC" or "Code" fuses. Interrupting rating not marked on label. Tested by UL on circuits of 10,000 amps. Rated 0-600 amps in both 250- and 600-volt range.

NEC 240-60 Cartridge Fuses and Fuseholders:
(a) 300-volt cartridge fuses and fuseholders not permitted where over 300 volts between conductors except on circuits of not over 300 volts to ground, i.e., 277/480-volt systems.

(b) Cartridge fuseholders (0-6000 amps) shall be so designed to make it difficult to put fuse of given class into fuseholder for current lower or voltage higher than intended. Current-limiting fuseholders shall permit insertion only of current-limiting fuses; (c) Fuses shall be marked to show: (1) amp rating, (2) volt rating, (3) interrupting rating if other than 10,000 amps, (4) current-limiting where applicable except where used as supplementary protection, (5) trademark of manufacturer.

NEC 250-100 Feeders over 600 volts shall have short-circuit protective device or comply with 230-208(d)(2) or (d)(3) and be able to detect and interrupt currents which can occur in excess of fuse setting or melting point. Fuse rating in continous amps shall not exceed 3 times the amp rating of conductor.

NEC 240-101 Fuses on branch circuits of over 600 volts shall be able to detect and interrupt available fault currents.

NEC 240-61 Cartridge fuses and fuseholders classified according to their amperage and voltage ranges.

Some common classifications for not over 600 volts —

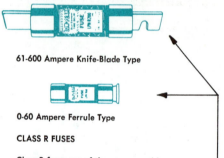

61-600 Ampere Knife-Blade Type

0-60 Ampere Ferrule Type

CLASS R FUSES

Class R fuses are of the nonrenewable, cartridge type and are the most recently listed fuse in the standards of UL. This fuse is the same size as Class H and K fuse except in 0-60 amps one end-cap ferrule will have an annular grooved ring and in 61-6000 amps one knife blade will have a slot in its side. Fuseholders to be designed to accept only Class R fuses, but rejection features in fuse would not prohibit the use of Class R fuse in a Class H or K fuseholder. Fuse rated to interrupt fault current up to 200,000 rms amps. If Class R to have UL label as time-delay, it would basically have a minimum of 10 seconds delay at 5 times fuse rating.

Amperes		
0-30	401-600	2001-2500
31-60	601-800	2501-3000
61-100	801-1200	3001-4000
101-200	1201-1600	4001-5000
201-400	1601-2000	5001-6000

Fuses of 600 volts or less permitted for use at voltages below their voltage ratings. (See 240-6 for standard ampere ratings of inverse time circuit breakers and fuses.)

NEC 240-3 Conductors shall be protected at ampacities as per Tables 310-16 thru 310-19 except where ampacity of conductor and standard rating of fuse do not correspond – then next larger fuse may be used only up to 800 amps. (Also see Ex 2 through 8).

NEC 240-2 Equipment shall be protected against overcurrent. (See 240-2 for list of equipment.)

Class G are available in sizes 0-60 amps for circuits of 300 volts or less to ground (i.e., 277/480-volt systems). They are a small dimension cartridge fuse (commonly referred to as Type SC fuses) and basically Type S cartridge fuses because they are noninterchangeable in that a 15-amp fuse will fit only into a 15-amp fuseholder, and a 20-amp fuse will fit only into a 20-amp fuseholder. They are available in sizes 15, 20, 30 and 60 amps. They are listed by UL at 100,000 symmetrical rms amps interrupting capacity.

OVERCURRENT PROTECTION-- CARTRIDGE FUSES

BUSS TYPE SC FUSES, CLASS G

(Illustrations, courtesy of Bussman, Cooper Industries)

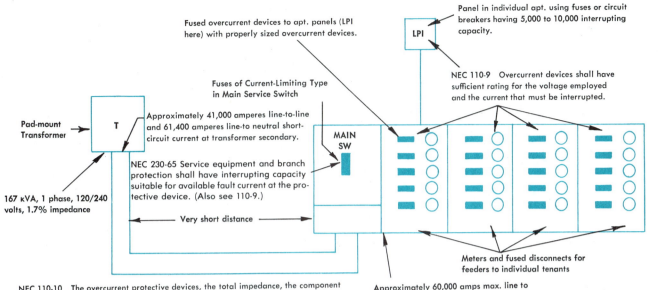

Fused overcurrent devices to apt. panels (LPI here) with properly sized overcurrent devices.

Panel in individual apt. using fuses or circuit breakers having 5,000 to 10,000 interrupting capacity.

LPI

Fuses of Current-Limiting Type in Main Service Switch

NEC 110-9 Overcurrent devices shall have sufficient rating for the voltage employed and the current that must be interrupted.

Pad-mount Transformer → T

Approximately 41,000 amperes line-to-line and 61,400 amperes line-to neutral short-circuit current at transformer secondary.

MAIN SW

NEC 230-65 Service equipment and branch protection shall have interrupting capacity suitable for available fault current at the protective device. (Also see 110-9.)

167 kVA, 1 phase, 120/240 volts, 1.7% impedance

Very short distance

Meters and fused disconnects for feeders to individual tenants

Approximately 60,000 amps max. line to neutral available short-circuit current

NEC 110-10   The overcurrent protective devices, the total impedance, the component short-circuit withstand ratings, and other characteristics of circuit to be protected shall be selected and coordinated to permit protective devices to clear fault without extensive damage to electrical components of circuit. In this diagram, the properly selected current-limiting fuses will protect the individual breakers in the panels rated 5,000 or 10,000 A.I.C.

THE ABOVE INSTALLATION MEETS REQUIREMENTS OF NEC 110-9, 110-10 AND 230-65.

Simple formula for calculating the short-circuit currents for above transformer:
Step 1. Calculate transformer full load current (single-phase):

$$I = \frac{kVA \times 1000}{E} = \frac{167 \times 1000}{240} = 696 \text{ amperes (FLA)}$$

Step 2. Find multiplier:

$$\frac{100}{\text{Trans \% Impedance}} = \frac{100}{1.7} = 58.8$$

Step 3. Find short-circuit current at transformer secondary:
Trans FLA × multiplier = 696 × 58.8 = 40,941 amperes short-circuit L-L (line-to-line).

Note 1. The L-N (line-to-neutral) fault current at transformer (single-phase center-topped) secondary is approximately 1.5 × L-L value, i.e., 40,941 × 1.5 = 61,411 amperes L-N. (Multiplier 1.5 varies from 1.33 to 1.67.)

Note 2. For three-phase systems, use the following formula in Step 1, then proceed with Steps 2 and 3.

$$I = \frac{kVA \times 1000}{E \times 1.73}$$

When a fault on an individual branch circuit not only opens the branch-circuit overcurrent device but also opens the feeder overcurrent device, it is referred to as "non-selective coordination." The NEC recognizes this problem and gives general coverage in NEC 110-10 and FPN (fine print note) of 517-65 recommends that an internal failure in one branch will not interrupt the supply to other branches. A selectively coordinated system is basically one where a fault on a branch circuit or feeder occurs and only the overcurrent device immediately ahead (upstream) from the fault opens, thereby a fault on a branch circuit would open only the branch-circuit device and not affect the feeder or service overcurrent device.

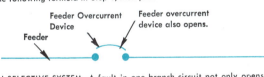

Feeder Overcurrent Device

Feeder overcurrent device also opens.

Feeder

Branch-Circuit Overcurrent Device

Branch-circuit overcurrent device opens.

Short on Branch Circuit

Branch Circuits

NON-SELECTIVE SYSTEM—A fault in one branch circuit not only opens the branch-circuit device but also opens the feeder overcurrent device.

NEC 517-14 (b) (c) In Health Care Facilities where ground-fault protection is provided for services and feeders, they shall be selective, coordinated, and performance tested when first installed.

NEC 240-12 (FPN) Coordination is properly localizing a fault condition to restrict outages to affected equipment by using proper selective fault-protective devices.

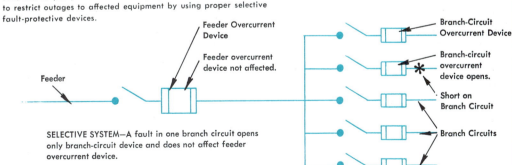

Feeder Overcurrent Device

Feeder overcurrent device not affected.

Feeder

Branch-Circuit Overcurrent Device

Branch-circuit overcurrent device opens.

Short on Branch Circuit

Branch Circuits

SELECTIVE SYSTEM—A fault in one branch circuit opens only branch-circuit device and does not affect feeder overcurrent device.

OVERCURRENT DEVICES AND AVAILABLE FAULT CURRENT

The 30A, 300V T-Tron fuse is no bigger than a nickel (interrupting rating is 200,000 Amps).

200 ampere Load Center with pull-out disconnect; Class T fuses bolted into panel board.

JJN (300 Volt) T-Tron fuses (all case sizes).

JJS (600 Volt) T-Tron fuses (all case sizes).

CLASS T FUSES

Class T fuses are listed by UL in sizes 1-600 amps in both 300 and 600 volt AC range. They are fast-acting with a high degree of current limitation and a high interrupting rating (200,000 amps). They provide overload and high performance short-circuit protection of main, feeder, and branch circuits, as well as general-purpose equipment. Class T fuses are the smallest and most compact of all UL-listed power distribution fuse devices. They comply with NEC 110-9 "Interrupting Rating" and 230-65 "available short-circuit current," which requires service equipment to be suitable for the short-circuit current available at its supply terminals. In addition, Class T fuses, when properly selected, can limit the short-circuit current for the protection of low interrupting capacity (IC) circuit breakers. (In a 10,000 amp, IC branch-circuit, the circuit breakers are protected by Class T fuses in the mains.)

Cable Limiters Available in Many Types of Terminals to Take Either Copper or Aluminum Cable.

NEC 240-11 Basic definition of current limiting device: an overcurrent device that when interrupting currents in current-limiting range, will reduce current in faulted circuit to a value less than obtainable in same circuit if replaced with a solid conductor having comparable impedance.

NEC 240-40 On circuits over 150 volts to ground, all fuses or thermal cutouts and all cartridge fuses of any voltage require disconnect means on supply side. Ex 1 Current limiters permitted on supply side of service disconnect as per 230-82.

NEC 230-82 (Ex 1) Equipment not permitted on supply side of service disconnect except cable limiters or other current-limiting devices.

NEC 230-91(a) Service overcurrent device permitted to be located immediately adjacent to the service disconnecting means.

NEC 240-60 (b) Current-limiting fuseholders shall only accept fuses that are current limiting.

Typical Application for Cable Limiters

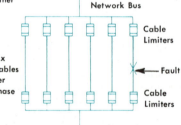
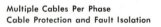

To Main Switchboard

Multiple Cables Per Phase
Cable Protection and Fault Isolation

Isolating faulted cables in multiple cable runs require cable limiters at each end of each cable.

Faulted cable would be isolated by the two cable limiters on each end of faulted cable.

Continuity of service would be maintained by the five remaining cables.

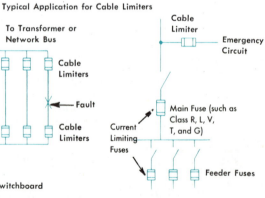

Emergency Circuit Taps
Ahead of Mains

Unprotected tapped conductors can result in violent conductor vaporization under short-circuit conditions initiating further switchboard short-circuit hazards. The protection of emergency circuit taps ahead of mains can be protected against short circuits by current-limiting cable limiters. The various types of cable limiter terminations makes adaptation to service switchboard construction relatively simple.

CABLE LIMITERS

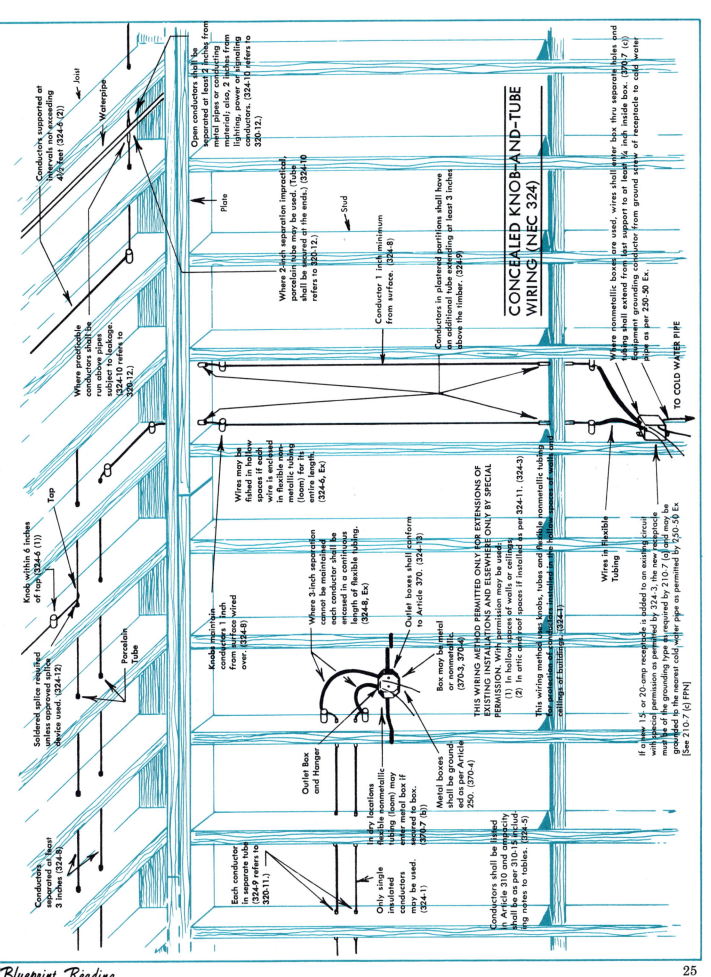

Conductors supported at intervals not exceeding 4½ feet (324-6 (2))

Joist

Waterpipe

Where practicable conductors shall be run above pipes subject to leakage. (324-10 refers to 320-12.)

Open conductors shall be separated at least 2 inches from metal pipes or conducting material; also, 2 inches from lighting, power or signaling conductors. (324-10 refers to 320-12.)

Plate

Where 2-inch separation impractical, porcelain tube may be used. (Tube shall be secured at the ends.) (324-10 refers to 320-12.)

Stud

Conductor 1 inch minimum from surface. (324-8)

Conductors in plastered partitions shall have an additional tube extending at least 3 inches above the timber. (324-9)

Where nonmetallic boxes are used, wires shall enter box thru separate holes and tubing shall extend from last support to at least ¼ inch inside box. (370-7 (c)) Equipment grounding conductor from ground screw of receptacle to cold water pipe as per 250-50 Ex.

CONCEALED KNOB-AND-TUBE WIRING (NEC 324)

TO COLD WATER PIPE

Knob within 6 inches of tap (324-6 (1))

Tap

Soldered splice required unless approved splice device used. (324-12)

Porcelain Tube

Conductors separated at least 3 inches (324-8)

Wires may be fished in hollow spaces if each wire is enclosed in flexible non-metallic tubing (loom) for its entire length. (324-6, Ex)

Knobs maintain conductors 1 inch from surface wired over. (324-8)

Where 3-inch separation cannot be maintained each conductor shall be encased in a continuous length of flexible tubing. (324-8, Ex)

Outlet boxes shall conform to Article 370. (324-13)

Box may be metal or nonmetallic. (370-3, 370-4)

Outlet Box and Hanger

Metal boxes shall be grounded as per Article 250. (370-4)

In dry locations flexible nonmetallic tubing (loom) may enter metal box if secured to box. (370-7 (b))

Only single insulated conductors may be used. (324-1)

Each conductor in separate tube (324-9 refers to 320-11.)

Conductors shall be listed in Article 310 and ampacity shall be as per 310-15 including notes to tables. (324-5)

THIS WIRING METHOD PERMITTED ONLY FOR EXTENSIONS OF EXISTING INSTALLATIONS AND ELSEWHERE ONLY BY SPECIAL PERMISSION. With permission may be used:
(1) In hollow spaces of walls or ceilings.
(2) In attic and roof spaces if installed as per 324-11. (324-3)

This wiring method uses knobs, tubes and flexible nonmetallic tubing for protection of conductors installed in the hollow spaces of walls and ceilings of buildings. (324-1)

Wires in Flexible Tubing

If a new 15- or 20-amp receptacle is added to an existing circuit with special permission as permitted by 324-3, the new receptacle must be of the grounding type as required by 210-7 (a) and may be grounded to the nearest cold water pipe as permitted by 250-50 Ex [See 210-7 (c) FPN]

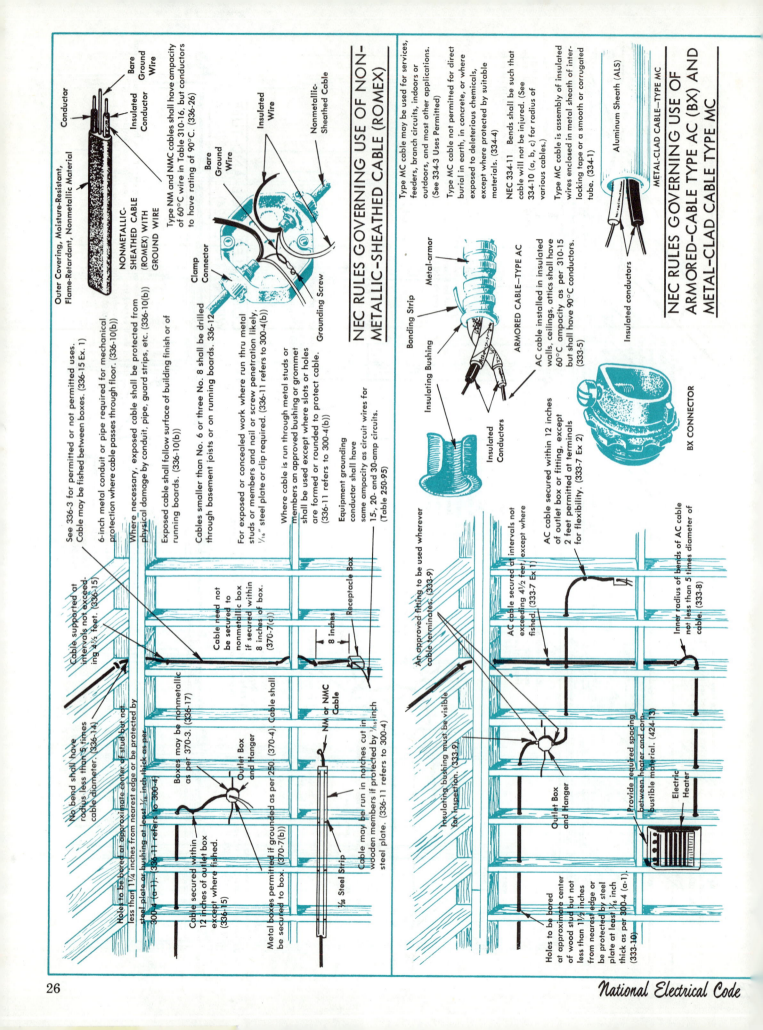

NEC RULES GOVERNING USE OF NON-METALLIC-SHEATHED CABLE (ROMEX)

NEC RULES GOVERNING USE OF ARMORED-CABLE TYPE AC (BX) AND METAL-CLAD CABLE TYPE MC

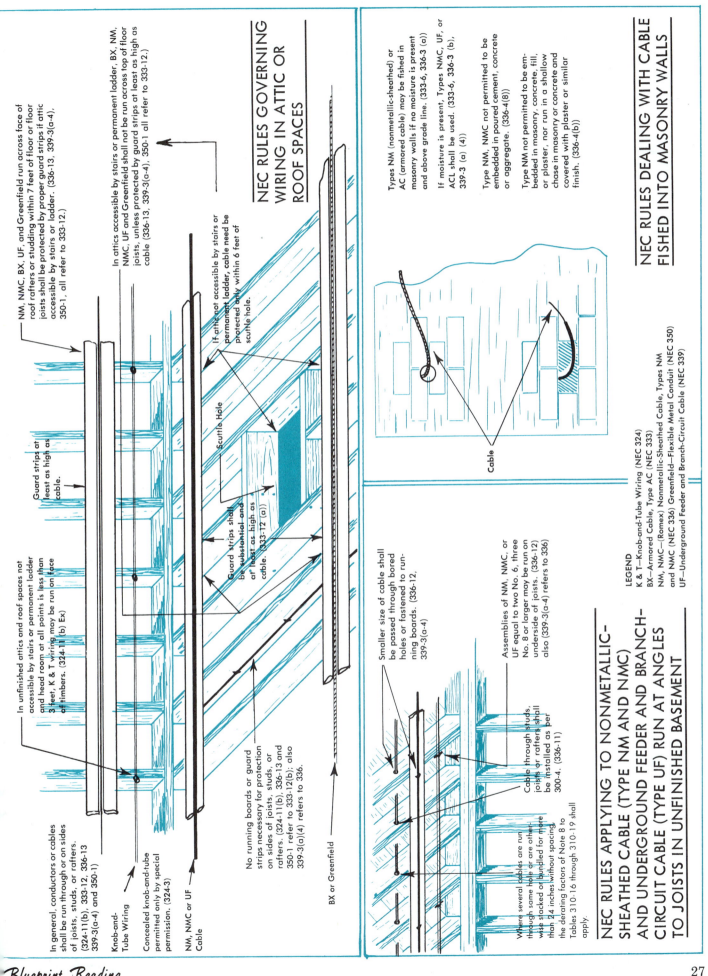

In general, conductors or cables shall be run through or on sides of joists, studs, or rafters. (324-11(b), 333-12, 336-13 339-3(a-4) and 350-1)

Knob-and-Tube Wiring

Concealed knob-and-tube permitted only by special permission. (324-3)

NM, NMC or UF Cable

In unfinished attics and roof spaces not accessible by stairs or permanent ladder and head room at all points is less than 3 feet, K & T wiring may be run on face of timbers. (324-11 (b) Ex)

NM, NMC, BX, UF, and Greenfield run across face of roof rafters or studding within 7 feet of floor or floor joists shall be protected by proper guard strips if attic accessible by stairs or ladder. (336-13, 339-3(a-4), 350-1, all refer to 333-12.)

In attics accessible by stairs or permanent ladder, BX, NM, NMC, UF and Greenfield shall not be run across top of floor joists, unless protected by guard strips at least as high as cable (336-13, 339-3(a-4), 350-1 all refer to 333-12.)

If attic not accessible by stairs or permanent ladder, cable need be protected only within 6 feet of scuttle hole.

Guard strips at least as high as cable.

Scuttle Hole

Guard strips shall be substantial and at least as high as cable. (333-12 (a))

NEC RULES GOVERNING WIRING IN ATTIC OR ROOF SPACES

Cable

NEC RULES DEALING WITH CABLE FISHED INTO MASONRY WALLS

Types NM (nonmetallic-sheathed) or AC (armored cable) may be fished in masonry walls if no moisture is present and above grade line. (333-6, 336-3 (a))

If moisture is present, Types NMC, UF, or ACL shall be used. (333-6, 336-3 (b), 339-3 (a) (4))

Type NM, NMC not permitted to be embedded in poured cement, concrete or aggregate. (336-4(8))

Type NM not permitted to be embedded in masonry, concrete, fill, or plaster, nor run in a shallow chase in masonry or concrete and covered with plaster or similar finish. (336-4 (b))

No running boards or guard strips necessary for protection on sides of joists, studs, or rafters. (324-11(b), 336-13 and 350-1 refer to 333-12(b); also 339-3(a)(4) refers to 336.

BX or Greenfield

Smaller size of cable shall be passed through bored holes or fastened to running boards. (336-12, 339-3(a-4)

Assemblies of NM, NMC, or UF equal to two No. 6, three No. 8 or larger may be run on underside of joists. (336-12) also (339-3(a-4) refers to 336)

Cable through studs, joists or rafters shall be installed as per 300-4. (336-11)

Where several cables are run through same hole or are otherwise stacked or bundled for more than 24 inches without spacing, the derating factors of Note 8 to Tables 310-16 through 310-19 shall apply.

NEC RULES APPLYING TO NONMETALLIC-SHEATHED CABLE (TYPE NM AND NMC) AND UNDERGROUND FEEDER AND BRANCH-CIRCUIT CABLE (TYPE UF) RUN AT ANGLES TO JOISTS IN UNFINISHED BASEMENT

LEGEND
K & T—Knob-and-Tube Wiring (NEC 324)
BX—Armored Cable, Type AC (NEC 333)
NM, NMC—(Romex) Nonmetallic-Sheathed Cable, Types NM and NMC (NEC 336) Greenfield—Flexible Metal Conduit (NEC 350) UF—Underground Feeder and Branch-Circuit Cable (NEC 339)

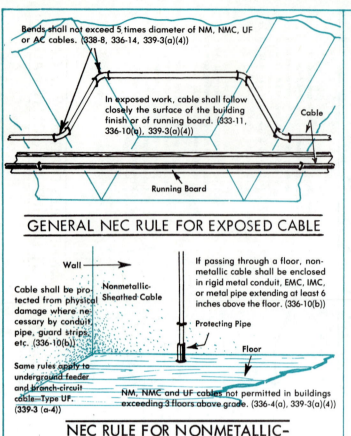

Bends shall not exceed 5 times diameter of NM, NMC, UF or AC cables. (338-8, 336-14, 339-3(a)(4))

In exposed work, cable shall follow closely the surface of the building finish or of running board. (333-11, 336-10(a), 339-3(a)(4))

Cable

Running Board

GENERAL NEC RULE FOR EXPOSED CABLE

Wall

Cable shall be protected from physical damage where necessary by conduit pipe, guard strips, etc. (336-10(b))

Same rules apply to underground feeder and branch-circuit cable—Type UF. (339-3 (a-4))

Nonmetallic-Sheathed Cable

If passing through a floor, nonmetallic cable shall be enclosed in rigid metal conduit, EMC, IMC, or metal pipe extending at least 6 inches above the floor. (336-10(b))

Protecting Pipe

Floor

NM, NMC and UF cables not permitted in buildings exceeding 3 floors above grade. (336-4(a), 339-3(a)(4))

NEC RULE FOR NONMETALLIC-SHEATED CABLE (ROMEX)

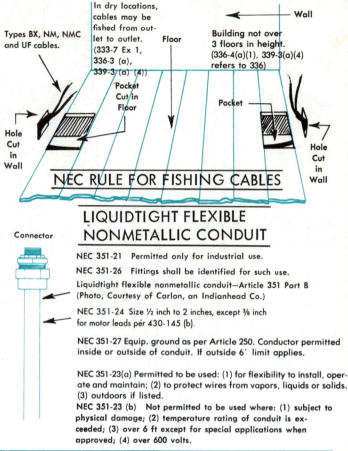

In dry locations, cables may be fished from outlet to outlet. (333-7 Ex 1, 336-3 (a), 339-3 (a) (4))

Floor

Types BX, NM, NMC and UF cables.

Building not over 3 floors in height. (336-4(a)(1), 339-3(a)(4) refers to 336)

Wall

Pocket Cut in Floor

Pocket

Hole Cut in Wall

Hole Cut in Wall

NEC RULE FOR FISHING CABLES

LIQUIDTIGHT FLEXIBLE NONMETALLIC CONDUIT

Connector

NEC 351-21   Permitted only for industrial use.

NEC 351-26   Fittings shall be identified for such use.

Liquidtight flexible nonmetallic conduit—Article 351 Part B (Photo, Courtesy of Carlon, an Indianhead Co.)

NEC 351-24 Size ½ inch to 2 inches, except ⅜ inch for motor leads per 430-145 (b).

NEC 351-27 Equip. ground as per Article 250. Conductor permitted inside or outside of conduit. If outside 6′ limit applies.

NEC 351-23(a) Permitted to be used: (1) for flexibility to install, operate and maintain; (2) to protect wires from vapors, liquids or solids. (3) outdoors if listed.

NEC 351-23 (b)   Not permitted to be used where: (1) subject to physical damage; (2) temperature rating of conduit is exceeded; (3) over 6 ft except for special applications when approved; (4) over 600 volts.

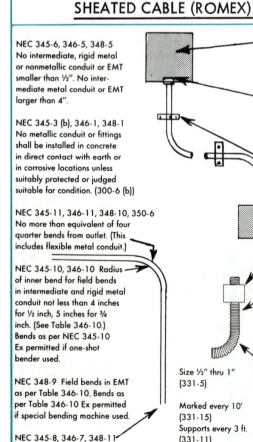

NEC 345-6, 346-5, 348-5 No intermediate, rigid metal or nonmetallic conduit or EMT smaller than ½". No intermediate metal conduit or EMT larger than 4".

NEC 345-3 (b), 346-1, 348-1 No metallic conduit or fittings shall be installed in concrete in direct contact with earth or in corrosive locations unless suitably protected or judged suitable for condition. (300-6 (b))

NEC 345-11, 346-11, 348-10, 350-6 No more than equivalent of four quarter bends from outlet. (This includes flexible metal conduit.)

NEC 345-10, 346-10 Radius of inner bend for field bends in intermediate and rigid metal conduit not less than 4 inches for ½ inch, 5 inches for ¾ inch. (See Table 346-10.) Bends as per NEC 345-10 Ex permitted if one-shot bender used.

NEC 348-9 Field bends in EMT as per Table 346-10. Bends as per Table 346-10 Ex permitted if special bending machine used.

NEC 345-8, 346-7, 348-11 Ends shall be reamed to remove rough edges.

NEC 345-15, 346-8, Bushing required where conduit enters box or fitting unless box or fitting affords conductor protection. See 373-6 (c).

NEC 250-76 Ex. (b)(c) Wherever over 250 volts to ground, requires double locknuts or fittings with shoulders that seat firmly against box.

NEC 345-12, 346-12, 348-12 Intermediate and rigid metal conduit and EMT shall be supported within 3 feet of outlet box, cabinet and fitting and at least every 10 feet of run.

Connector

Electrical Nonmetallic Tubing Made of nonmetallic corrugated, moisture chemical and flame-retardant material. (331-1)

Limit of 4, 90° Bends (331-10)

Size ½" thru 1" (331-5)

Marked every 10′ (331-15) Supports every 3 ft. (331-11)

Connector 349-18 approved fittings required.

Size ½" thru ¾" (See 349-10 Ex for ⅜") 349-4 Limited use and never over 6 ft. 349-16 Grounding per 250-91 (b) Ex.

Flexible Metallic Tubing 349-1 Similar to sealtight without nonmetallic jacket.

Bends per Table 349-20 (a) (b)

NEC 350-2   Flexible metal conduit shall not be used: (1) In wet location unless lead-covered or other special conductors used: (2) In hoistways except as per 620-21; (3) In storage battery rooms; (4) In hazardous area except as per 501-4 (b); (5) Rubber-covered wire used in areas where gas, oil, etc., will have deteriorating effect on rubber; (6) Underground or in concrete or aggregate.

NEC 230-43(13) six ft. of greenfield permitted for service if bonded as per 250-79.

NEC 350-3 Flexible metal conduit shall not be smaller than ½ inch except: (1) Underplaster extension as per 344-2; (2) To enclose motor leads as per 430-145 (b); (3) Part of approved fixture assembly not over 6 ft. as per 410-67(c); (4) Mfg. wiring systems as per 604-6(a).

NEC 350-4 Flexible metallic conduit to be supported within 12 inches of outlet box and every 4½ feet. Exception 1: Where fish... Exception 2: Not over 3... at terminals where flexib... is necessary; Exception 3: Not over 6... from box to fixture as re... by 410-67 (c).

See Table 350-3 for conductors allowed in ⅜" flexible conduit.

INTERMEDIATE METAL CONDUIT, RIGID METAL CONDUIT, ELECTRICAL METALLIC TUBING, FLEXIBLE METAL CONDUIT, AND FLEXIBLE METALLIC TUBING

CARLON P&C DUCT FITTINGS AND ACCESSORIES

 Couplings	 Male Adapters	 Female Adapters
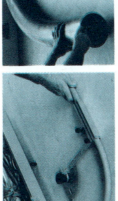 Plugs	 Bell Ends	Flexible Coupling

Expansion Fittings

(Photos courtesy of Carlon, an Indian Head Co.)

NEC 347-9 Expansion joints shall be provided to compensate for thermal expansion and contraction.

Type EPC-40 (electrical plastic conduit) a semi-rigid conduit for virtually every application requiring RIGID conduit. Heavy Wall PV-Duit for underground encased and many above-ground installations. See NEC Article 347. UL listed for above ground and underground use.

Type A-EPT (electrical plastic tubing) Thin Wall PV-Duit. UL listed for underground concrete encasement only.

NEC 300-5 (d) PVC Schedule 80 or equivalent may be used for physical protection of conductors from ground up poles. Conduit shall extend 8 feet up pole.

Type EPC-80 (electrical plastic conduit) Extra Heavy Wall for special applications requiring additional mechanical protection such as pole risers and bridge crossings.

NEC 347-10 Minimum size 1/2 inch—no limit on maximum size.

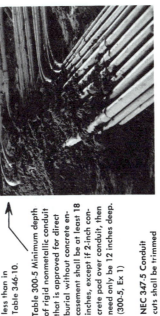

NEC 347-14 Not over four quarter bends between pulling points.

Table 300-5 Minimum depth of rigid nonmetallic conduit that is approved for direct burial without concrete encasement shall be at least 18 inches, except if 2-inch concrete pad over conduit, then need only be 12 inches deep. (300-5, Ex 1)

NEC 347-5 Conduit cuts shall be trimmed inside and outside to remove rough edges.

NEC 347-8 Conduit supported as per Table 347-8 and in addition a support within 3 feet of each box, conduit or other conduit termination.

NEC 347-1 DESCRIPTION Shall be resistant to moisture and chemical atmospheres. For use above ground shall be flame retardant, resistant to impact and crushing, resistant to heat encountered, resistant to low temperature and sunlight. For use underground shall be resistant to moisture and corrosive agents, shall have strength to withstand abuse during and after installation.

NEC 347-17 CONSTRUCTION SPECIFICATIONS
(a) Each length clearly and durably marked at least every 10 ft as per 110-21 including type of material unless visually identifiable. Markings listed aboveground use to be permanent and for underground use only legible until installed.

THREE EASY STEPS FOR FIELD BENDING

Small Diameters (1/2"-1 1/2")   Large Diameters (2"-6")

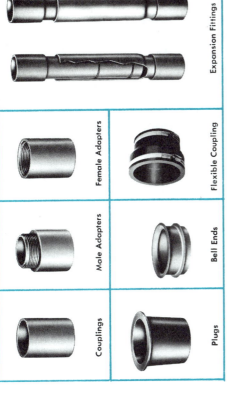

NEC 347-6 Joints shall be made by specially approved methods.

NEC 347-13 Field bends made only with special bending equipment and radius of inner bend not less than in Table 346-10.

1. HEATING. Rotate the conduit to provide even heating. Time varies with conduit size. In large sizes insert bending plugs as shown before heating conduit.

2. FORMING THE BEND. Use bend guide for small diameters, or a jig, or a line drawn on the work surface. Large sizes can be formed using a corner for an "extra set of hands." Plugs must remain tight during handling.

3. COOLING. The bend will "set" when cooled. Use a wet rag or sponge to speed up the cooling process. When "set", remove plugs. The bend is now ready to install.

NEC 347-2 USES PERMITTED:
(a) Concealed in ceilings, walls, and floors; (b) in corrosive areas as per 300-6 and where exposed to chemicals; (c) in cinder fill; (d) in wet locations where walls are washed, such as dairies or canneries; (e) in dry or damp areas where not prohibited by 347-3; (f) for exposed work if identified for such use and not exposed to physical damage; (g) underground if as per 300-5 and 710-3 (b).

NEC 347-3 USES NOT PERMITTED:
(a) In hazardous areas except as per 514-8, 515-5, 501-4 (b) Ex;
(b) To support fixtures or other equipment;
(c) Where exposed to physical damage unless identified for such use;
(d) Where ambient temperatures exceed those of conduit;
(e) Where conductor insulation temperature limitations exceed conduit insulation.

RIGID NONMETALLIC CONDUIT (NEC 347)

(See page 51 in this text for "Made" and Other Electrodes.)

Grounding Electrode Conductor

Cold Water Pipe

NEC 250-115 Ground clamps suitable for conductor and electrode materials, and suitable for direct soil burial where buried. One conductor only unless approved for multiple conductors. (See 250-115 (a, b, c, d) for connection methods.)

Electrically continuous metal underground gas pipe (as per 250-83 (a)) permitted only if NONE of the electrodes required by 250-81 (such as underground metal water pipe) are available at building and then gas pipe permitted ONLY when acceptable and expressly permitted by gas supplier and authority enforcing code. (250-83 (a))

Approved Ground Clamp

GROUNDING ELECTRODE SYSTEM (NEC 250-81) If available at building or structure, the following grounding electrodes SHALL be bonded together and used as the "Grounding Electrode System": (Sized as per 250-94, connected as per 250-115).
 a. Metal water pipe with at least 10 feet in contact with earth (including well casing if bonded to pipe). Water pipe to be supplemented by additional electrode as per 250-81 or 250-83.
 b. Effectively grounded metal frame of building.
 c. Electrode consisting of re-bar or rod at least ½ inch diameter and 20 feet long or at least 20 feet of bare copper wire not smaller than No. 4 and encased in at least 2 inches of concrete at bottom of footing.
 d. Endless ring encircling building, consisting of at least 20 feet of bare copper wire not smaller than No. 2 and in direct contact with earth at least 2½ feet below surface.

Connection of grounding electrode conductor to grounding electrode shall be permanent, effective and accessible, except the connections to concrete encased or buried electrodes need not be accessible. Where water pipe is used, connection required on street side of water meter or bonding around water meter, valves, unions, etc., required. (250-112)

Aluminum or copper-clad aluminum grounding electrode conductors not permitted in direct contact with masonry, earth or where subject to corrosive conditions. (250-92 (a))

Metallic enclosures for grounding conductors shall be continuous from cabinet to grounding electrode. Enclosures which are not continuous from cabinet to electrode can be bonded at each end to grounding conductor. (250-92 (a))

Aluminum or copper-clad aluminum conductors shall not be used within 18 inches of earth when used outside. (250-92 (a))

GROUNDING ELECTRODE SYSTEM -- NEC 250-81

3-Wire Circuit

Black

White

Red

Ungrounded conductors shall be clearly distinguishable from grounded and grounding conductors by colors or combination of color plus distinguishing marking, and in either case the color shall be other than white, natural gray, or green. (310-12 (c))

Branch circuits supplying medium-base lampholders of the screw-shell type or 125-volt devices or receptacles shall not exceed 125 volts between conductors or to ground. (210-6(a))

NEC 210-19 (a) (Fine Print Note) Voltage drop should not exceed 3 percent to farthest outlet on branch circuit. Maximum voltage drop for feeders plus branch circuits should not exceed 5 percent. (Also see 215-2 (b) (FPN).)

Grounded conductor, on premises, shall be identified as per 200-6. Grounded conductor, when insulated, to have same insulation as ungrounded conductor. (200-2)

Conductors shall be insulated except where covered, or bare conductors are specifically permitted in Code. (310-2)
For covered grounded circuit conductors for electric ranges and dryers see 250-60.
For bare service entrance grounded conductors see 230-41.
For bare or covered overhead service grounded conductors see 230-22.
For bare or covered underground service grounded conductors see 230-30.

Conductors having a white or natural gray covering shall be used only as identified conductors. (200-7) (See Exceptions.)
Where necessary to use these conductors otherwise than above, they may be painted to render unidentified. (200-7, Ex 1)
White wire in cable may be used as supply to S_1, S_3, or S_4 switch but not permitted as return wire to switched outlet. (200-7, Ex 2)

Continuous loads shall not exceed 80 percent of branch-circuit rating, except: (1) Motor loads that have been computed as per Article 430; (2) Circuits supplied by assembly listed for continuous operation at 100 percent of rating. (210-22(c)(Ex. 1,2))

CONDUCTORS -- GENERAL RULES

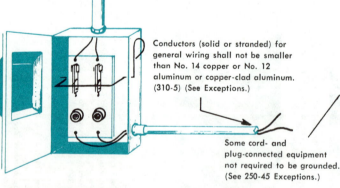

Conductors (solid or stranded) for general wiring shall not be smaller than No. 14 copper or No. 12 aluminum or copper-clad aluminum. (310-5) (See Exceptions.)

Some cord- and plug-connected equipment not required to be grounded. (See 250-45 Exceptions.)

Flexible cord approved for specific appliances considered protected by branch-circuit overcurrent device conforming to Article 210 as per following:
 20-amp circuit – Tinsel cord No. 18 or larger
 30-amp circuit – No. 16 or larger cord
 40-amp circuit – 20 amp or more rated cord
 50-amp circuit – 20 amp or more rated cord. (240-4)

Plug fuses and fuseholders shall not be used in circuits exceeding 125 volts between conductors except in systems having a grounded neutral and of which no conductor operates at more than 150 volts to ground. (240-50 (a))

Plug fuses of the Edison-base type shall be used only as a replacement item in existing installations where no evidence of overfusing or tampering. (240-51 (b))

NEC 310-4 Conductors 1/0 and larger may be run in parallel if of same length, same conductor material, same size, same insulation type, and terminated in the same manner. If in separate raceways or cables, these must have the same physical characteristics. (See Exceptions.)

NEC RULES GOVERNING CONDUCTORS AND CIRCUITS

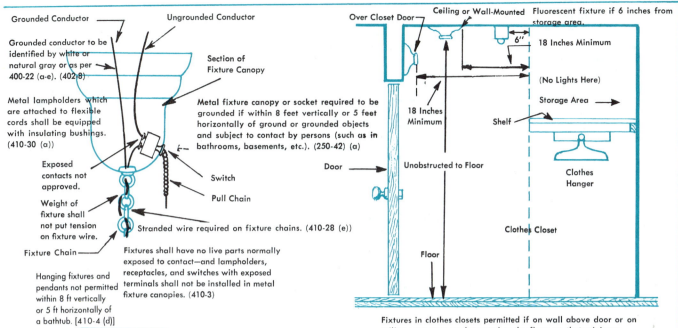

Grounded Conductor

Ungrounded Conductor

Grounded conductor to be identified by white or natural gray or as per 400-22 (a-e). (402-8)

Section of Fixture Canopy

Metal lampholders which are attached to flexible cords shall be equipped with insulating bushings. (410-30 (a))

Metal fixture canopy or socket required to be grounded if within 8 feet vertically or 5 feet horizontally of ground or grounded objects and subject to contact by persons (such as in bathrooms, basements, etc.). (250-42) (a)

Exposed contacts not approved.

Switch

Pull Chain

Weight of fixture shall not put tension on fixture wire.

Stranded wire required on fixture chains. (410-28 (e))

Fixture Chain

Fixtures shall have no live parts normally exposed to contact—and lampholders, receptacles, and switches with exposed terminals shall not be installed in metal fixture canopies. (410-3)

Hanging fixtures and pendants not permitted within 8 ft vertically or 5 ft horizontally of a bathtub. [410-4 (d)]

Over Closet Door

Ceiling or Wall-Mounted

Fluorescent fixture if 6 inches from storage area.

18 Inches Minimum

6"

(No Lights Here)

Storage Area

18 Inches Minimum

Door

Unobstructed to Floor

Shelf

Clothes Hanger

Clothes Closet

Floor

Fixtures in clothes closets permitted if on wall above door or on ceiling over area unobstructed to the floor, so that minimum horizontal clearance is at least 18 inches between fixture and areas where combustibles may be stored. A flush recessed solid-lens fixture or fluorescent fixture on ceiling if a 6-inch horizontal clearance betwen fixture and storage area. Pendants not permitted. (410-8 (a) (b))

Cast Outlet Box

Threads at Upper Ends of Enclosing Globe

Fixtures installed in damp or wet locations shall be so constructed or installed so water cannot enter or accumulate. Fixtures in wet locations to be marked "Suitable for Wet Locations." Fixtures in damp locations shall be marked "Suitable for Wet Locations" or "Suitable for Damp Locations." (410-4 (a))

Hinged Cover

Metal Plate

Lampholders and receptacles in damp or wet locations shall be of the weatherproof type. (410-49), (410-57)

FIXTURES IN CLOTHES CLOSETS

Socket with Attached Leads

White or Natural Gray

NEC 200-10 (d) For devices with attached leads, the conductor attached to the screw-shell shall have white or natural gray finish.

NEC 410-47 Grounded conductor to be connected to shell of screw-shell type lampholder. (See 410-23)

Screw-Shell

NEC 200-10 (c) In devices with screw-shells—except fuseholders—shall have the grounded conductor attached to the screw-shell.

NEC 240-50 (e) The screw-shell of plug type fuseholders shall be connected to the load side of the circuit.

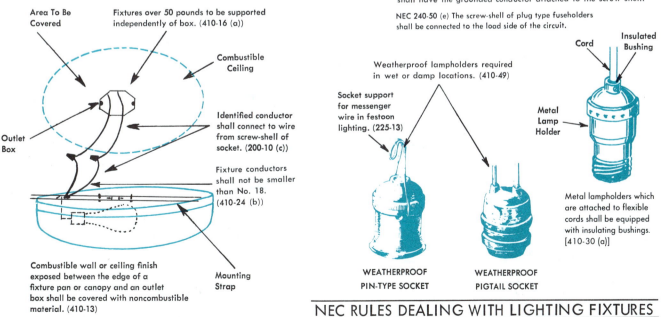

Area To Be Covered

Fixtures over 50 pounds to be supported independently of box. (410-16 (a))

Combustible Ceiling

Identified conductor shall connect to wire from screw-shell of socket. (200-10 (c))

Outlet Box

Fixture conductors shall not be smaller than No. 18. (410-24 (b))

Weatherproof lampholders required in wet or damp locations. (410-49)

Socket support for messenger wire in festoon lighting. (225-13)

Cord

Insulated Bushing

Metal Lamp Holder

Metal lampholders which are attached to flexible cords shall be equipped with insulating bushings. [410-30 (a)]

Combustible wall or ceiling finish exposed between the edge of a fixture pan or canopy and an outlet box shall be covered with noncombustible material. (410-13)

Mounting Strap

WEATHERPROOF PIN-TYPE SOCKET

WEATHERPROOF PIGTAIL SOCKET

NEC RULES DEALING WITH LIGHTING FIXTURES

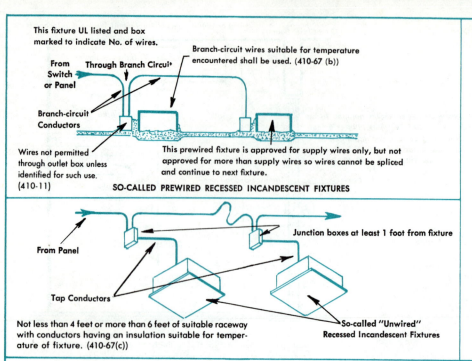

This fixture UL listed and box marked to indicate No. of wires.

From Switch or Panel

Through Branch Circuit

Branch-circuit wires suitable for temperature encountered shall be used. (410-67 (b))

Branch-circuit Conductors

Wires not permitted through outlet box unless identified for such use. (410-11)

This prewired fixture is approved for supply wires only, but not approved for more than supply wires so wires cannot be spliced and continue to next fixture.

SO-CALLED PREWIRED RECESSED INCANDESCENT FIXTURES

From Panel

Tap Conductors

Junction boxes at least 1 foot from fixture

Not less than 4 feet or more than 6 feet of suitable raceway with conductors having an insulation suitable for temperature of fixture. (410-67(c))

So-called "Unwired" Recessed Incandescent Fixtures

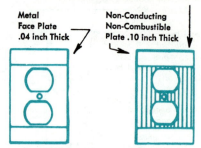

Metal face plates shall be grounded. (410-56 (c))

Shall be installed to completely cover wall opening and seat against wall surface.

Metal Face Plate .04 inch Thick

Non-Conducting Non-Combustible Plate .10 inch Thick

Face plates of nonferrous metal shall be not less than .04 inch thick, ferrous .03 inch thick, and composition .10 inch thick, unless formed or reinforced to provide adequate mechanical strength. (410-56 (c))

Face plates to completely cover opening and seat against mounting surface. (410-56 (d))

FACE PLATES NEC 410-56 (c & d)

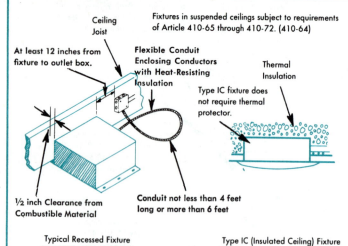

Ceiling Joist

Fixtures in suspended ceilings subject to requirements of Article 410-65 through 410-72. (410-64)

At least 12 inches from fixture to outlet box.

Flexible Conduit Enclosing Conductors with Heat-Resisting Insulation

Thermal Insulation

Type IC fixture does not require thermal protector.

½ inch Clearance from Combustible Material

Conduit not less than 4 feet long or more than 6 feet

Typical Recessed Fixture

Type IC (Insulated Ceiling) Fixture

GENERAL RULE: Wires to flush and recessed fixtures shall have insulation suitable for temperature encountered. (410-67 (a))

RECESSED INCANDESCENT FIXTURES TO HAVE THERMAL PROTECTION AND BE SO IDENTIFIED - Ex. 1 if installed in poured concrete and identified for such use. (410-65(c))

Recessed portions of enclosures, other than at points of support, shall be spaced at least ½ inch from combustible material. [410-66 (a)]

Thermal insulation not permitted within 3 inches of fixture and not above fixture so as to entrap heat and prevent free air circulation except when identified for the purpose. [410-66 (b)]

Branch-circuit wires permitted to fixture if insulation suitable for fixture temperature. (410-67 (b))

Tap wires with insulation suitable for fixture temperature permitted from fixture terminal to outlet box if in suitable raceway. Outlet box at least 1 foot from fixture and raceway shall be at least 4 feet long but not longer than 6 feet. (410-67(c))

RECESSED FIXTURES

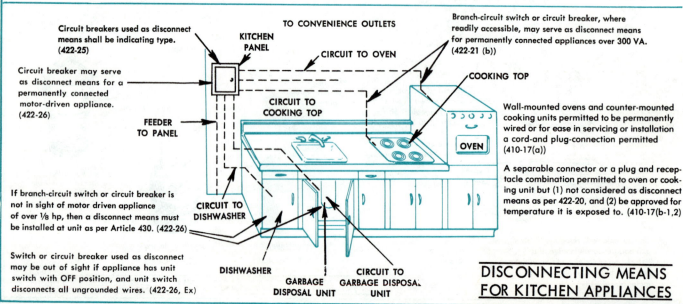

Circuit breakers used as disconnect means shall be indicating type. (422-25)

Circuit breaker may serve as disconnect means for a permanently connected motor-driven appliance. (422-26)

KITCHEN PANEL

TO CONVENIENCE OUTLETS

CIRCUIT TO OVEN

CIRCUIT TO COOKING TOP

FEEDER TO PANEL

If branch-circuit switch or circuit breaker is not in sight of motor driven appliance of over ⅛ hp, then a disconnect means must be installed at unit as per Article 430. (422-26)

Switch or circuit breaker used as disconnect may be out of sight if appliance has unit switch with OFF position, and unit switch disconnects all ungrounded wires. (422-26, Ex)

CIRCUIT TO DISHWASHER

DISHWASHER

GARBAGE DISPOSAL UNIT

CIRCUIT TO GARBAGE DISPOSAL UNIT

Branch-circuit switch or circuit breaker, where readily accessible, may serve as disconnect means for permanently connected appliances over 300 VA. (422-21 (b))

COOKING TOP

OVEN

Wall-mounted ovens and counter-mounted cooking units permitted to be permanently wired or for ease in servicing or installation a cord-and plug-connection permitted (410-17(a))

A separable connector or a plug and receptacle combination permitted to oven or cooking unit but (1) not considered as disconnect means as per 422-20, and (2) be approved for temperature it is exposed to. (410-17(b-1,2)

DISCONNECTING MEANS FOR KITCHEN APPLIANCES

National Electrical Code

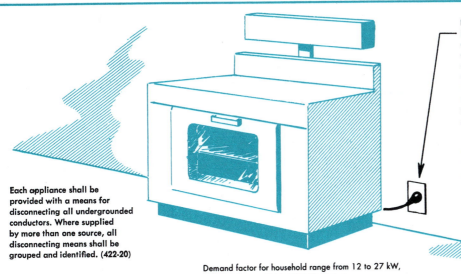

For cord- and plug-connected appliances, including free-standing household ranges and clothes dryers, a separable connector or attachment plug and receptacle may serve as the disconnecting means where accessible, otherwise shall be as per 422-21(a) or (b). (422-22(a))

For residential ranges, plug and receptacle if accessible by removal of a drawer is acceptable. (422-22 (b))

NEC 210-50 (c) Outlets for special appliances, such as laundry equipment, shall be within 6 feet of the appliance.

Use Table 220-20 for demand factors for commercial cooking equipment such as dishwashers, booster heaters, water heaters and other kitchen equipment.

Each appliance shall be provided with a means for disconnecting all undergrounded conductors. Where supplied by more than one source, all disconnecting means shall be grouped and identified. (422-20)

Demand factor for household range from 12 to 27 kW, calculated as per Table 220-19, Note 1.
Demand factor for household ranges from 12 to 27 kW of unequal ratings calculated per Table 220-19, Note 2.

DISCONNECT MEANS FOR ELECTRIC DRYERS AND FREE-STANDING RANGES

Jar Which Contains Driven Element

NEC 422-20 Each appliance shall have a disconnect.

For cord and plug and permanently connected appliances rated at not more than 300 volt-amperes or ⅛ hp, the branch-circuit overcurrent device may serve as the disconnecting means. (422-21 (a))

For permanently connected appliances over 300 VA or ⅛ hp, the branch-circuit switch or circuit breaker may serve as the disconnect means, if readily accessible to user of appliance. (422-21 (b))

Disconnect means for motor-driven appliances over ⅛ hp shall be within sight of motor controller. (422-26)

Disconnect may be out of sight if unit switch on appliance has OFF position and disconnects all ungrounded conductors. (422-26, Ex)

Motor-Driven Impeller of Built-In Juice Extractor

OFF ON

General Rule: Switches which are a part of the appliance shall not be considered as taking the place of the single disconnecting means when such is required by this section. (422-24)

Exception to the General Rule: In a single-family dwelling the service switch is acceptable as the single disconnecting means if the unit switch on the appliance has an OFF position and disconnects all ungrounded conductors. (422-24 (c)) (Also see 422-24 (a, b, d) for other occupancies.)

DISCONNECTING MEANS FOR APPLIANCES

No wiring permitted in duct of commercial-type cooking equipment. (300-22 (a))

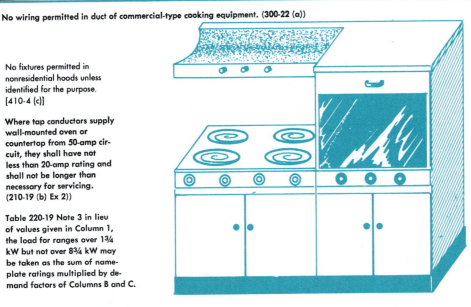

No fixtures permitted in nonresidential hoods unless identified for the purpose. [410-4 (c)]

Where tap conductors supply wall-mounted oven or countertop from 50-amp circuit, they shall have not less than 20-amp rating and shall not be longer than necessary for servicing. (210-19 (b) Ex 2))

Table 220-19 Note 3 in lieu of values given in Column 1, the load for ranges over 1¾ kW but not over 8¾ kW may be taken as the sum of nameplate ratings multiplied by demand factors of Columns B and C.

Table 220-19 Note 4 The branch-circuit load for one wall-mounted oven or countertop unit shall be taken as the nameplate rating. Where countertop and not more than two wall-mounted ovens are in the same room, the nameplate ratings shall be added and the load treated as that of a single range.

Separable connector or plug and receptacle used for servicing, not acceptable as disconnecting means. (422-17 (b-1))

WALL-MOUNTED OVEN AND COUNTERTOP RANGES

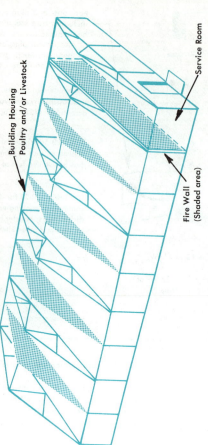

NEC 547-1 (a) This Article applies to agricultural buildings where dust and dust with water are present. Includes all totally enclosed areas and environmentally controlled poultry and livestock confinement systems, where feed dust and litter dust may accumulate, and other similar enclosed areas.

NEC 547-1 (b) This Article also applies to agricultural buildings where corrosive fumes exist, such as totally enclosed and environmentally controlled areas where: (1) animal and poultry excrement causes corrosive vapors, (2) corrosive particles mix with water, (3) periodic washing and sanitizing causes area to be damp or wet, (4) other similar conditions exist.

NEC 547-2 For buildings other than 547-1 above applicable code rules shall be used.

NEC 547-3 Electrical equipment and devices to be installed to function at full rating without overheating.

NEC 547-4 Wiring methods to be UF, NMC, SNM or other cables or raceways suitable for location with identified and listed fittings.

NEC 547-4(a) All boxes containing devices, taps, joints, or terminal connections shall be dust- and watertight and made of corrosion-resistant material (FPN See 300-7 and 347-9 for seals).

NEC 547-4(b) Flexible connections to be made with dust-tight flexible connectors, liquid-tight flexible metal conduit, or hard usage flexible cord, with approved fittings.

NEC 547-5 Devices, such as switches, fuses, circuit breakers, controller, pushbuttons, relays, etc., shall be installed in weatherproof, corrosion-resistant enclosure designed to keep out dust, water, and corrosive elements, and shall have telescoping or close-fitting cover.

NEC 547-6 Rotating electrical machinery to be designed to limit entrance of dust, moisture, or corrosive particles, or to be totally enclosed.

NEC 547-7(a) Lighting fixtures installed to minimize entrance of dust, foreign matter, moisture, and corrosive material.

NEC 547-7(b) Light fixture exposed to physical damage to have suitable guard.

NEC 547-7(c) Light fixture to be water-tight if exposed to water from condensation and/or cleansing water or solution.

NEC 547-8 Grounding to comply with NEC 250 and bonding to be as per 547-8 and EX.

AGRICULTURAL BUILDINGS

Diagram labels: Service Room; Building Housing Poultry and/or Livestock; Fire Wall (Shaded area)

NEC 545-3 Definition of a Manufactured Building: A Manufactured Building is of closed construction, normally factory-mode or assembled for installation, or assembly and installation on building site. (Does not include mobile home or recreational vehicles.)

Manufactured buildings not restricted to dwellings; could be Industrial or Commercial.

NEC 545-6 Service-entrance conductors to be installed at building site, except where point of attachment is known before manufactured.

NEC 545-4 (a) All wiring methods in code permitted for use, and other wiring systems and fittings specifically approved for use with manufactured buildings also permitted. Wiring devices with integral enclosures permitted if sufficient length of conductor provided for replacement of devices.

NEC 545-4 (b) Concealed cables of No. 10 and smaller permitted to be installed without supports, such as staples or straps, as required in normal construction, if cables are secured at cabinets, boxes, and fittings and protected from physical damage as per 300-4.

NEC 545-5 Service-entrance conductors to be installed as per 230. (See FPN for temperature limitations)

NEC 545-7 Service equipment located at nearest, readily accessible point, either inside or outside of building.

NEC 545-8 Protection of equipment and conductors that are exposed during manufacturing, transit, and erection at building site is required.

NEC 545-9(a) Boxes other than those in Table 370-6(a) permitted if listed and tested. (b) Boxes under 100 cu. in. for mounting in closed construction shall have anchors or clamps for rigid and secure installation.

NEC 545-10 Switch or receptacle with integral enclosure permitted when tested and listed.

NEC 545-11 Bonding and grounding to conform to 250, Parts E, F, and G.

NEC 545-12 Means to be provided to route grounding electrode conductor from service equipment to grounding electrode and to conform to 250, Part J.

NEC 545-13 Where on-site interconnection of modules or other building components is necessary, fittings and connectors when tested, identified, and listed are permitted to be used. These fittings to be equal to wiring method and may be concealed at time of on-site assembly.

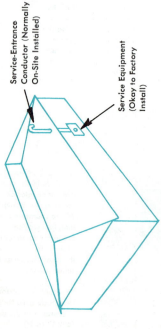

Diagram labels: Service-Entrance Conductor (Normally On-Site Installed); Service Equipment (Okay to Factory Install)

MANUFACTURED BUILDINGS

TRADE COMPETENCY TEST NO. 1A

Student
Number_____

Instructor's
Name_____

INSTRUCTIONS: The following statements are either True or False. Draw a circle around the **T** if the statement is True, or around the **F** if it is False. The questions are to be answered, as nearly as possible, in accordance with the specific provisions of the NEC.

• True or False •

Service drops must clear sidewalks by not less than 15 ft.   T   (F)

1. The term "service-entrance conductors" includes the term "service drops."   T   F

2. The smallest permissible size of hard-drawn copper service drop is No. 10.   T   F

3. Service drops must clear the ground by not less than 10 feet.   T   F

4. The service disconnecting means may consist of not more than eight switches or circuit breakers.   T   F

5. The disconnecting means for a lighting service shall disconnect all ungrounded conductors simultaneously.   T   F

6. Service conductor connections for single-phase services may be made with the aid of solder lugs.   T.   F

7. Cable limiters are permitted on the supply side of the service disconnect means.   T   F

8. An insulated copper grounding electrode conductor used with a single service disconnect means may be spliced under the proper circumstances.   T   F

9. A No. 6 grounding electrode conductor which is secured to the building surface and is exposed to moderate physical damage requires metal covering or protection.   T   F

10. The sheet-metal-straps type of ground clamp constructed solely of strong sheet metal is not permitted for connecting the grounding electrode conductor to the grounding electrode.   T   F

11. A plug receptacle in the hallway outside of the kitchen shall not be connected to the small appliance circuit.   T   F

12. Additional plug receptacles in bedrooms, above the number required by the Code, must be assessed at ½ amp each.   T   F

Continued on Next Page

TEAR OUT HERE

13. In a bedroom, if a point on any available wall is 7 feet from the nearest plug receptacle, an additional receptacle is needed.   T   F

14. A receptacle in the middle of the dining room floor is not counted toward fulfilling minimum Code requirements.   T   F

15. A general purpose branch circuit supplies outlets for both lighting and appliances.   T   F

16. Wooden supports for boxes in concealed work shall not be less than nominal 1 inch thickness.   T   F

17. Junction boxes shall be readily accessible.   T   F

18. When two switches are mounted on the same strap, the number of conductors allowed in the box shall be reduced by one.   T   F

19. One of the standard circuit breaker ratings is 45 amperes.   T   F

20. The screw shell of a plug type fuseholder shall be connected to the load side of the circuit.   T   F

21. When the neutrals of two branch circuits which are derived from two different systems enter the same enclosure, one neutral may be white and the other natural gray.   T   F

22. All 125-volt, single-phase, 20-ampere receptacles installed in bathrooms of dwelling units shall have ground-fault circuit-interrupter protection for personnel.   T   F

23. Where not permanently accessible, nonmetallic-sheathed cable within 5 feet of a scuttle hole does not require protection.   T   F

24. Standard type receptacle boxes may be used in elevated floors of show windows.   T   F

25. The maximum number of quarter bends permitted in an exposed or concealed run of ½-inch flexible metal conduit is five.   T   F

TEAR OUT HERE

TRADE COMPETENCY TEST NO. 1B

INSTRUCTIONS: In the space provided at the right of each of the following questions, write **a, b, c,** or **d** to indicate which of the alternatives given makes it most nearly correct. The questions are to be answered, as nearly as possible, in accordance with the specific provisions of the NEC.

• Multiple-Choice •

The minimum clearance of service drops over a flat roof is:
(a) 8 ft., (b) 6 ft., (c) 10 ft., (d) 12 ft.   1. *(a)*

1. A bare copper neutral service-entrance conductor is permitted only if the voltage is not over:

   (a) 150 volts to ground, (b) 300 volts to ground, (c) 600 volts to ground, (d) a bare copper neutral may be used on all voltages   1. _____

2. The service disconnect rating for a single-family dwelling shall be not less than 100 amps when the net computed load is:

   (a) 7.5 kW, (b) 10 kW, (c) 12.5 kW, (d) 15 kW   2. _____

3. The minimum clearance of service drops above a public driveway on other than residential property is:

   (a) 10 feet, (b) 12 feet, (c) 16 feet, (d) 18 feet   3. _____

4. In general, the minimum ampacity of the ungrounded service-entrance conductors of a service is:

   (a) 50 amps, (b) 60 amps, (c) 80 amps, (d) 100 amps   4. _____

5. The general rule with respect to the service disconnecting means states that a grounding connection shall not be made to any grounded circuit on:

   (a) the supply side, (b) the load side, (c) either supply or load side, as desired, (d) both supply and load sides   5. _____

6. An equipment bonding jumper is permitted outside a raceway:

   (a) when not over 10 feet long, (b) when not over 8 feet long, (c) when not over 6 feet long, (d) is not permitted.   6. _____

Continued on Next Page

Blueprint Reading

TEAR OUT HERE

7. According to the Code, metal enclosures for grounding electrode conductors shall be:

   (a) not permitted, (b) isolated, (c) electrically continuous, (d) rigid conduit

7. _____

8. The unit lighting load for a dwelling unit, expressed in "watts-per-square-foot", shall be:

   (a) 1-½ watts (b) 2 watts (c) 3 watts (d) 5 watts

8. _____

9. For the kitchen small appliance load in dwelling occupancies, the Code requires not less than:

   (a) two 20-amp circuits, (b) one 15-amp circuit, (c) two 15-amp circuits, (d) one 20-amp circuit

9. _____

10. To provide for this small appliance load, the feeder and service allowance is required to be not less than:

   (a) 1800 watts, (b) 3000 watts, (c) 2400 watts, (d) 3600 watts

10. _____

11. The service or feeder capacity necessary to provide for six 5 kW household electric clothes dryers is:

   (a) 24 kW, (b) 30 kW, (c) 19.5 kW, (d) 21 kW

11. _____

12. When using the optional calculation method for a dwelling unit service, all other load above the initial 10 kW is to be assessed at:

   (a) 40%, (b) 50%, (c) 60%, (d) 75%

12. _____

13. The maximum rating of the overcurrent device on a 17-amp noncontinuous, single nonmotor appliance circuit is:

   (a) 14 amps, (b) 15 amps, (c) 20 amps, (d) 25 amps

13. _____

14. The rating of a cord-and-plug appliance used on a 20-amp branch circuit having two or more outlets shall not exceed:

   (a) 16 amps, (b) 15 amps, (c) 18 amps, (d) 12 amps

14. _____

15. The demand load for a 16 kW electric range should be assessed at:

   (a) 8200 watts, (b) 9600 watts, (c) 12600 watts, (d) 14400 watts

15. _____

16. The internal depth of outlet boxes intended to enclose flush devices shall be at least:

   (a) ½ inch, (b) ⅞ inch, (c) $^{15}/_{16}$ inch, (d) 1½ inches

16. _____

National Electrical Code

17. Threaded boxes and fittings which are threaded into two or more properly supported conduits from two or more sides and do not support fixtures or contain devices need not be further secured provided their volume does not exceed:

(a) 100 cubic inches, (b) 125 cubic inches, (c) 150 cubic inches, (d) 200 cubic inches

17. _____

18. The maximum number of No. 14 conductors permitted in a 1½ x 4-inch octagonal outlet box is:

(a) 5, (b) 9, (c) 8, (d) 7

18. _____

19. If this box contains a fixture stud and a cable clamp, the number of such wires must be reduced to:

(a) 6, (b) 4, (c) 7, (d) 5

19. _____

20. In combustible walls or ceilings, the front edge of an outlet box or fitting may set back of the finished surface:

(a) ¼ inch, (b) ⅛ inch, (c) ½ inch, (d) not at all

20. _____

21. Plug fuses are never permitted in circuits which exceed:

(a) 125 volts between conductors, (b) 125 volts to ground, (c) 150 volts between conductors, (d) 150 volts to ground

21. _____

22. When 100 feet of No. 16 fixture wire is connected to a standard 120-volt circuit, the branch-circuit overcurrent device shall not exceed:

(a) 15 amps, (b) 30 amps, (c) 20 amps, (d) 40 amps

22. _____

23. Under ordinary circumstances, the Code requires that the distance between supports for open wiring on insulators shall not exceed:

(a) 4½ feet, (b) 3 feet, (c) 7½ feet, (d) 4 feet

23. _____

24. The Code requires that recessed portions of fixture enclosures shall be spaced from combustible material by at least:

(a) ⅜ inch, (b) ½ inch, (c) ⅝ inch, (d) ¾ inch

24. _____

25. The metal raceway enclosing tap conductors for recessed fixtures shall not be:

(a) over 2 feet long, (b) less than 6 feet long, (c) over 3 feet long, (d) less than 4 feet long

25. _____

Blueprint Reading

TRADE COMPETENCY TEST NO. 1C

Student
Number_____

Instructor's
Name_____

INSTRUCTIONS: Complete each of the following statements, by writing the missing word or words in the space provided at the right of each question, to make a true statement. The questions are to be answered, as nearly as possible, in accordance with the specific provisions of the NEC.

• Completion •

Outlet boxes may be recessed _____ inch or less.   1.____¼____

1. In general, the rating of a service disconnecting means shall not be less than _____ amps.

   1._____

2. The service disconnecting means need not be greater than 30 amps where there are not more than _____ 2-wire branch circuits.

   2._____

3. The clearance of service drops above a flat roof shall be not less than _____ feet.

   3._____

4. The clearance of service drops from the side or bottom of windows shall not be less than _____ feet.

   4._____

5. Not more than six switches at one service location may be used as the service disconnecting means if they are _____ marked.

   5._____

6. The grounding electrode conductor for an AC ungrounded system need not be larger than No. _____ copper when connected to a made electrode.

   6._____

7. In residential occupancies the code requires one 15-amp lighting circuit for every _____ square feet.

   7._____

8. In determining "watts-per-sq-ft" area, the _____ dimensions of the building shall be used.

   8._____

9. Under the optional calculation for a single-family residence, air conditioning load is assessed at _____ percent.

   9._____

10. Under the code, receptacles in living areas are spaced so that no point along the available floor line perimeter shall be more than _____ feet from a receptacle outlet.

    10._____

11. A new receptacle outlet installed on a 15-amp or 20-amp branch circuit shall be of the _____ type.

    11._____

12. Conduit shall not be connected to _____ of round boxes.

    12._____

Continued on Next Page

13. Where necessary, a shallow outlet box not less than _____ inch(es) deep may be used.

13. _____

14. A metal fixture shall be grounded if located within _____ feet vertically or 5 feet horizontally of a kitchen sink.

14. _____

15. Unused openings in boxes and fittings shall be effectively _____ .

15. _____

16. Nonmetallic boxes may have _____ or _____ covers.

16. _____

17. The front edge of an outlet box in a non-combustible ceiling may be _____ inches back from the surface.

17. _____

18. Pull boxes shall be installed so they are _____ .

18. _____

19. The free length of conductor at a switch outlet shall be not less than _____ inches.

19. _____

20. The smallest stranded copper conductor permitted for general wiring is No. _____ .

20. _____

21. In general the voltage to ground on a branch circuit supplying 15-amp receptacles shall be not greater than _____ volts.

21. _____

22. In dwelling occupancies the voltage between conductors supplying screw shell lampholders shall not exceed _____ volts.

22. _____

23. Nonmetallic sheathed cable shall be installed so no bend will have a radius less than ___ times the diameter of the cable.

23. _____

24. In general Type AC armored cable shall be supported within _____ inches of an outlet box.

24. _____

25. The branch-circuit switch or circuit breaker may serve as the disconnecting means for a permanently connected appliance of over 300 VA or 1/8 h.p., where _____ accessible to the user.

25. _____

National Electrical Code

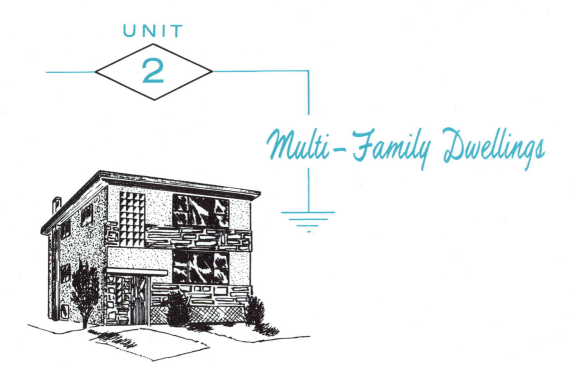

Multi-Family Dwellings

CONDUCTORS

Allowable Current-Carrying Capacities

Table 310-16 lists current-carrying values for copper and aluminum wires when not more than three conductors are grouped in raceway, cable, or buried in the ground. Table 310-17 lists values for copper and aluminum conductors in free air. Tables 310-18 and 310-19 list ampacities of conductor insulation for specialized application.

Table 310-16

The upper portion of this table deals with ampacities of wires that are used at a nominal room temperature of 30 degrees Centigrade (C), or 86 degrees Fahrenheit (F). The lower part of Tables 310-16 through 19 supplies correction factors for limiting ampacities of conductors used at room temperatures greater than 30 degrees C or 86 degrees F.

Column 1 in the main part of the table lists AWG sizes of conductors from No. 18 to 2000 MCM inclusive. Succeeding columns show ampacities for each size of conductor according to the type of insulation it bears. The columns relating to 60°C, 75°C, and 90°C insulation in Tables 310-16 and 310-17 are the ones most commonly used by inside wiremen.

With size No. 8, for example, Column 2 shows that with 60° insulation, the conductor will carry 40 amps without subjecting insulation to damaging heat. If the No. 8 conductor has 75° insulation, Column 3, it may be loaded to 50 amps without harmful result. This size conductor may carry 55 amps with the insulations in Columns 4 and 5.

Suppose that room temperature for this particular application is 50 degrees C (122 degrees F) instead of 30 degrees C (86 degrees F). Referring to lower part of Table 310-16, *Correction Factors*, a multiplying factor of .58 must be used to determine the ampacity of a conductor having any of the Column 2 insulations, .75 for Column 3 insulations, .80 for Column 4 and .82 for Column 5 insulations. The absence of any multiplying factor in some columns indicates that these insulations shall not be used for the listed ambient temperatures.

Using the indicated factors, the ampacity of No. 8 Type TW wire used at a room tem-

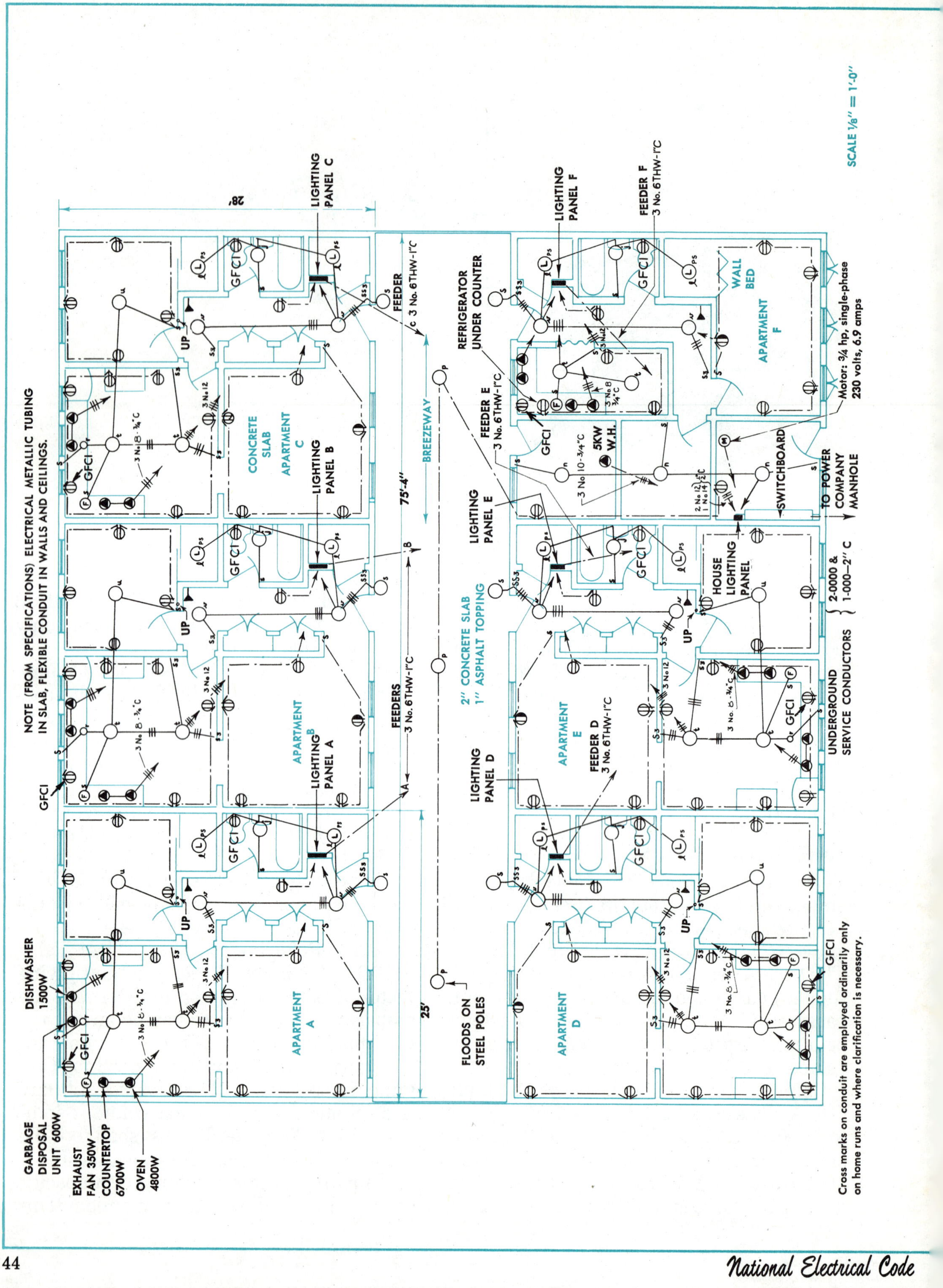
NOTE (FROM SPECIFICATIONS) ELECTRICAL METALLIC TUBING IN SLAB, FLEXIBLE CONDUIT IN WALLS AND CEILINGS.
LIGHTING PANEL C
28'
FEEDER C 3 No. 6THW-1"C
CONCRETE SLAB APARTMENT C
LIGHTING PANEL B
BREEZEWAY
75'-4"
APARTMENT B
LIGHTING PANEL A
FEEDERS 3 No. 6THW-1"C
25'
APARTMENT A
DISHWASHER 1500W
GARBAGE DISPOSAL UNIT 600W
EXHAUST FAN 350W
COUNTERTOP 6700W
OVEN 4800W
FLOODS ON STEEL POLES
GFCI
3 No. 8-3/4"C
3 No 12
UP
LIGHTING PANEL F
FEEDER F 3 No. 6THW-1"C
REFRIGERATOR UNDER COUNTER
FEEDER E 3 No. 6THW-1"C
3 No 8 3/4"C
WALL BED
APARTMENT F
5KW W.H.
3 No 10-3/4"C
2 No 12 1 No 14 1/2"C
SWITCHBOARD
TO POWER COMPANY MANHOLE
Motor: 3/4 hp, single-phase 230 volts, 6.9 amps
LIGHTING PANEL E
2" CONCRETE SLAB 1" ASPHALT TOPPING
HOUSE LIGHTING PANEL
2-0000 & 1-000-2" C
UNDERGROUND SERVICE CONDUCTORS
APARTMENT E
FEEDER D 3 No. 6THW-1"C
LIGHTING PANEL D
APARTMENT D
SCALE 1/8" = 1'-0"
Cross marks on conduit are employed ordinarily only on home runs and where clarification is necessary.

perature of 50 degrees C, 122 degrees F, is equal to .58 × 40 amps, or 23 amps. That of a No. 8 Type THWN conductor is equal to .75 × 50 amps, or 37.5 amps.

The table shows that Column 2 insulations are not to be used at room temperatures greater than 131 degrees F or column 5 insulations at temperatures greater than 176 degrees F. The notes below the table indicate the rating of No. 14, 12 and 10 copper is 15, 20 and 30 amps and No. 12 and 10 aluminum is 15 and 25 amps.

Table 310-17

This table is similar except that it deals with the ampacities of conductors in free air. Ampacities of conductors in free air are greater than those of conductors in raceways, cables, or earth.

Tables 310-18 and 310-19

These tables are similar in construction and they deal with allowable ampacities of conductors rated 110 to 250°C. The correction factor tables are also similar, but they also relate to higher temperatures.

Notes to the Tables

There are eight notes connected with Tables 310-16 through 310-19, which state specific applications. These notes should be carefully evaluated. Although the facts are clearly stated, it may be well to call attention here to No. 5 which states that bare conductors have the same ampacity as insulated conductors with which they are used, and to No. 8, which lists derating factors where more than three conductors are installed in a raceway or cable.

Table 310-13 "Conductor Application and Insulations"

This table lists conductor applications and the physical characteristics of the various types of insulated conductors. It should be noted that 310-13 *Conductor Application* states that the conductors listed in this table may be installed for any of the wiring methods recognized by Chapter 3 as specified in the tables. It should also be

noted that Section 310-13 *Conductor Construction* states that insulated conductors used for 600 volts or less shall comply with Table 310-13. There are several types of wire that are given two maximum operating temperatures depending upon how they are used, such as Type THW. Type THW has a 75°C rating when used as a general purpose wire and installed in a wet or dry location, but when it is installed within three inches of a ballast it is given a 90°C rating as required in 410-31.

Table 310-13 also lists several types of conductors for specialized applications as noted, such as: MI cable which has an outer metal sheath, Types TA, TBS and SIS which are limited to switchboard wiring only and Types TFE, A, AA, AI and AIA which are limited to leads within apparatus in a dry location. Types A, AA and AI are limited to 300 volts.

Table 402-3 "Fixture Wire"

The operating temperature of fixture wires shall not exceed the temperature listed in Table 402-3.

Table 402-5 "Ampacity of Fixture Wires"

Table 402-5 lists the allowable ampacities of fixture wires in sizes 18 through 10 AWG. The ampacities of the fixture wires listed in this table are based on an ambient temperature of 30°C, 86°F.

Flexible Cords

Table 400-5 shows ampacities of flexible cords where not more than three conductors are present. If there are more than three conductors in the group, ampacities given here are to be reduced to 80 percent of listed values. It may be noted that ordinary portable cord size No. 18 will carry 7 amps and size No. 16 will carry 10 amps. Notes 1 through 9 of Table 400-4 should be read carefully.

Table 400-4 "Flexible Cords and Cables"

This table sets forth constructional features of the various cords and the conditions of use and application. Section 400-4

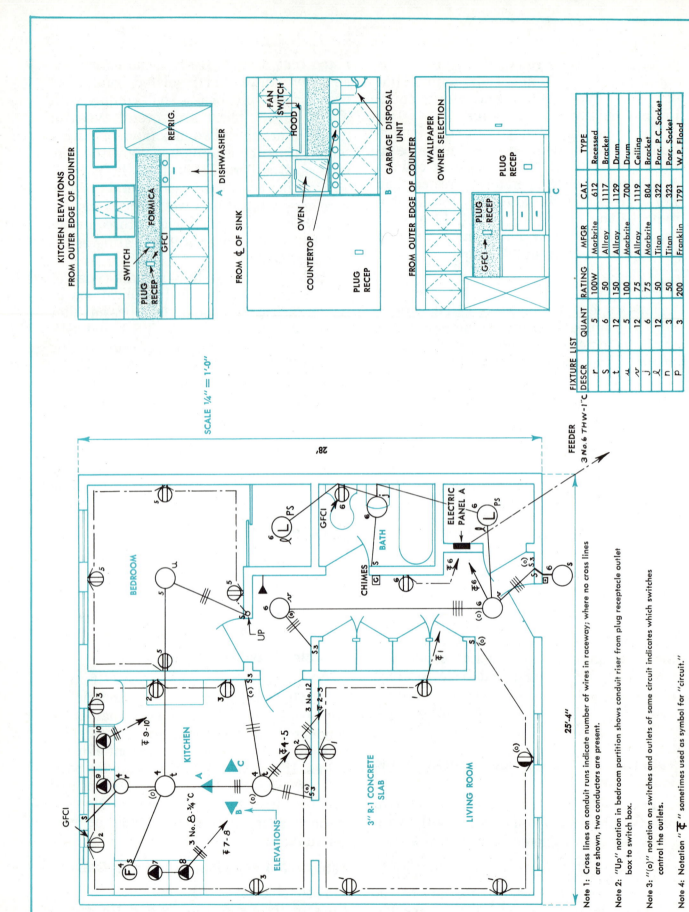

KITCHEN ELEVATIONS
FROM OUTER EDGE OF COUNTER

SWITCH
PLUG RECEP
FORMICA
GFCI
REFRIG.
DISHWASHER
A

FROM ℄ OF SINK

FAN SWITCH
HOOD
OVEN
COUNTERTOP
PLUG RECEP
GARBAGE DISPOSAL UNIT
B

FROM OUTER EDGE OF COUNTER

WALLPAPER
OWNER SELECTION
PLUG RECEP
GFCI
C

SCALE ¼″ = 1′-0″

FIXTURE LIST

DESCR	QUANT	RATING	MFGR	CAT.	TYPE
r	5	100W	Morbrite	612	Recessed
s	6	50	Allray	1117	Bracket
t	12	150	Allray	1129	Drum
u	5	100	Morbrite	700	Drum
v	12	75	Allray	1119	Ceiling
j	6	75	Morbrite	804	Bracket
l	12	50	Titan	322	Porc. P.C. Socket
n	3	50	Titan	323	Porc. Socket
p	3	200	Franklin	1791	W.P. Flood

28′

25′-4″

BEDROOM

KITCHEN

GFCI

3 No. 8-¾″C

ELEVATIONS

3″ R-1 CONCRETE SLAB

LIVING ROOM

BATH

CHIMES

GFCI

ELECTRIC PANEL A

FEEDER
3 No.6 THW-1″C

Note 1: Cross lines on conduit runs indicate number of wires in raceway; where no cross lines are shown, two conductors are present.

Note 2: "Up" notation in bedroom partition shows conduit riser from plug receptacle outlet box to switch box.

Note 3: "(o)" notation on switches and outlets of same circuit indicates which switches control the outlets.

Note 4: Notation "℄" sometimes used as symbol for "circuit."

46   National Electrical Code

states that flexible cords and cables shall conform to the description and shall be the subject of special investigation and shall not be used before being approved.

General Drawing, Page 44

This drawing, based originally upon a ⅛″ scale, shows a six-unit, one-story apartment house. Five of the apartments are identical, having living room, kitchen-dinette, bedroom and bath. The lighting panel, a flush type, is placed in the entrance storeroom. The kitchen has a wall-mounted oven, a counter-mounted hotplate or cooking top, a hood fan, a garbage-disposal unit and a dishwasher.

Caretaker's apartment, *F*, has a living room, kitchen-dinette and bath. It has no separate bedroom, a wall bed being used in the living room. Kitchen equipment is similar to that in the larger apartments.

There is a washroom, a storeroom and a combination heating equipment and switchboard room. There is a house plug circuit, an electric water heater circuit and a pump motor. A house lighting circuit supplies the three work rooms and three night-light standards. All these circuits originate in a special house panel which is adjacent to service equipment in the switchboard room.

Arrows pointing in the direction of the lighting panel indicate circuit runs. The number of cross marks on the arrow shows the number of wires in that particular *home* run. Where no cross marks are found, it is understood that the circuit consists of two wires. Conductors are assumed to be No. 14 unless referred to in the specifications or marked adjacent to the arrow.

Dot-and-dash lines denote branch-circuit raceways concealed in the floor; solid black lines mark branch-circuit conduit concealed in wall or ceiling. Designating signs for various types of outlets including lights, plug receptacles, recessed units, special outlets, telephones and fans are those recommended by the list of standard symbols. Feeders are shown by arrows originating at the various panels and pointing in the direction of the switchboard.

It should be observed that a note taken from the electrical specifications states that electrical metallic tubing is to be used in concrete floor slabs, flexible conduit in walls and ceilings.

LOAD CALCULATIONS

Unit load specified for dwelling occupancies in NEC Table 220-2(b) is 3 watts per square foot. Allowance for two small appliance or utility circuits as per NEC 220-16(a) is 3000 watts. Load required for two cooking units as per Column C of Table 220-19, is 65 percent of the sum of nameplate ratings. On the basis of above Code sections, the load for each of the five apartments, *A* to *E* inclusive, is determined as follows:

Area of each apartment, 710 sq ft
Lighting requirement equals:
710 × 3 watts, or	(2 circuits)	2130 watts
Allowance for two utility circuits		3000 watts
	Total	5130 watts

Feeder Demand Factors (NEC Table 220-11):
1st 3000 watts at 100%	3000 watts
Remainder, 2130 watts at 35%	745 watts
Total	3745 watts

Feeder load for oven and countertop:
Oven rated at	4800 watts	
Cooktop rated	6700 watts	
	Total 11500 watts	

Feeder capacity .65 × 11500 watts = 7475 watts
Feeder load:
Lighting and utility	3745 watts
Oven and countertop	7475 watts
Dishwasher	1500 watts
Garbage disposal unit	600 watts
Exhaust fan	350 watts
Total	13670 watts

FEEDER SIZE (115-230 volt service):
13670 watts divided by 230 volts = 59 amps
Table 310-16 shows that a No. 6 Type THW, copper is needed.

Table 3A, Chapter 9, shows that a 1-inch conduit is required for 3 No. 6 Type THW conductors.

Load calculations for Apartment F:
Area 448 sq ft, requiring:
448 × 3 watts, or	(1 circuit)	1344 watts
Two appliance circuits		3000 watts
	Total	4344 watts

Feeder load, lighting and appliance:
1st 3000 watts at 100%		3000 watts
Remainder, 1344 watts at 35%		470 watts
	Total	3470 watts

Feeder load:
Lighting and appliance		3470 watts
Oven and countertop		7475 watts
Dishwasher		1500 watts
Garbage disposal		600 watts
Exhaust fan		350 watts
	Total	13395 watts

Feeder size: 13395 watts divided by 230 volts = 58 amps. Here, too, 3 No. 6 Type THW copper conductors and 1-inch conduit are indicated.

In some cases, the neutral feeder may be smaller than the outer conductors. NEC 220-22 allows a 70 percent demand factor on that portion of neutral feeder load which supplies an electric range or equivalent cooking devices. Although the specifications and drawings here state that three No. 6 Type THW copper conductors shall be installed to each panel, it may be well to perform necessary calculations to see if a smaller neutral conductor might have been otherwise admissible. Calculations for the "A" type apartment proceed as follows:

Load on wall-mounted oven and countertop: 7475 watts.
.7 X 7475 watts = 5235 watts (to nearest 5 watts)

Assumed feeder load:
Lighting and appliance		3745 watts
Oven and countertop		5235 watts
Dishwasher		1500 watts
Garbage disposal unit		600 watts
Exhaust fan		350 watts
	Total	11430 watts

Neutral feeder size:
11430 watts divided by 230 volts = 50 amps (approx)
Table 310-16 indicates a No. 8 THW copper conductor is acceptable for the neutral. (Maximum unbalance of 50 amps as per 220-22.)

Another factor sometimes important when the load appears close to carrying capacity of a particular size wire is that of balancing circuits. In the "A" type apartment, for example, a three-wire lighting circuit, a three-wire utility circuit, and a three-wire cooking circuit are balanced across the ungrounded conductors and the neutral wire. But the 1500-watt, 115-volt

dishwasher which is connected between one ungrounded conductor and the neutral, does not balance the 950-watt garbage disposal and exhaust fan load which is connected between the other ungrounded conductor and the neutral.

In such case, an imaginary load may be assumed in parallel with the smaller load in order to gain balance, its value here being equal to 1500 watts minus 950 watts, or 550 watts. The listing then becomes:

Lighting and appliance	3745 watts
Oven and countertop	7475 watts
Dishwasher	1500 watts
Garbage disposal unit	600 watts
Exhaust fan	350 watts
Artificial load	550 watts
Total	14220 watts

Adjusted load:

14220 watts divided by 230 volts = 62 amps (approx) permitting No. 6 THW copper conductors (Table 310-16).

Although the calculation has not altered the result here, the method is worth noting. Some inspection authorities insist upon balancing loads; others do not.

SERVICE CALCULATIONS

NEC 220-3(c)(5) provides that a load of 180 volt-amperes per outlet shall be allowed for outlets other than for general illumination. The two plug receptacles in the "work" area come within such classification. NEC 220-13 permits this type of load to be included with general lighting load in applying demand factors. These plug receptacles and the three 200-watt night-light standards come within scope of the section. NEC 220-17 states that where four or more fixed appliances other than ranges, clothes dryers, air conditioning equipment or space heating units are connected to the same feeder or service in a multifamily dwelling, a demand factor of 75 percent may be applied to the load. The six dishwashers, six garbage disposals, six exhaust fans and electric water heater qualify here. The service conductor sizes are governed by NEC 230-42 which refers us to Article 220, so that service conductor sizes are calculated as follows:

Area to be lighted, approx 4220 sq ft
Lighting requirement equals:
4220 × 3 watts, or		12660 watts
Appliance circuits: 6 × 3000 watts =		18000 watts
Laundry circuit: 1500 watts		1500 watts
Three night-light standards:		
3 × 200 watts =		600 watts
Two receptacle outlets in work area:		
2 × 180 volt-amps =		360 watts
	Total	33120 watts

Load for lighting and appliance (NEC 220-11):
1st 3000 watts at 100%		3000 watts
Remainder, 30120 watts at 35%		10542 watts
	Total	13542 watts

Load for ovens and countertops:
6 × 4800 watts = 28800 watts		
6 × 6700 watts = 40200 watts		
	Total	69000 watts

Table 220-19, Column C, permits a demand factor of 32% for 12 ranges
.32 × 69000 watts =	22080 watts

Loads for dishwashers, garbage disposals, exhaust fans and water heater:
Dishwashers	6 × 1500 =	9000 watts
Disposals	6 × 600 =	3600 watts
Exhaust fans	6 × 350 =	2100 watts
Water heater	1 × 5000 =	5000 watts
	Total	19700 watts

.75 × 19700 watts =	14775 watts

Motor located in switchboard room
Load of motor (NEC 430-25 (a))
¾ hp, 1 phase 230 V × 6.9 amps = 1600 watts (approx)	
1600 watts × 125% =	2000 watts

Service load:
Lighting and appliance	13542 watts
Cooking units	22080 watts
Fixed appliance	14755 watts
Motor	2000 watts
Total	52397 watts

Service entrance conductors:
52397 watts divided by 230 volts = 228 amps
Table 310-16 shows that 4/0 Type THW copper conductors are acceptable.
Neutral conductor:
Lighting and appliance	13542 watts
Cooking units	
.7 × 22080 watts (NEC 220-22) =	15455 watts
Fixed appliances	14775 watts
Total	43772 watts

43772 watts divided by 230 volts = 190 amps
Table 310-16 shows that 3/0 Type THW copper conductor is acceptable as the neutral.
Service Size:
Two 4/0 and one 3/0 Type THW conductors, all in a 2-inch conduit are acceptable.

Note: Method of determining conduit size for different sizes of conductor in the same raceway will be taken up in a later assignment.

DETAILS AND ELEVATIONS

The larger drawing, made to ¼-inch scale, shows one of the five similar apartments. The plan view brings out details to better advantage than is possible on the general, or ⅛-inch, drawing, so that the position of equipment may be determined with greater accuracy.

Conduit in the floor slab is indicated by dot-and-dash line, that in wall or ceiling by solid black line. All wires are No. 14 except the three-wire No. 8 circuit to the cooking units and the three-wire No. 12 circuit to kitchen appliance receptacles.

Circuits may be traced out to learn the reason for three or four wires, as the case may be, in various conduits. Certain runs that might otherwise have been combined are kept separate to avoid the penalty of derating where more than three wires are installed in one raceway.

One point that should not be overlooked is the arrangement for connecting bedroom plug receptacles. They are supplied by conduit laid in the floor slab. This run continues up to the switch outlet box in the wall at the side of the door and then to the ceiling outlet. A small circle and the word *up* indicate this procedure. It should also be noted that one half of the symbol designating the duplex receptacle outlet in each of the living rooms is a solid color. This indicates that one half of this receptacle outlet is controlled by a wall switch.

Triangles *A*, *B*, and *C* marked in a semi-circle on the kitchen floor represent one scheme for directing attention to elevations seen at the right of the drawing. From these elevations it is possible to determine exact locations of outlets, such as those for plug receptacles and for cooking units. Location of the garbage disposal device and its method of connection are shown in elevation *B*.

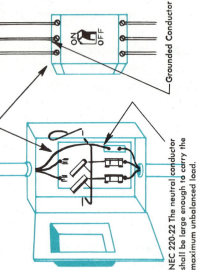

Grounded Conductor

4-Wire 3-Phase Circuit

ON OFF

Grounded Conductor

NEC 230-90 (b) No overcurrent device shall be inserted in a grounded service conductor except a circuit breaker which simultaneously opens all conductors.

NEC 220-22 The neutral conductor shall be large enough to carry the maximum unbalanced load.

NEC 230-42 (c) The grounded (neutral) conductors shall not be smaller than required by 250-23 (b)

NEC 250-23 (b) Service grounded conductor not smaller than grounding electrode conductor as per Table 250-94 and if phase conductors are larger than 1100 MCM copper or 1750 MCM aluminum, the grounded conductor shall not be smaller than 12½ percent of area of largest phase conductor except need never be larger than largest service ungrounded conductor.

NEC 240-20 An overcurrent device shall be placed in each ungrounded conductor. Individual single-pole circuit breakers may be inserted in each ungrounded conductor provided there is a grounded neutral, and voltage to ground is not over 150 volts. (See 210-6.)

NEC 110-10 Impedance and other circuit characteristics must be considered so that selected overcurrent device will clear the fault before serious damage results.

(See APPENDIX, page 187 for typical Fault Current Form.)

NEC RULES CONCERNING SERVICES AND FEEDERS

Approved terminal bar shall be provided inside cabinet and secured to cabinet for connection of all feeder and branch-circuit equipment grounding conductors, when such equipment grounding conductors are used. (384-27)

Terminal bar for equipment grounding conductors shall be secured to cabinet. (384-27)

Connection to neutral bar required in separate building as per 250-24 (a). (See 384-27 Ex 2)

NEC 373-6 (b) Conductors at terminals on entering or leaving cabinet shall have bending space as per Tables 373-6 (a) and 373-6 (b).

Ungrounded conductors No. 4 and larger must be protected by a substantial insulating material, securely fastened in place. (373-6 (c))

Insulating bushing shall have temperature rating not less than temperature rating of conductors. (373-6 (c))

NEC 373-11 (c) Cabinets and cutout boxes which contain devices or apparatus connected to more than 8 conductors, not including the supply circuit or a continuation thereof shall have back or side wiring spaces.

Panelboard protected on supply side by overcurrent devices not exceeding rating of panelboard. (384-16 (a), Ex 1)

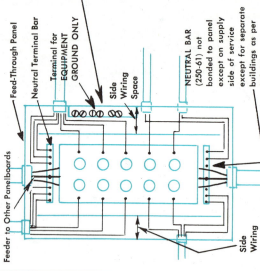

Feed-Through Panel

Neutral Terminal Bar

Terminal for EQUIPMENT GROUND ONLY

Side Wiring Space

NEUTRAL BAR (250-61) not bonded to panel except on supply side of service except for separate buildings as per 250-24.

Feeder to Other Panelboards

Side Wiring Space

Feeder to Panelboard

DISTRIBUTION PANELBOARDS

One service drop or lateral to a single-occupant building may feed 2 to 6 services that supply separate loads. (230-40 Ex. 2)

For the purpose of 230-40 only, 1/0 and larger multiple lateral services may be run to the same location, if connected together at supply end, but not connected together at other end. (230-2, Ex 7)

NEC 230-40 Each lateral or service drop to supply one set of service-entrance conductors except; (1) one set of service-entrance conductors permitted for each occupancy or group of occupancies in multiple occupancy bldgs.; (2) where 2 to 6 disconnects are grouped and in separate enclosures in one location and supply separate loads.

NEC 230-2 A building shall be served by one service. Where exceptions permit more than one service, a permanent plaque or directory required at each, indicating location of all others and area served by each.

NEC 230-84 (b) Ex In garages or outbuildings on residential property, the disconnecting means may consist of a snap switch.

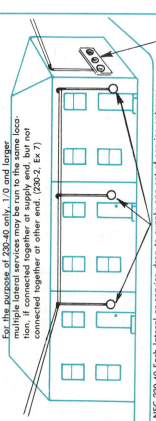

SERVICE—MULTIPLE OCCUPANCY BUILDING

NEC 230-43 Wiring methods for service conductors are limited to: (1) open wiring on insulators; (2) rigid metal conduit; (3) IMC; (4) EMT; (5) SE cable; (6) wireways; (7) busways; (8) auxiliary gutters; (9) PVC; (10) cablebus; (11) MC cable; (12) MI cable; (13) 6 ft. greenfield bonded as per 250-79

NEC 373-5 (c) Where cable is used, each cable required to be connected to cabinet. Two cables from one connector not permitted unless connector approved for two cables.

NEC 384-27 One conductor per terminal in equipment grounding terminal bar, unless terminal approved for two conductors

(See page 20 in this text for fused lighting and appliance panelboard.)

To Electric Utilities Meter

Service Entrance Conduit

Terminal Bar Mounting Screws

Equipment Grounding Conductor Terminal Bar

Branch circuits

Insulated Bushing

Service Entrance Conductor

Wiring Space

Neutral Terminal Bar

Service Disconnect Means

Branch circuits (240 volts)

Provisions to mount terminal bar in panelboard.

Branch circuits (120 volt-2 wire, or multi wire 120/240 volts)

Side Wiring Space Connections to Equipment Grounding Conductors

(Square D Company)

SERVICE—LIGHTING AND APPLIANCE PANELBOARD

Pump

Well Casing

Grounding Electrode Conductors

Grounding Terminal

NEC 250-83 (b) Underground systems or structures, such as piping systems or tanks, also permitted as "Other Electrodes".

Underground Tank.

NEC 250-83 Made and Other Electrodes:
(a) Electrically continuous underground gas piping system, without nonconductive coating permitted only when expressly permitted by gas supplier and authority enforcing the Code.
(b) Underground systems or tanks.
(c) Rods and pipes at least 8 ft long: (1) Galvanized pipe at 3/4 inch size; (2) Rod electrode at least 5/8 inch iron or 1/2 inch nonferrous; (3) Pipes and rods driven to depth of at least 8 ft unless rock, then buried in trench at least 2 1/2 ft deep or driven at oblique angle not over 45 degrees. (d) Buried plate of iron or steel at least 1/4 inch thick or 0.06 inch thick if of nonferrous metal. Plate electrode required to expose 2 square foot surface area to soil.

NEC 250-112 Grounding conductor connection should be on street side of meter. When not on street side, bonding is required around all meters, valves, service unions, etc.

(See page 30 in this text for "Grounding Electrode System".)

No. 10 or Smaller Conductor (110-14 (a-Ex))

Terminal Plates with Upturned Lugs

Conductors of dissimilar metals not to be intermixed unless terminal, lug, or connector suitable for purpose. (Some wire nuts suitable to splice copper to aluminum if in dry location.) (110-14)

Terminals shall be so identified when used for more than one conductor or used to connect aluminum wire. (110-14 (a))

SOLDERLESS LUGS

Crimp-lug (not for service conductors) (230-81)

Pressure Connectors

Terminals for aluminum shall be identified for such use.

NEC 110-14 (a) Connections to terminals shall insure good connection without damaging conductors.

CONNECTIONS TO TERMINALS

Grounding Terminal

Grounding Conductor

Gas Meter

Ground Clamp

To Street

NEC 250-83 (a) Electrically continuous underground gas piping system, without nonconductive coating, and then gas pipe permitted only when expressly permitted by gas supplier and authority enforcing the Code.

NEC 250-112 Attachment made in a manner to assure a permanent and effective ground.

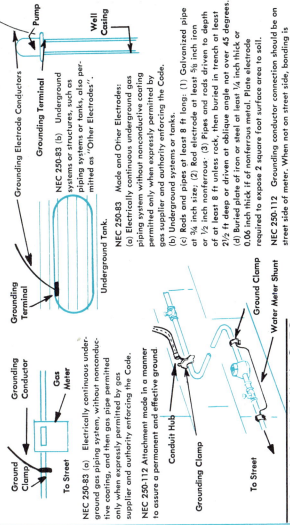

Conduit Hub

Grounding Clamp

Ground Clamp

Water Meter Shunt

To Street

NEC RULES PERTAINING TO GROUNDING ELECTRODES

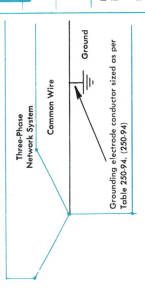

Operating NEC 380-8 (a) Center of grip of operating handle when in highest position not more than 6 1/2 feet from floor or platform.

NEC 373-8 Ex These conductors permitted if not more than 40 percent of cross-sectional area is used. For splices and taps, not over 75 percent of area may be used.

Operating Handle.

NEC 373-8 Switch enclosures are not to be used as auxiliary gutters or junction boxes unless adequate space is provided.

NEC 110-12 Equipment shall be installed in a neat, workmanlike manner.

NEC 110-13 (a) Electrical equipment to be securely fastened to mounting surface. Wooden plugs driven into masonry or concrete are not acceptable for support.

NEC 110-3 (a) (1) Evidence of suitability of equipment may be by listing or labeling.

SWITCH ENCLOSURES

NEC 250-81 (c) At least 20 feet of bare copper not smaller than No. 4 or steel reinforcing rod or bar encased by at least 2 inches of concrete within bottom of concrete footing in contact with the earth.

NEC 250-81 Where none of the electrodes listed in 250-81 are available at a building, one or more of the "Made and Other Electrodes" listed in 250-83 shall be used. (See page 30 in this text for "Grounding Electrode System.")

NEC 250-83 (d) .25 inch thick if iron or steel. .06 inch thick if nonferrous.

Buried metallic plate, surface area 2 square feet.

NEC 250-83 (c) (3) Driven rod at least 8 ft except where rock encountered, then rod driven at not more than 45° angle or buried in a trench at least 2 1/2 ft deep. Upper end of rod to be flush or below ground [with suitaibe clamp (250-115)] level unless rod and clamp are protected from damage as per 250-117.

NEC 250-83 (c-2) Rod electrode 5/8 inch iron, or 1/2 inch nonferrous.

Grounding Electrode Conductors

Approved Ground Clamps (NEC 250-115)

Aluminum wire permitted if used with proper UL listed ground clamp. Not permitted outside within 18 inches of earth. (250-92 (a) and 250-115))

Only one wire per clamp unless approved for purpose.

NEC 250-83—Rod or pipe electrodes of different systems shall be separated at least 6 feet. Lightning rod electrode may be bonded to other rods as permitted by 250-86. (Also see 250-46, 800-31 (b-7) and 820-22 (h).)

NEC 250-84 If resistance of made electrode exceeds 25 ohms, one additional electrode required and it shall be at least 6 ft away.

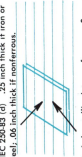

Trench

NEC 250-83 (c) (1) Galvanized iron pipe, 3/4 inch trade size.

NEC 250-83 Where more than one electrode system is used, they shall be separated at least 6 ft. (Two or more bonded electrodes treated as one system. (FPN))

NEC 250-83 "MADE" ELECTRODES

Three-Phase Network System

Common Wire

Ground

Grounding electrode conductor sized as per Table 250-94. (250-94)

NEC 250-25 (3) For multiphase systems having a conductor common to all phases, the common conductor shall be grounded.

Same conductor to be grounded for separately derived AC systems as per NEC 250-26.

NEC RULE PERTAINING TO GROUNDING IN MULTIPHASE SYSTEM

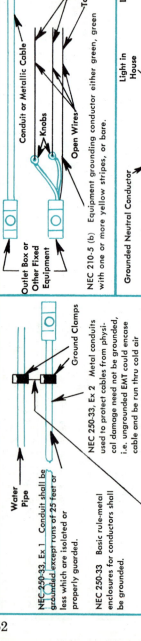

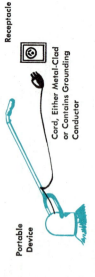

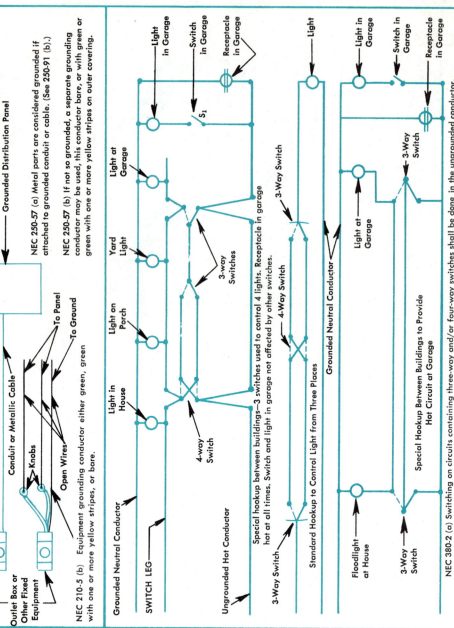

GROUNDING -- METAL ENCLOSURES

Water Pipe

Ground Clamps

NEC 250-33, Ex 1 Conduit shall be grounded except runs of 25 feet or less which are isolated or properly guarded.

NEC 250-33 Basic rule-metal enclosures for conductors shall be grounded.

NEC 250-33, Ex 2 Metal conduits used to protect cables from physical damage need not be grounded, i.e. ungrounded EMT could encase cable and be run thru cold air return duct and comply with 300-22 (b).

NEC Table 250-95 Equipment grounding conductor for 15, 20- and 30-amp circuits shall have ampacity at least equal to overcurrent protection of circuit. For circuits over 30 amps, use Table 250-95.

Grounded Distribution Panel

NEC 250-57 (a) Metal parts are considered grounded if attached to grounded conduit or cable. (See 250-91 (b).).

NEC 250-57 (b) If not so grounded, a separate grounding conductor may be used, this conductor bare, or with green or green with one or more yellow stripes on outer covering.

Conduit or Metallic Cable

To Panel

To Ground

Knobs

Open Wires

NEC 210-5 (b) Equipment grounding conductor either green, green with one or more yellow stripes, or bare.

Outlet Box or Other Fixed Equipment

GROUNDING CORD-AND-PLUG CONNECTED EQUIPMENT

Receptacle

Cord, Either Metal-Clad or Contains Grounding Conductor

Portable Device

NEC 250-59 Noncurrent-carrying metal parts of cord- and plug-connected equipment which must be grounded may be grounded: (a) by means of metal enclosure of conductors or (b) by a grounding conductor which is part of the cable or flexible cord, and which is either bare or has green or green with one or more yellow stripes on outer covering.

NEC 250-95 For grounding cord connected equipment protected at not more than 20 amps, No.18 wire may be used if part of an approved cord assembly. Where a separate grounding conductor is used, Table 250-95 shall be used to determine size. (See 250-94 Ex. 1)

NEC 250-45 (c) In residential occupancies the following cord- and plug connected equipment shall be grounded: clothes washers and dryers, refrigerators, freezers, air conditioners, dishwashers, electrical aquarium equipment, sump pumps, portable hand-held power tools and motor-operated appliances, such as saws, sanders, drills, lawn mowers, wet scrubbers, snow blowers, portable handlamps, and hedge clippers. Exception: Such tools and appliances protected by approved double insulation and so marked need not be grounded.

NEC 250-59 (a), Ex and (b), Ex The grounding member of grounding type attachment plugs on the supply cord of hand-held portable tools or appliances may be of the self-restoring type.

NEC 410-58 (c) A grounding terminal or grounding device shall be used only for grounding purpose.

Light in Garage

Switch in Garage

Receptacle in Garage

Light

Light in Garage

Switch in Garage

Receptacle in Garage

S1

Light at Garage

Yard Light

3-Way Switch

3-way Switches

4-Way Switch

3-Way Switch

Light on Porch

Light in House

4-way Switch

Grounded Neutral Conductor

SWITCH LEG

Ungrounded Hot Conductor

3-Way Switch

Special hookup between buildings-3 switches used to control 4 lights. Receptacle in garage hot at all times. Switch and light in garage not affected by other switches.

Grounded Neutral Conductor

Standard Hookup to Control Light from Three Places

Light at Garage

3-Way Switch

Floodlight at House

3-Way Switch

Special Hookup Between Buildings to Provide Hot Circuit at Garage

NEC 380-2 (a) Switching on circuits containing three-way and/or four-way switches shall be done in the ungrounded conductor. Wiring between switches and lights, if in metal enclosures, shall be run with both polarities in the same enclosure.

NEC 410-56 (a) Receptacles for portable cords shall be rated not less than 15 amps for 125 volts or 250 volts. (See Ex for 10 amps, 250 volts nonresidential. Also see 210-21 (b).)

NEC 210-7 (a), (b) Receptacles installed on 15-amp and 20-amp circuits shall be of grounding type.

NEC 210-7 (b) (c) Receptacles having grounded contacts must be grounded. Branch circuits must include a grounding conductor unless the raceway provides grounding. (Fine Print Note): For extension to non-grounding existing circuits, an adjacent water pipe may be used for grounding. (See 250-50 (a, b, Ex).)

NEC 210-7 (d) Grounding-type receptacles to be used as replacements for existing nongrounding types. Except where receptacle box does not have grounding means, a GFCI or nongrounding type receptacle is permitted.

RECEPTACLE GROUNDING

NONGROUNDING TYPE RECEPTACLE

"Hot" Wire Screws (Brass)

Equipment Grounding Screw (Green and Hexagonal)

GROUNDING TYPE RECEPTACLE

Neutral Wire Screw (Silver Color On Side Not Shown)

NEC 250-74, Ex 2 Support hole in yoke, if designed and listed for purpose, may be used to bond receptacle to flush-type boxes.

Neon Tubing

NEC 410-73 (e) Integral ballast protection required for indoor fluorescents. Also required for replacement ballasts. Ex Simple reactance ballasts are exempt.

NEC 410-76 (b) Fluorescent fixtures mounted on combustible low-density cellulose fiber-board required to have 1½-inch air space or be approved for direct mounting.

ELECTRIC-DISCHARGE LIGHTING

604-6(a): (1) Cable to be AC or MC with No. 12 copper and bare No. 12 copper grounding conductor or; (2) flex metal conduit with No. 12 copper and No. 12 insulated or bare copper grounding wire; (3) each section marked as per type of cable or conduit.

NEC 604-7 Unused outlets to be capped to close connector openings.

NEC 604-6 (c) Component parts to be listed for appropriate system.

NEC 604-6 (b) Receptacles and connectors to be locking type, uniquely polarized and identified for purpose and part of listed assembly for appropriate system.

NEC 604-4 Not permitted where conductors or cables limited as per Articles 333, 334.

NEC 604-5 Installations to conform to Articles 110, 200, 210, 220, 250, 300, 310, 333, 334, 350, 410, 545, 640, 700, 725, and 800 plus other applicable sections. This includes securing Type AC cable every 4½ ft and within 12 in. from boxes or fittings. (See 333-7 Ex. 1, 2, 3)

High-Voltage Terminals

SIGN TRANSFORMER

Transformer open-circuit potential not to exceed 1,000 volts. (410-73 (a))

Low-Voltage Leads

FLUORESCENT FIXTURE

NEC 410-75 (a), (b) Equipment having an open-circuit voltage of more than 1000 volts shall not be installed in dwelling occupancies. Equipment having an open-circuit voltage of more than 300 volts shall not be installed in dwelling occupancies unless so designed that no live parts are exposed when lamps are being handled.

NEC 410-76 (b) (Fine Print Note) Combustible low-density cellulose fiberboard includes sheets, panels and tiles formed of bonded plant fiber and has a density of 20 pounds or less per cubic foot. Does not include solid or laminated wood or material integrally treated with fire-retardang chemicals to limit flame spread to not over 25 as per ANSI A2.5-1977 test methods.

SWITCHES

NEC 380-14 (a) AC general-use snap switch, suitable for use on AC only, may be used to control resistive and inductive loads not to exceed the ampere rating of the switch or the voltage of the circuit, or tungsten-filament loads of rating of switch at 120 volts, or motor loads not to exceed 80% of rating of switch at rated voltage. Snap switches rated 20 amps or less shall be marked CO/ALR when used with aluminum wire.

NEC 380-8 Switches and circuit breakers shall be readily accessible and not over 6½ feet from floor or platform except: (1) busway switches permitted at same level as busway, (2) switches located adjacent to motors, appliances or other equipment, (3) hookstick operated isolating switches.

NEC 380-8(b) Voltage between adjacent snap switches in same box shall not exceed 300 volts. If voltage between adjacent switches exceeds 300 volts then permanent barriers must be installed.

NEC 380-14 (b-1, 2, 3) AC-DC general-use snap switches suitable for AC or DC may be used to control resistive loads not exceeding the ampere rating of the switch or the voltage of the circuit, or inductive loads not exceeding 50% of the ampere rating of the switch at applied voltage, or when horsepower rated may control motor leads not exceeding voltage applied, or tungsten-filament loads not exceeding the ampere rating of the switch at applied voltage, when "T" rated. (See 430-83, 430-109, 430-110 for motors, and 600-2 for signs.)

Grounding Screw

(LEVITON)

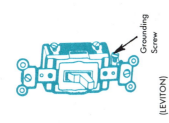

MANUFACTURED WIRING SYSTEM--ARTICLE 604

NEC 604-1 applies to field-installed wiring using factory manufactured subassemblies for branch circuits, signaling and communication circuits in accessible areas. (Drop ceiling with lift-out panels considered accessible—See Definitions.)

NEC 604-3 Permitted in accessible dry areas and plenums when listed and installed as per 300-22. Ex Permitted to extend into hollow walls for termination of switch and outlets.

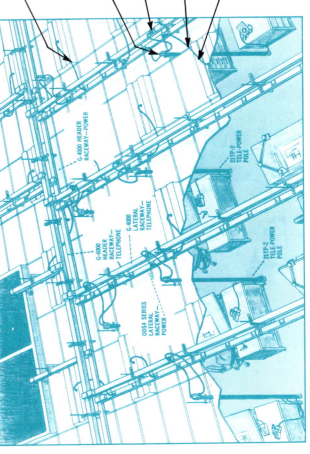

G-4000 HEADER RACEWAY—POWER

21TP-2 TELE-POWER POLE

G-6000 HEADER RACEWAY—TELEPHONE

G-4000 LATERAL RACEWAY—TELEPHONE

ODS4 SERIES LATERAL RACEWAY—POWER

Illustration, courtesy; The Wiremold Co. ODS® (Overhead Distribution System)

NEC 342-7 (a-2) Supported at intervals not over 8 inches apart except where fed by attachment cord may be 12 inches. Extension to be free from contact with any conductive material.

NEC 342-4 (e) The extension shall not run outside the room where it originates.

NEC 342-5 Must be unbroken lengths except for taps.

NEC 342-7 (a-1) Not on floor or within 2 inches of floor.

Tap

Existing Outlet

Supply Connection

NEC 342-3 Nonmetallic surface extensions permitted: (a) extensions from an existing outlet on a 15- or 20-amp circuit, (b) in an exposed dry location, (c) where building is occupied for residential or office purposes and not over 3 floors in height as per 336-4(a).

NEC 342-4 They are not permitted: (a) as aerial cable to substitute for one of the general wiring methods, (b) in unfinished basements, attics, or roof spaces, (c) if the voltage between conductors exceeds 150 volts, (d) if subject to corrosive vapors, or (e) if extended outside room where they originate.

NEC 342 NONMETALLIC SURFACE EXTENSIONS

NEC 210-23 Individual branch circuits may supply any loads.

NEC 210-21 (b-1) A single receptacle on an individual branch circuit shall have rating not less than that of the branch circuit.

NEC 210-23 Individual branch circuits may supply any loads. Branch circuits having two or more outlets may supply only loads as follows: (a) 15- and 20-amp circuits may supply lighting units and/or appliances, the rating of any cord-and-plug connected appliance not exceeding 80 percent of circuit, and total rating of fixed appliances not over 50 percent; (b) 30-amp circuits may supply heavy-duty lampholders in other than dwellings, or appliances anywhere, the rating of a cord-and-plug connected appliance not exceeding 24 amps; (c) 40- or 50-amp branch circuits may supply fixed cooking appliances in any occupancy or fixed lighting units with heavy-duty lampholders or infrared heating units or other utilization equip. in other than dwellings; (d) circuits over 50 amps to supply only non-lighting outlet loads.

BRANCH CIRCUITS -- PERMITTED LOADS

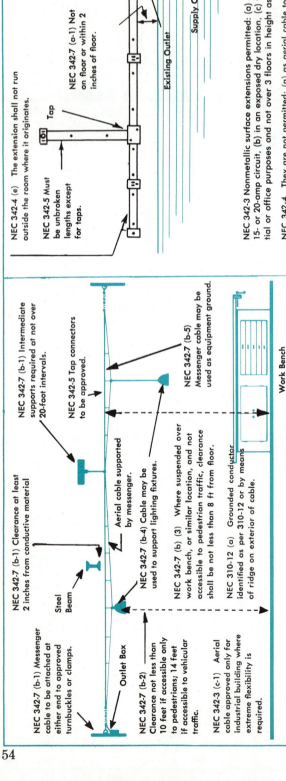

NEC 342-7 (b-1) Clearance at least 2 inches from conductive material

NEC 342-7 (b-1) Intermediate supports required at not over 20-foot intervals.

NEC 342-5 Tap connectors to be approved.

NEC 342-7 (b-5) Messenger cable may be used as equipment ground.

Aerial cable supported by messenger.

NEC 342-7 (b-4) Cable may be used to support lighting fixtures.

NEC 342-7 (b) (3) Where suspended over work bench, or similar location, and not accessible to pedestrian traffic, clearance shall be not less than 8 ft from floor.

NEC 310-12 (a) Grounded conductor identified as per 310-12 or by means of ridge on exterior of cable.

Work Bench

NEC 342-7 (b-1) Messenger cable to be attached at either end to approved turnbuckles or clamps.

Steel Beam

NEC 342-7 (b-2) Clearance not less than 10 feet if accessible only to pedestrians; 14 feet if accessible to vehicular traffic.

NEC 342-3 (c-1) Aerial cable approved only for industrial building where extreme flexibility is required.

Outlet Box

NEC 342-4 (c) Not approved where voltage between conductors exceeds 300 volts.

NEC 342 NONMETALLIC EXTENSIONS--AERIAL CABLE

Ungrounded Conductors.

Grounded Conductor.

Four-Wire Circuit

Grounding Conductor

Black
Red
Blue
White or Natural Gray
Green

NEC 310-12 (c) Ungrounded conductors shall be identified by colors, or a combination of colors plus a distinguishing marking, other than white, natural gray, or green. These colors shall clearly distinguish the ungrounded conductor from the grounded or grounding conductor and shall not conflict with markings required by 310-11 (b-1).

NEC 210-5(a) The grounded conductor of a branch circuit shall be white or natural gray. Each other system in same raceway or enclosures shall have the grounded conductor (if required) identified by outer covering of white with colored stripe (not green).

NEC 310-12 Grounded conductors No. 6 or smaller shall be white or natural gray. For sizes over No. 6, see 200-6. (Also see 210-5.)

NEC 210-5 (b) Any conductor intended for grounding purposes shall be green, green with one or more yellow stripes, or bare.

CONDUCTOR IDENTIFICATION

NEC RULES DEALING WITH BRANCH CIRCUITS

(See page 86 this text.)

NEC 384-16 (a) Lighting and appliance panelboards shall be protected on supply side not to exceed rating of panelboard. (See Exceptions.)

NEC 240-30 Overcurrent devices should be completely enclosed. Exception: The operating handle of a circuit breaker may project from a cabinet.

NEC 240-51 (b) Plug fuses of Edison-base type may be used as replacement in existing installations where no evidence of overfusing.

Telescoping Cover on Cutout Box (See Article 100)

Fuse Panel

PANELBOARD -- LIGHTING AND APPLIANCE

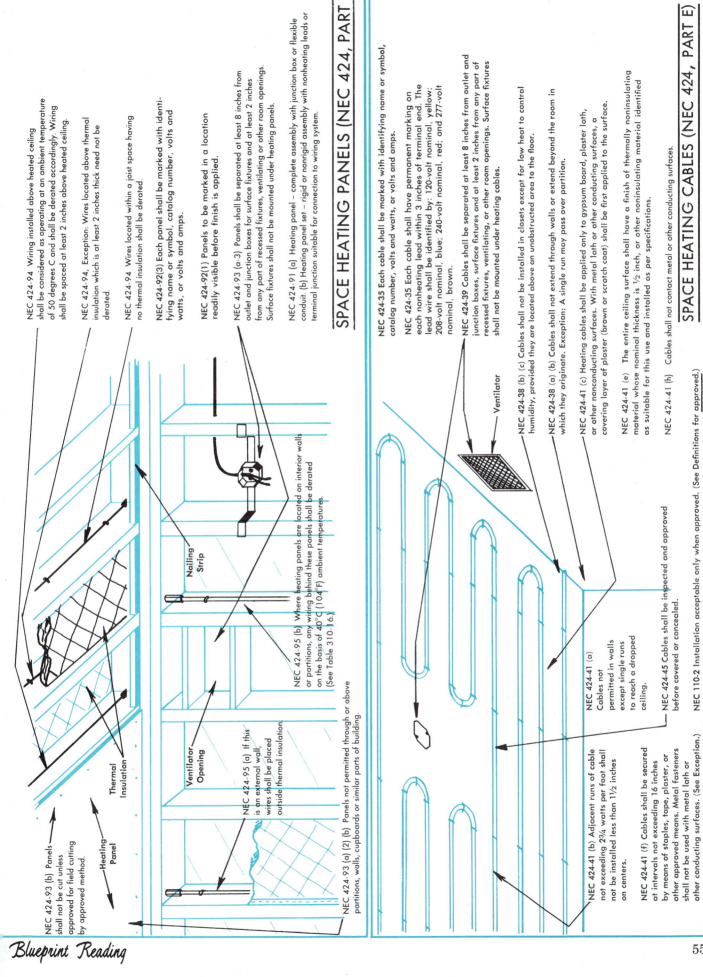

SPACE HEATING PANELS (NEC 424, PART I)

NEC 424-94 Wiring installed above heated ceiling shall be considered as operating at an ambient temperature of 50 degrees C and shall be derated accordingly. Wiring shall be spaced at least 2 inches above heated ceiling.

NEC 424-94, Exception: Wires located above thermal insulation which is at least 2 inches thick need not be derated.

NEC 424-94 Wires located within a joist space having no thermal insulation shall be derated.

NEC 424-92(3) Each panel shall be marked with identifying name or symbol, catalog number, volts and watts, or volts and amps.

NEC 424-92(1) Panels to be marked in a location readily visible before finish is applied.

NEC 424-93 (a-3) Panels shall be separated at least 8 inches from outlet and junction boxes for surface fixtures and at least 2 inches from any part of recessed fixtures, ventilating or other room openings. Surface fixtures shall not be mounted under heating panels.

NEC 424-91 (a) Heating panel – complete assembly with junction box or flexible conduit. (b) Heating panel set – rigid or nonrigid assembly with nonheating leads or terminal junction suitable for connection to wiring system.

NEC 424-95 (b) Where heating panels are located on interior walls or partitions, any wiring behind these panels shall be derated on the basis of 40°C (104°F) ambient temperatures. (See Table 310-16.)

Nailing Strip

NEC 424-95 (a) If this is on external wall, wires shall be placed outside thermal insulation.

Ventilator Opening

Thermal Insulation

Heating Panel

NEC 424-93 (b) Panels shall not be cut unless approved for field cutting by approved method.

NEC 424-93 (a) (2) (b) Panels not permitted through or above partitions, walls, cupboards or similar parts of building.

SPACE HEATING CABLES (NEC 424, PART E)

NEC 424-35 Each cable shall be marked with identifying name or symbol, catalog number, volts and watts, or volts and amps.

NEC 424-35 Each cable shall have permanent marking on each nonheating lead within 3 inches of terminal end. The lead wire shall be identified by: 120-volt nominal, yellow; 208-volt nominal, blue; 240-volt nominal, red; and 277-volt nominal, brown.

NEC 424-39 Cables shall be separated at least 8 inches from outlet and junction boxes, surface fixtures and at least 2 inches from any part of recessed fixtures, ventilating, or other room openings. Surface fixtures shall not be mounted under heating cables.

Ventilator

NEC 424-38 (b) (c) Cables shall not be installed in closets except for low heat to control humidity, provided they are located above an unobstructed area to the floor.

NEC 424-38 (a) (b) Cables shall not extend through walls or extend beyond the room in which they originate. Exception: A single run may pass over partition.

NEC 424-41 (c) Heating cables shall be applied only to gypsum board, plaster lath, or other nonconducting surfaces. With metal lath or other conducting surfaces, a covering layer of plaster (brown or scratch coat) shall be first applied to the surface.

NEC 424-41 (e)   The entire ceiling surface shall have a finish of thermally noninsulating material whose nominal thickness is ½ inch, or other noninsulating material identified as suitable for this use and installed as per specifications.

NEC 424-41 (h)   Cables shall not contact metal or other conducting surfaces.

NEC 424-41 (a) Cables not permitted in walls except single runs to reach a dropped ceiling.

NEC 424-45 Cables shall be inspected and approved before covered or concealed.

NEC 110-2 Installation acceptable only when approved. (See Definitions for approved.)

NEC 424-41 (b) Adjacent runs of cable not exceeding 2¾ watts per foot shall not be installed less than 1½ inches on centers.

NEC 424-41 (f) Cables shall be secured at intervals not exceeding 16 inches by means of staples, tape, plaster, or other approved means. Metal fasteners shall not be used with metal lath or other conducting surfaces. (See Exception.)

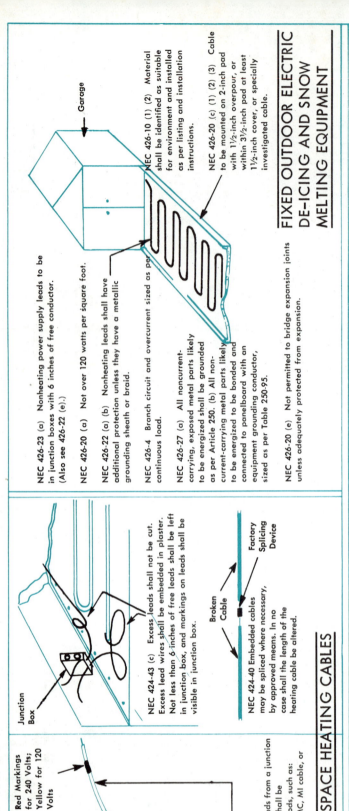

Garage

NEC 426-23 (a)   Nonheating power supply leads to be in junction boxes with 6 inches of free conductor. (Also see 426-22 (e).)

NEC 426-20 (a)   Not over 120 watts per square foot.

NEC 426-22 (a) (b)   Nonheating leads shall have additional protection unless they have a metallic grounding sheath or braid.

NEC 426-4   Branch circuit and overcurrent sized as per continuous load.

NEC 426-27 (a)   All noncurrent-carrying, exposed metal parts likely to be energized shall be grounded as per Article 250. (b) All non-current-carrying metal parts likely to be energized to be bonded and connected to panelboard with an equipment grounding conductor, sized as per Table 250-95.

NEC 426-10 (1) (2)   Material shall be identified as suitable for environment and installed as per listing and installation instructions.

NEC 426-20 (c) (1) (2) (3)   Cable to be mounted on 2-inch pad with 1½-inch overpour, or within 3½-inch pad at least 1½-inch cover, or specially investigated cable.

NEC 426-20 (e)   Not permitted to bridge expansion joints unless adequately protected from expansion.

FIXED OUTDOOR ELECTRIC DE-ICING AND SNOW MELTING EQUIPMENT

Junction Box

NEC 424-43 (c)   Excess leads shall not be cut. Excess lead wires shall be embedded in plaster. Not less than 6 inches of free leads shall be left in junction box, and markings on leads shall be visible in junction box.

Broken Cable

Factory Splicing Device

NEC 424-40   Embedded cables may be spliced where necessary, by approved means. In no case shall the length of the heating cable be altered.

SPACE HEATING CABLES

Red Markings for 240 Volts; Yellow for 120 Volts

Coil of Heating Cable

NEC 424-34   Heating cables shall have factory-assembled nonheating, general-use wire leads at least 7 feet in length.

NEC 424-35   Nonheating leads of heating cable shall be marked and color coded: 120-volt yellow, 208-volt blue, 240-volt red and 277-volt brown.

NEC 424-43 (a)   Free nonheating leads from a junction box to a location within the ceiling shall be installed with approved wiring methods, such as: conductors in raceways, Type UF, NMC, MI cable, or other approved conductors.

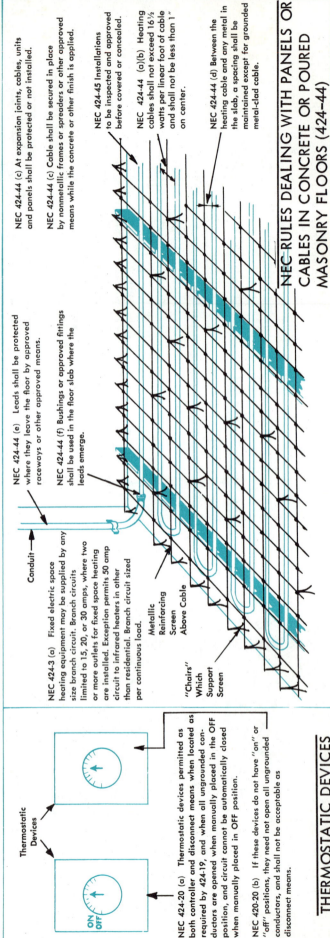

NEC 424-44 (c)   At expansion joints, cables, units and panels shall be protected or not installed.

NEC 424-44 (c)   Cable shall be secured in place by nonmetallic frames or spreaders or other approved means while the concrete or other finish is applied.

NEC 424-45   Installations to be inspected and approved before covered or concealed.

NEC 424-44 (a)(b)   Heating cables shall not exceed 16½ watts per linear foot of cable and shall not be less than 1" on center.

NEC 424-44 (d)   Between the heating cable and any metal in the slab, a spacing shall be maintained except for grounded metal-clad cable.

NEC RULES DEALING WITH PANELS OR CABLES IN CONCRETE OR POURED MASONRY FLOORS (424-44)

NEC 424-44 (e)   Leads shall be protected where they leave the floor by approved raceways or other approved means.

NEC 424-44 (f)   Bushings or approved fittings shall be used in the floor slab where the leads emerge.

Conduit

Metallic Reinforcing Screen Above Cable

"Chairs" Which Support Screen

NEC 424-3 (a)   Fixed electric space heating equipment may be supplied by any size branch circuit. Branch circuits limited to 15, 20, or 30 amps, where two or more outlets for fixed space heating are installed. Exception permits 50 amp circuit to infrared heaters in other than residential. Branch circuit sized per continuous load.

Thermostatic Devices

ON OFF

NEC 424-20 (a)   Thermostatic devices permitted as both controller and disconnect means when located as required by 424-19, and when all ungrounded conductors are opened when manually placed in the OFF position, and circuit cannot be automatically closed when manually placed in OFF position.

NEC 420-20 (b)   If these devices do not have "on" or "off" positions, they need not open all ungrounded conductors, and shall not be acceptable as disconnect means.

THERMOSTATIC DEVICES

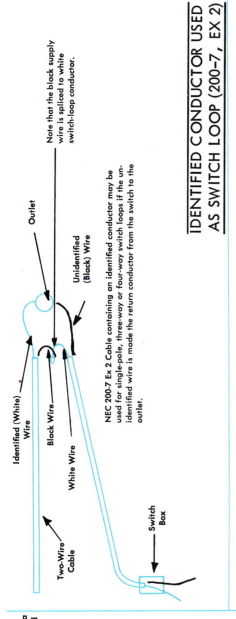

Note that the black supply wire is spliced to white switch-loop conductor.

Outlet

Unidentified (Black) Wire

Identified (White) Wire

Black Wire

White Wire

Two-Wire Cable

Switch Box

IDENTIFIED CONDUCTOR USED AS SWITCH LOOP (200-7, EX 2)

NEC 200-7 Ex 2 Cable containing an identified conductor may be used for single-pole, three-way or four-way switch loops if the un-identified wire is made the return conductor from the switch to the outlet.

Control Switch For Food Waste Disposer

Food Waste Disposer

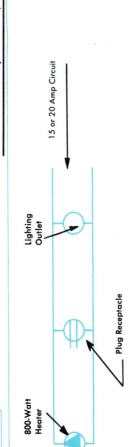

Disconnect in circuit-breaker panel only acceptable if in sight of motor controller.

NEC 422-26 The disconnecting means for a motor-driven appliance of more than 1/8 hp shall be located within sight of the motor controller and shall comply with Part H of Article 430.

NEC 422-26, Ex   Where appliance unit switch disconnects all ungrounded conductors, and switch is marked with "off" position, then switch or circuit breaker permitted to be out of sight, only if conforming to 422-24 (a, b, c, d).

DISCONNECT MEANS -- MOTOR APPLIANCE (OVER 1/8 H.P.) (422-26)

---

15 or 20 Amp Circuit

Lighting Outlet

Plug Receptacle

800-Watt Heater

NEC 422-27 (a) Appliances (other than motors) shall be considered as protected against overcurrent when supplied by branch circuits of 422-5, 422-6, and 422-27 (e). Exception: For motors see NEC 430 and 440.

APPLIANCES -- BRANCH CIRCUIT PROTECTION (422-27)

Original Building

DWELLING OCCUPANCIES   New Addition or Area Not Previously Wired

NEC 220-3 (d-1) When calculating loads in existing dwellings for previously unwired portions or new additions, either of which exceeds 500 square feet, the watts-per-square-foot basis shall be used. For new circuits or extended circuits, the watts-per-square-foot, or the actual total load served (as per 220-3 (c)) shall be used.

NEC 220-3(d-2) In other than dwellings, adding new or extending existing circuits to be calculated as per 220-3(b, c).

CALCULATING LOADS -- BUILDING ADDITIONS

NEC 424-65   Duct heater controller equipment shall be accessible with disconnect at or within sight of controller. (Ex see 424-19 (a).)

NEC 424-58   Heaters installed in air ducts must be identified as suitable for such use.

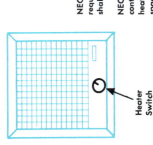

NEC 424-11 Fixed electric space heaters requiring supply conductors over 60°C shall be clearly and permanently marked.

NEC 424-13 Unless approved for direct contact with combustible material, fixed heating equipment must have required spacing.

Heater Switch

NEC 424-12 (a) (b)   Fixed electric space heating equipment shall not be used: (a) where exposed to severe mechanical injury, unless adequately protected, nor (b) in damp or wet locations, unless approved for such locations.

ELECTRIC HEATERS -- INSTALLATION

Blueprint Reading

NEC 422-15 (b) Infrared lamps over 300 watts shall not use screw-shell lampholders (See Ex.)

NOT OVER 300 WATTS

NEC 422-15 (c) Infrared lamps permitted in series, and each group considered one appliance.

NEC 422-15 (a) Infrared heating lamps rated at 300 watts or less may be used with lampholders of the medium-base unswitched type, or other types approved for the purpose.

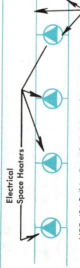

Medium Base

Infrared Heat Lamp Used in Industrial Heating Appliances

INFRARED LAMPS -- INDUSTRIAL HEATING

NEC 220-19 Thirty 12 kW ranges supplied by a 3φ, 4-wire, 120/208-volt feeder, equal number of ranges on each feeder.

10 12 kW Ranges

Neutral

NEC 220-19 Where single-phase ranges are supplied by a 3-phase, 4-wire feeder, the load shall be computed (as per Table 220-19) on a demand basis of twice the maximum number between any two phases.

Method No. 1:
10 12 kW ranges per phase $\times$ 2 $=$ 20 ranges
Table 220-19 gives 35 kW.
35 kW $\div$ 2 $=$ 17,500 $\div$ 120 $=$ 146 amps

Method No. 2
10 $\times$ 2 $=$ 20 ranges
Table 220-19 gives 35 kW.
35 kW $\div$ 2 $=$ 17,500 $\times$ 3 $=$ 52,500 watts
$$\frac{52,500}{208 \times 1.73} = 146 \text{ amps}$$

ELECTRIC RANGES -- 3 Ø FEEDER DEMAND

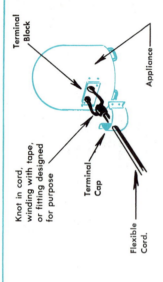

Terminal Block

Appliance

Knot in cord, winding with tape, or fitting designed for purpose

Terminal Cap

Flexible Cord.

NEC 400-10 Flexible cords shall be so connected to devices and fittings that tension will not be transmitted to joints or terminal screws.

NEC 422-8 (c), 400-7 (a-6) Flexible cord permitted for connection to appliances to facilitate frequent interchange.

NEC 210-23(a) One cord-and plug appliance on a 15- or 20-amp ckt. not to exceed 80% of ckt. rating. Total fastened in place appliances not to exceed 50% of circuit rating if lights and other portable equip., or both are supplied.

CORD CONNECTED APPLIANCES

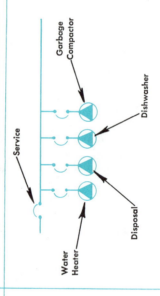

Garbage Compactor

Dishwasher

Service

Disposal

Water Heater

NEC 220-17 Where four or more fixed appliances, other than electrical ranges, clothes dryers, air conditioning equipment, or space heating equipment, are connected to the same feeder in a single-family or multifamily dwelling, a demand factor of 75 percent may be applied to the fixed appliance load.

NEC 220-21 In making load calculations, the smaller of two dissimilar loads may be omitted from the total where it is unlikely that both loads will be served simultaneously.

FIXED APPLIANCE AND NONCOINCIDENT LOADS -- DWELLINGS

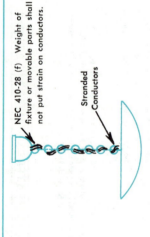

Stranded Conductors

NEC 410-28 (f) Weight of fixture or movable parts shall not put strain on conductors.

NEC 410-28 (e) Stranded conductors shall be used on chain fixtures and other movable parts.

NEC 410-17 Fixtures required to be grounded.

NEC 410-20 All fixtures with exposed metal parts shall provide means to connect equipment grounding conductor to the fixture.

NEC 410-4 (d) No parts of pendants or hanging fixtures permitted within 8 feet vertically or 5 feet horizontally from top of bathtub.

CHAIN SUPPORTED FIXTURES (410-26)

Electrical Space Heaters

NEC 424-3 (b) Branch circuit and overcurrent device sized 125% of heaters.

NEC 220-15 The computed load of a feeder supplying fixed electrical space heating equipment shall be the total load connected thereto unless duty-cycle features are present.

NOTE: NEC 220-15 does not apply when feeder capacity is calculated in accordance with optional methods in NEC 220-30, 220-31 for single dwelling or single apartment in multifamily dwelling or in NEC 220-32 for multifamily dwellings.

NEC 424-22 (b) When resistance-type heating elements of over 48 amps are used in space heating equipment, they shall be subdivided into smaller circuits of not over 48 amps and protected at not more than 60 amps. The rating of the overcurrent device shall be as per 424-3 (b) where subdivided load is less than 48 amps. See 424-72 (a) for resistance type boilers.

SPACE HEATING EQUIPMENT OVERCURRENT PROTECTION

GROUNDING -- RANGES AND SIMILAR APPLIANCES

NEC 250-60 Frames of electric ranges, cook-tops, ovens, dryers, and their associated outlet or junction boxes shall be grounded the same as other electrical equipment except the grounded circuit conductor (neutral) may be used as a grounding means if all of the following conditions are met: (a) The circuit is 120/240-volt, single-phase, 3-wire, or 120/208-volt derived from a 3-phase, 4-wire system; (b) The grounded conductor is No. 10 or larger; (c) The grounded conductor is insulated, or is uninsulated and part of a SE Cable and is fed from service equipment; (d) The grounded contacts of receptacles are bonded to the equipment. (The above does not apply to mobile homes or travel trailers.)

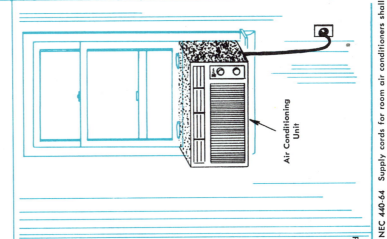

Air Conditioning Unit

NEC 440-64 Supply cords for room air conditioners shall not exceed 10 ft at 120 volts or 6 ft at 208 or 240 volts.

NEC 440-60 Generally a room air conditioner means a room air conditioner for the wall of room being cooled, incorporate a hermetic-refrigerant motor-compressor, and operate at not over 250 volts, single-phase. A 3-phase unit rated over 250 volts shall be directly wired.

(Dryer diagram labels)
- 3-Wire Cap
- Grounding Strap from Neutral Terminal to Frame of Dryer
- Cord so connected so tension not transmitted to terminals. Accomplished here by cord connector. (NEC 410-10)
- 3-Wire Cord Grounded Circuit Conductor Not Smaller Than No. 10
- Cord so connected so tension not transmitted to terminals. Accomplished here by cord connector.
- Electric Dryer
- Cover Plate

NEC 250-61 Grounding of room air conditioning units shall be in accordance with 250-42, 250-43, and 250-45.

NEC 250-42 (a) (b) (c) (d) (e) (f) Exposed noncurrent-carrying metal parts of fixed equipment shall be grounded if: (a) within 8 feet vertically or 5 feet horizontally of ground or grounded objects and possible contact by persons, (b) in damp or wet location and not isolated, (c) in electrical contact with metal, (d) in hazardous location, (e) if fed by metal enclosed wiring (except for 25 feet of isolated raceway as per 250-33), (f) general when voltage exceeds 150 volts to ground.

NEC 250-45 Exposed metal noncurrent-carrying parts of cord- and plug-connected equipment shall be grounded if: (a) in hazardous location, (b) over 150 volts to ground, (c) in residential occupancies: refrigerators, freezers, air conditioners, clothes washers, clothes dryers, dishwashers, sump pumps, and portable tools unless double insulated. (See 250-45 (d) for grounding of cord-and-plug equipment in other than residential occupancies.)

NEC 440-62 (c) The total load of cord- and plug-connected room air conditioning equipment shall not exceed 50% of the rating of a branch circuit which also supplies lighting units or appliances.

NEC 440-62 (a) In most cases a cord-and-plug connected room air conditioner is considered as a single motor.

NEC 440-63 A plug and receptacle may serve as disconnecting means for a single-phase air conditioner rated 250 volts or less when the manual controls are readily accessible and within 6 feet of floor, or a manually operable switch is readily accessible and within sight of unit.

NEC 440-14 Disconnect within sight and readily accessible for air conditioning or refrigerating equipment except for appliances connected by cord and plug.

NEC 440-62 (b) The rating of a cord- and plug-connected room air conditioner shall not exceed 80% of its individual branch circuit rating.

AIR CONDITIONING EQUIPMENT

(Range diagram labels)
- Range 8¾ kW or Over
- 3-Wire Range Circuit
- No. 8
- No. 10 Neutral
- No. 8
- R

NEC 210-19 (b) Conductors for ranges 8¾ kW and larger shall not be rated less than 40 amps.
Exception: The neutral conductor for a household electric range may have an ampacity equal to 70% of that of the branch-circuit rating, but shall not be smaller than No. 10 in any case.

NEC 110-3 (b) Equipment that is listed or labeled shall be installed or used as per instructions included in the listing or labeling. (For additional information, see UL standard for Safety, Central Cooling Air Conditioners, No. 465-Paragraph 56.25.)

NEC 440-14 Disconnect means within sight and readily accessible from unit. Permitted on or within equip. See 430-107 for readily accessible. Ex. cord-and plug-connected appliances.

Where fuses are indicated in installation instructions and on nameplate, then fuses shall be used somewhere in circuit.

Where proper fuse is used in panel, then non-fused disconnect or circuit breaker can be used. When nameplate states that unit is to be protected by "overcurrent device" or "circuit breaker or fuse", then a HACR (Heating, Air-Conditioning and Refrigeration) listed breaker is OK.

(Central A/C diagram labels)
- Central Air Conditioner (Listed for Outdoor Use)
- Nameplate (NEC 440-4)
- Fused Disconnect at Unit
- Circuit Breaker at Service Panel

TYPICAL NAMEPLATE INFORMATION

	HP	VOLTS	PHASE	HZ	FLA	LRA
COMPRESSOR		230	1	60	15.3	71
FAN	1/4	230	1	60	2.0	4.7

Fuse Size—Time Delay 25 Amp Max 35A

ALSO on nameplate "Listed Section of UL 465N, Central Cooling Air Conditioner." ALSO: Installation and operating instructions list fuses for overcurrent protection.

NEC 440-12 (a) Disconnect selected on basis of nameplate rated-load current, or branch-circuit selection current (whichever is greater) and locked-rotor current: (a-1) At least 115% of FLA where non-horsepower disconnect (such as circuit breaker) is used. (a-2) To determine size of h.p. rated disconnect, as required by 430-109; the h.p. corresponding to the FLA in Tables 430-148, 149, 150; the h.p. corresponding to the LRA in Table 430-151, where different h.p. ratings are obtained, the largest h.p. shall be used.

VOLTAGE DROP FORMULAS

The K factor (constant) is the resistance of a circular-mil-foot of wire (a wire 1 foot long and having a cross sectional area of 1 cir mil). The resistance increases with the temperature from 10.4 at 20°C (68°F) to 10.75 at 25°C (77°F). The American Electricians Handbook, The Standard Handbook for Electrical Engineers and other engineering data (also refer to 1981 NEC Chapter 9, Table 8) lists the resistance of copper and aluminum conductors in ohms per 1000 ft. at 25°C (77°F). Based on this temperature we will use a K (constant) of 10.7 for copper and 17.4 for aluminum conductors.

K = 10.7
VD = voltage drop
L = length of circuit in feet (one way)
I = current
CM = circular mil area of conductor

$$VD = \frac{2KLI}{CM} \quad \text{(single phase circuit)}$$

$$VD = \frac{KLI\ 1.73}{CM} \quad \text{or} \quad VD = \frac{2KLI\ 0.866}{CM}$$

(3-wire, 3-phase circuit or 4-wire, 3-phase wye circuit with balanced load)

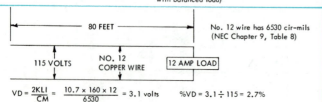

No. 12 wire has 6530 cir-mils (NEC Chapter 9, Table 8)

$$VD = \frac{2KLI}{CM} = \frac{10.7 \times 160 \times 12}{6530} = 3.1 \text{ volts} \qquad \%VD = 3.1 \div 115 = 2.7\%$$

Check by using a resistance of 1.62 ohms per M/ft.

VD = IR   R = 0.00162 Ω × 160 ft = 0.2592

VD = 12 × 0.2592 = 3.1 volts

> THIS CIRCUIT HAS A VOLT DROP OF 3.1 VOLTS OR 2.7% AT 115 VOLTS WITH A 12 AMP LOAD.

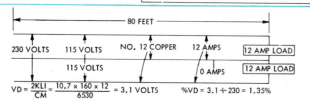

$$VD = \frac{2KLI}{CM} = \frac{10.7 \times 160 \times 12}{6530} = 3.1 \text{ VOLTS} \qquad \%VD = 3.1 \div 230 = 1.35\%$$

> THIS CIRCUIT HAS A VOLT DROP OF 3.1 VOLTS OR 1.35% AT 115/230 VOLTS WITH A 12 AMP, 115/230 VOLT BALANCED LOAD. THE NEUTRAL CARRIES NO CURRENT.

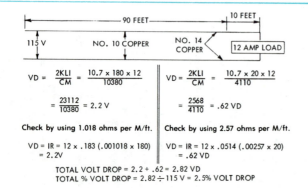

$$VD = \frac{2KLI}{CM} = \frac{10.7 \times 180 \times 12}{10380} \qquad\qquad VD = \frac{2KLI}{CM} = \frac{10.7 \times 20 \times 12}{4110}$$

$$= \frac{23112}{10380} = 2.2 \text{ V} \qquad\qquad\qquad\qquad = \frac{2568}{4110} = .62 \text{ VD}$$

Check by using 1.018 ohms per M/ft.   Check by using 2.57 ohms per M/ft.

VD = IR = 12 × .183 (.001018 × 180)   VD = IR = 12 × .0514 (.00257 × 20)
    = 2.2V   = .62 VD

TOTAL VOLT DROP 2.2 + .62 = 2.82 VD
TOTAL % VOLT DROP 2.82 ÷ 115 V = 2.5% VOLT DROP

Three-phase, 120/208 Volt (wye), 4-wire feeder with 45 amps per phase with balanced load.

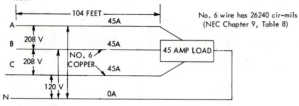

No. 6 wire has 26240 cir-mils (NEC Chapter 9, Table 8)

See page 186 this book for formulas of unbalanced load.

Any of the following formulas may be used.

$$VD = \frac{KLI\ 1.73}{CM} = \frac{86631}{26240} = 3.3 \text{ VD}$$

$$\text{or } VD = \frac{2KLI \times 0.866}{CM} = \frac{86732}{26240} = 3.3 \text{ VD}$$

Voltage drop is 3.3 Volts or 3.3 ÷ 208 = 1.6% VD

Check by using .410 ohms per M/ft.

VD = I × R × 1.73   where R = 0.00041 × 104 = 0.043 Ω
VD = 45 × 0.043 × 1.73 = 3.3 VD or 3.3 ÷ 208 = 1.6% VD

or, by using 120 volts (phase to neutral):
VD = IR = 45 amps × 0.043 = 1.94 VD or 1.94 ÷ 120 V = 1.6% VD

Voltage drop in all of above is 1.6% of circuit voltage.

TO FIND CONDUCTOR SIZE when the voltage drop, length of circuit and current are known:

$$CM = \frac{2 \times K \times L \times I}{VD} \quad \text{(single-phase, 2-wire circuit)}$$

$$CM = \frac{K \times L \times I \times 1.73}{VD} \quad \text{(three-phase circuit)}$$

$$\text{or } CM = \frac{2 \times K \times L \times I \times 0.866}{VD}$$

TO FIND CURRENT when the voltage drop, length of circuit and conductor are known:

$$I = \frac{VD \times CM}{2 \times K \times L} \quad \text{(single-phase, 2-wire circuit)}$$

$$I = \frac{VD \times CM}{K \times L \times 1.73} \quad \text{(three-phase circuit)}$$

$$\text{or } I = \frac{VD \times CM}{2 \times K \times L \times .866}$$

TO FIND LENGTH OF CIRCUIT when the voltage drop, conductor size and current are known:

$$L = \frac{VD \times CM}{2 \times K \times I} \quad \text{(single-phase circuit)}$$

$$L = \frac{VD \times CM}{K \times I \times 1.73} \quad \text{(three-phase circuit)}$$

$$\text{or } L = \frac{VD \times CM}{2 \times K \times I \times .866}$$

TO FIND THE RESISTANCE of a conductor

L = length of conductor
R = resistance
K = constant (10.7 for copper, 17.4 for aluminum)

$$R = \frac{KL}{CM} \quad ; \qquad L = \frac{R \times CM}{K} \quad ; \qquad CM = \frac{K \times L}{R}$$

Student
Number _____

Instructor's
Name _____

INSTRUCTIONS: The following statements are either True or False. Draw a circle around the **T** if the statement is True, or around the **F** if it is False. The questions are to be answered, as nearly as possible, in accordance with the specific provisions of the NEC.

• True or False •

1. A ground rod used to ground an electrical system shall be at least 8 feet from a ground rod used to ground a lightning protection system. T F

2. If the resistance of a driven ground rod is too high, a larger rod shall be used. T F

3. Iron wire shall not be employed for bonding jumpers. T F

4. In general, switch enclosures may not be used for junction boxes. T F

5. Stranded wires, No. 10 and smaller, may be connected to equipment by means of screws used with terminal plates having upturned lugs. T F

6. A single 15-amp receptacle may be installed on an individual 20-amp branch circuit. T F

7. The grounding conductor in a cable assembly for cord-and-plug connected equipment need not be insulated. T F

8. A separate flexible wire or strap is never allowed for grounding cord-and-plug connected equipment. T F

9. Soldered connections are not permissible for grounding equipment. T F

10. Not less than 10 feet of No. 4 bare copper conductor encased under the concrete footing in contact with the earth is an approved made electrode. T F

11. An AC general-use snap switch shall not be used to control a tungsten-filament lamp load. T F

12. Time switches need not be of the externally-operable type. T F

Continued on Next Page

Blueprint Reading

13. Double-throw knife switches may be mounted either vertically or horizontally. T F

14. Type NM nonmetallic-sheathed cable may be used for under-plaster extensions. T F

15. Wiring located 2 inches above electrically-heated ceilings and over 1 inch of thermal insulation must be derated. T F

16. Heating panels shall not be extended beyond partitions. T F

17. Heating cable shall not be used with metal lath ceilings. T F

18. Heating cable may be spliced if necessary. T F

19. Heating cables in a concrete floor must be placed on at least 2-inch centers. T F

20. The nonheating leads of heating cables installed in concrete may be protected by rigid nonmetallic conduit. T F

21. Branch circuit conductors for fixed resistance space heaters shall be rated at 125% of the total load of heaters and motors. T F

22. In a 1000 square foot addition to a dwelling unit, the "watts-per-square-foot" method need not be adhered to. T F

23. A 20-amp branch circuit supplying air conditioning equipment shall not supply any lighting outlets. T F

24. For a motor-driven permanently connected appliance rated not over ⅛ h.p., the branch circuit overcurrent device may serve as the disconnecting means. T F

25. A separable connector will satisfy the Code requirements with respect to a disconnecting means for a wall-mounted oven. T F

TEAR OUT HERE

TRADE COMPETENCY TEST NO. 2B

Student
Number _____

Instructor's
Name _____

INSTRUCTIONS: In the space provided at the right of each of the following questions, write **a, b, c,** or **d** to indicate which of the alternatives given makes it most nearly correct. The questions are to be answered, as nearly as possible, in accordance with the specific provisions of the NEC.

• Multiple-Choice •

1. The three wires of a single-phase 115/230 volt service are designated as *A*, *N*, and *B*. A 100-amp 230-volt load is connected across wire *A* and *B*. A 50-amp 115-volt load is connected across *A* and *N*. A copper, TW, neutral conductor shall be:

   (a) No. 10,   (b) No. 8,   (c) No. 6,   (d) No. 4   1. _____

2. The minimum diameter of a steel ground rod shall be:

   (a) ⅜ inch,   (b) ½ inch,   (c) ⅝ inch,   (d) ¾ inch   2. _____

3. The Code requires that the minimum area of exposed surface offered by a plate electrode shall be:

   (a) 1 sq ft,   (b) 2 sq ft,   (c) 3 sq ft,   (d) 4 sq ft   3. _____

4. A single grounding electrode is permitted when the resistance to ground does not exceed:

   (a) 5 ohms,   (b) 10 ohms,   (c) 15 ohms,   (d) 25 ohms   4. _____

5. The minimum size of a copper equipment grounding conductor required for equipment connected to a 40-amp circuit is:

   (a) No. 12,   (b) No. 14,   (c) No. 8,   (d) No. 10   5. _____

6. Non-current-carrying metal parts of cord-and-plug connected equipment may be grounded:

   (a) through the neutral conductor,   (b) by length of No. 14 solid wire run to a water pipe,   (c) by a green wire run as part of the flexible cable,   (d) by all three of these   6. _____

7. The maximum permissible open-circuit voltage of electric discharge lighting equipment used in a dwelling occupancy is:

   (a) 1,000 volts,   (b) 500 volts,   (c) 2,000 volts,   (d) 1,500 volts   7. _____

Continued on Next Page

Blueprint Reading   63

8. Nonmetallic surface extension may be run:

   (a) over 2 inches from the floor,   (b) upon the floor if not subject to damage,   (c) through a dry wall,   (d) through a floor if protected by a kick plate

   8. _____

9. Side or back-wiring spaces are required in cabinets when, in addition to the supply circuit or a continuation thereof, there are devices connected to more than:

   (a) 10 conductors,   (b) 8 conductors,   (c) 12 conductors, (d) 6 conductors

   9. _____

10. Wiring located above electric heating panels shall be spaced above the heated ceiling not less than:

    (a) 5 inches,   (b) 4 inches,   (c) 3 inches,   (d) 2 inches

    10. _____

11. The Code requires that heating panels be separated from outlet boxes that are to be used for mounting fixtures not less than:

    (a) 12 inches,   (b) 8 inches,   (c) 6 inches,   (d) 10 inches

    11. _____

12. Adjacent runs of heating cable in plaster ceilings not exceeding 2¾ watts per foot shall be installed on centers of not less than:

    (a) 3 inches,   (b) 1 inch,   (c) 2 inches,   (d) 1½ inches

    12. _____

13. Ground-fault circuit protection for personnel is required for all 125-volt, single-phase, 15- and 20-ampere receptacles that are installed in a dwelling unit:

    (a) attic,   (b) bathroom,   (c) laundry,   (d) utility room

    13. _____

14. According to the Code, receptacles in a dwelling unit dining area are required:

    (a) to be on a 15-amp circuit,   (b) to be on a separate 20-amp circuit,   (c) for each counter space wider than 12 inches, (d) to be on the 20-amp kitchen appliance circuit.

    14. _____

15. The rating of any one cord- and plug-connected appliance used on a 30-amp branch circuit shall not exceed:

    (a) 30 amps,   (b) 24 amps,   (c) 27 amps,   (d) 21 amps

    15. _____

16. The total rating of appliances fastened in place connected to a 20-amp branch circuit which also supplies lighting units shall not exceed:

    (a) 20 amps,   (b) 16 amps,   (c) 12 amps,   (d) 10 amps

    16. _____

TEAR OUT HERE

National Electrical Code

17. No. 8 Type TW conductors supply a central air conditioner, consisting of a sealed (hermetic-type) motor-compressor. The branch circuit may be loaded to a maximum value of:

 (a) 40 amps, (b) 32 amps, (c) 36 amps, (d) 30 amps

 17. _____

18. A 20-amp circuit in a residence supplies only fixed resistance type space heaters. This circuit may be loaded to a maximum value of:

 (a) 16 amps, (b) 20 amps, (c) 12 amps, (d) 18 amps

 18. _____

19. The smallest Type TW neutral feeder conductor supplying two 12 kW ranges shall be:

 (a) No. 6, (b) No. 8, (c) No. 10, (d) No. 4

 19. _____

20. The feeder load for three electric ranges, one of 10 kW, one of 16 kW, and one of 17 kW, should be assessed at:

 (a) 14.5 kW, (b) 15 kW, (c) 15.6 kW, (d) 16.1 kW

 20. _____

21. A feeder supplying a 5-kW wall-mounted oven and a 7-kW counter-mounted cooking unit must have capacity for:

 (a) 12 kW, (b) 9.2 kW, (c) 7.8 kW, (d) 7 kW

 21. _____

22. Frames of electric clothes dryers may be grounded by connection to the grounded circuit conductor only when this condutor is not smaller than:

 (a) No. 12, (b) No. 8, (c) No. 10, (d) the outer conductors

 22. _____

23. The neutral conductor of a 3-wire branch circuit to a household electric range shall not be smaller than:

 (a) No. 4 THW, (b) No. 6 THW, (c) No. 8 THW, (d) No. 10 THW

 23. _____

24. A heavy-duty lampholder of the admedium type shall have a rating of not less than:

 (a) 150 watts, (b) 300 watts, (c) 600 watts (d) 660 watts

 24. _____

25. In multifamily dwellings, the disconnecting means for a fixed appliance:

 (a) must be only in the same apartment, (b) may be on the same floor, (c) may be on the floor below, (d) may be on the service switch for the occupancy

 25. _____

TRADE COMPETENCY TEST NO. 2C

Student
Number _____

Instructor's
Name _____

INSTRUCTIONS: Complete each of the following statements, by writing the missing word or words in the space provided at the right of each question, to make a true statement. The questions are to be answered, as nearly as possible, in accordance with the specific provisions of the NEC.

• Completion •

1. The maximum standard ampere rating of a fuse or inverse time circuit breaker shall be considered to be _____ amps.

1. _____

2. Where a change occurs in the size of the _____ conductor, a similar change may be made in the size of the grounded conductor.

2. _____

3. The neutral feeder conductor must be capable of carrying the maximum _____ load.

3. _____

4. _____ _____ are required around service conduits entering concentric or eccentric knockouts.

4. _____

5. Where a service run of rigid metal raceway is interrupted by an expansion joint the sections of rigid conduit shall be bonded by an _____ _____ jumper or other approved means.

5. _____

6. In a single-phase, 3-wire AC system the _____ _____ shall be grounded.

6. _____

7. The smallest permissible size of a bronze ground rod is _____ inch(es).

7. _____

8. Where rock bottom is encountered, ground rods are permitted to be buried if the trench is at least _____ deep.

8. _____

9. Double locknuts are required where rigid metal raceways in other than service runs enter a box or cabinet, if the raceway contains any wire of more than _____ volts to ground.

9. _____

10. Splices which are to be soldered shall first be made mechanically and _____ secure.

10. _____

Continued on Next Page

TEAR OUT HERE

11. Unless special precautions are taken, exposed live parts operating at _____ volts or more shall be guarded against accidental contact.

11. _____

12. Receptacles connected to a 30-amp branch circuit supplying two or more outlets shall be rated not less than _____ amps.

12. _____

13. In dwelling occupancies, equipment in which live parts are exposed while changing lamps shall have an open-circuit voltage not exceeding _____ volts.

13. _____

14. When installing Type AC armored cable other than the lead type, _____ bushings must be used.

14. _____

15. Wires run above heated ceilings and within thermal insulation, shall be derated on the basis of _____ degrees C.

15. _____

16. Heating cables on ceilings shall be kept free from contact with metal or other _____ surfaces.

16. _____

17. Leads of 240-volt heating cables are marked with _____ .

17. _____

18. Ungrounded conductors shall be clearly distinguishable from _____ and _____ conductors by color or other marking.

18. _____

19. Flexible cord shall not be used as a substitute for _____ wiring.

19. _____

20. The minimum branch circuit rating to a single electric range of over 8¾ kW shall be _____ amps.

20. _____

21. Flexible cords shall be connected to devices so that no _____ will be transmitted to terminal screws.

21. _____

22. In general flexible cords other than tinsel cords shall be not smaller than No. _____ .

22. _____

23. _____ conductors must be used with chain fixtures.

23. _____

24. The smaller of two dissimilar loads may be omitted in calculating the load on a feeder if it is unlikely that both loads will be connected _____ .

24. _____

25. Screw-shell lampholders shall not be used with infrared lamps of over _____ watts except when identified as suitable for such use.

25. _____

Commercial Locations

WIRING PLANS FOR A STORE BUILDING

FIRST STEPS — In preparing to handle a large installation, the electrical foreman glances through the complete set of plans to obtain a bird's-eye view of the entire project. He surveys electrical plans to note the extent of the work, kinds of circuits and equipment, locations of main elements. He reads electrical specifications, referring to the plans from time to time as becomes necessary. The electrical foreman must refer to the plot, architectural, structural, framing, mechanical, and electrical plans to observe all the details essential to his work. By this time, he is ready to arrange for an orderly sequence of operations. Excerpts from electrical specifications for a store building are presented below, along with notes that might be taken from the architectural plans.

Excerpts From The Specifications

Note: The Article number indicates where the main topic will be found in the National Electrical Code. The manufacturer's directions should always be followed.

29-1 Main Service (See NEC Art. 230)

(a) The main service shall consist of a 4-wire, 3-phase, 120-208-volt secondary system from utility company's manhole on Sheppard Avenue to main switchboard.

(b) All underground conduit outside the building shall be buried at least 18 in. below grade and painted with bitumastic paint.

(c) All service conductors shall be insulated, including the neutral conductor.

29-4 Conductors (See NEC Art. 310)

(a) Conductors shall be Type THW unless otherwise specified or required by applicable provisions of the NEC. All conductors to be copper unless otherwise noted.

(b) No conductors for lighting, plug receptacles, or power shall be smaller than No. 12 copper AWG.

(c) No block and tackle or other mechanical means shall be used to draw conductors smaller than No. 2 into raceways,

and powdered soapstone will be the only lubricant permitted.

(d) Feeder voltage drop from service location to any distribution panel shall not exceed 1%.

29-5 Wiring Methods and Materials (See NEC Arts. 300 to 390)

(c) Approximate locations of all outlets are marked on the drawings. The Contractor shall check all such measurements and, in case of doubt, consult the Architect as to exact locations so that boxes will be properly centered with acoustical tile, interior trim, paneling, etc.

29-6 Equipment for General Use (See NEC 400 to 480)

(b) Lock type wall switches shall be installed at points where a switch is marked Sk on drawing.

(m) Two special plug receptacles on the East wall of the Alterations Department shall be equipped with signal lights. These outlets . . .

29-7 Telephone and Loud Speaker Systems (See NEC Arts. 640, 800, 810)

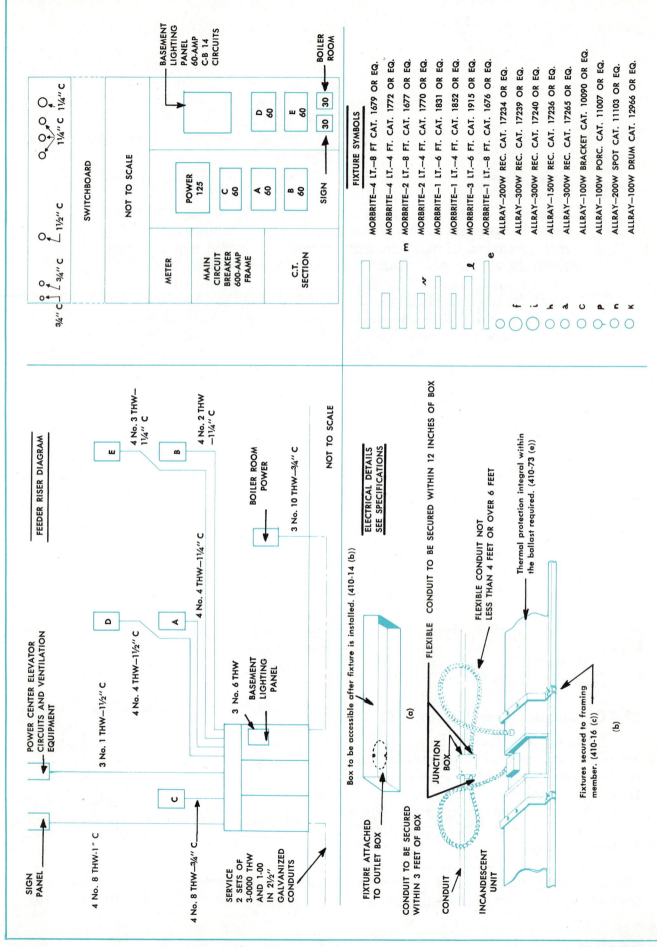

SWITCHBOARD

NOT TO SCALE

BASEMENT LIGHTING PANEL 60-AMP C-B 14 CIRCUITS

BOILER ROOM

¾" C
¾" C
1½" C
1¼" C
1¼" C

METER

MAIN CIRCUIT BREAKER 600-AMP FRAME

C.T. SECTION

POWER 125

C 60

A 60

B 60

D 60

E 60

SIGN

30

30

FIXTURE SYMBOLS

MORBRITE—4 LT.—8 FT CAT. 1679 OR EQ.
MORBRITE—4 FT.—4 FT. CAT. 1772 OR EQ.
MORBRITE—2 LT.—8 FT. CAT. 1677 OR EQ.
MORBRITE—2 LT.—4 FT. CAT. 1770 OR EQ.
MORBRITE—1 LT.—6 FT. CAT. 1831 OR EQ.
MORBRITE—1 LT.—4 FT. CAT. 1852 OR EQ.
MORBRITE—3 LT.—6 FT. CAT. 1915 OR EQ.
MORBRITE—1 LT.—8 FT. CAT. 1676 OR EQ.
ALLRAY—200W REC. CAT. 17234 OR EQ.
ALLRAY—300W REC. CAT. 17239 OR EQ.
ALLRAY—300W REC. CAT. 17240 OR EQ.
ALLRAY—150W REC. CAT. 17236 OR EQ.
ALLRAY—300W REC. CAT. 17265 OR EQ.
ALLRAY—100W BRACKET CAT. 10090 OR EQ.
ALLRAY—100W PORC. CAT. 11007 OR EQ.
ALLRAY—200W SPOT CAT. 11103 OR EQ.
ALLRAY—100W DRUM CAT. 12966 OR EQ.

m
r
l
e

f
i
h
a
c
p
n
k

FEEDER RISER DIAGRAM

E
4 No. 3 THW—1¼" C

B
4 No. 2 THW—1¼" C

POWER CENTER ELEVATOR CIRCUITS AND VENTILATION EQUIPMENT

3 No. 1 THW—1½" C

D
4 No. 4 THW—1½" C

A
4 No. 4 THW—1¼" C

C
4 No. 8 THW—¾" C

3 No. 6 THW

BASEMENT LIGHTING PANEL

BOILER ROOM POWER
3 No. 10 THW—¾" C

NOT TO SCALE

SIGN PANEL
4 No. 8 THW—1" C

SERVICE 2 SETS OF 3-0000 THW AND 1-00 IN 2½" GALVANIZED CONDUITS

ELECTRICAL DETAILS SEE SPECIFICATIONS

FIXTURE ATTACHED TO OUTLET BOX

Box to be accessible after fixture is installed. (410-14 (b))

CONDUIT TO BE SECURED WITHIN 3 FEET OF BOX

(a)

CONDUIT

INCANDESCENT UNIT

FLEXIBLE CONDUIT TO BE SECURED WITHIN 12 INCHES OF BOX

FLEXIBLE CONDUIT NOT LESS THAN 4 FEET OR OVER 6 FEET

Thermal protection integral within the ballast required. (410-73 (e))

JUNCTION BOX

Fixtures secured to framing member. (410-16 (c))

(b)

(a) Telephone outlets shall be located approximately as shown . . .

29-8 Miscellaneous Equipment
(See NEC Art. 422)

(j) Toilet exhaust fans shall be Nuway, 130-watt, Type C2, or . . .

NOTES TAKEN FROM THE ARCHITECTURAL BLUEPRINTS

First floor, 8-inch poured, reinforced concrete slab.

Second floor, 6-inch poured, reinforced concrete slab.

Basement floor, 4-inch poured, reinforced concrete slab.

Partition details . . . (toilets, offices, fitting rooms, Alteration Department, bin and shelf details in stock area)

Hung ceilings first and second floors, acoustical tile.

Basement ceiling unfinished concrete.

Canopy, entrance, show-window details . . .

Details, basement passageway, toilets to elevator . . .

Types of finished floors . . .

Kinds of walls . . .

Details of stair ceilings . . .

Trap door to roof, built-in iron-rung ladder, wall of freight elevator shaft, leads to trap door.

Dimensions . . . (lengths, widths, thicknesses, ceiling heights)

ELECTRICAL PLANS

Before taking up detailed analysis of the store building installation, notice that four drawings are included here. There are three floor plans: Basement, First Floor, and Second Floor. Also, there is a special drawing which shows a riser diagram, switchboard elevation, electrical details, and a list of fixture symbols. Discussion could be started with the Basement plan, then carried on to the First Floor and Second Floor plans in turn. Or, the First Floor might be dealt with, then Second Floor, and finally Basement.

A different order is followed, however, starting with the Second Floor, working downward to the First Floor, and lastly the Basement. This procedure is dictated by the wish to simplify explanation. Since the least complicated area is the second floor, it has been chosen as the starting point. The more involved First Floor plan is then studied. The third step, examination of the Basement plan, leads naturally to the consideration of service requirements.

SECOND FLOOR PLAN

Two lighting panels are used for this floor, panel E at the north wall, and panel D at the south wall. Conduit runs between lighting outlets and panelboards are indicated by solid lines showing them to be concealed in the hung ceiling. Plug receptacle runs are indicated by broken lines showing that they are to be run in the poured concrete slab. Telephone runs are indicated by lines marked —— T —— T ——, speaker system lines marked —— S —— S ——. Telephone outlets are indicated by standard symbols, speaker outlets by the symbol S. The runs, in both cases, are to be installed as outlined in the specifications.

Conduit *riser* indications are shown in the south wall. At the left or west side the first such riser is marked *down* to show that a conduit run from the stair outlet leads to the floor below. The next two risers are marked *up*, and since no conduit runs lead to them from the second floor, it is apparent that they originate below and continue upward toward the roof. The final riser is marked *down*, a run of conduit leading downward from a lighting outlet at the head of the stairs. Two risers in the east wall of the passenger elevator shaft are both shown as *down*, one marked T to indicate a telephone run, the other S to indicate a speaker run. The PBX telephone board and the sound amplifier are both located in the small room to the right of the elevator shaft.

In determining load on the watts-per-square-foot basis, NEC 220-3(b) requires that the floor area shall be computed from outside dimensions of the building. The measurements here are 72 ft by 96 ft, so the area of the whole floor is 6912 sq ft. From this amount, it is permissible to deduct 520

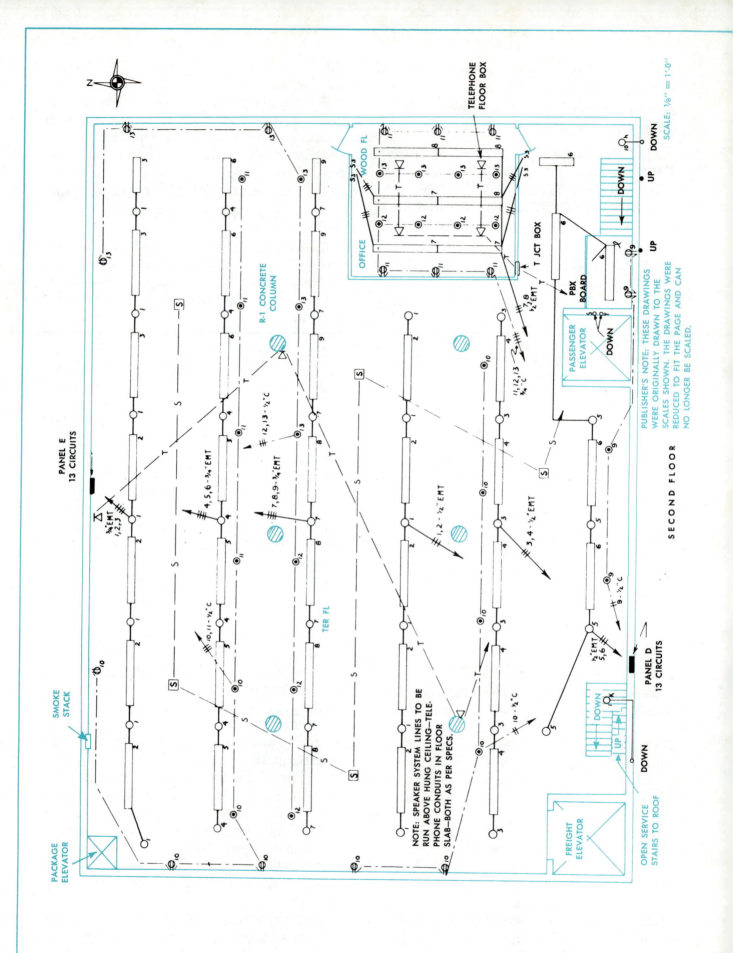

National Electrical Code

sq ft which is the area of the office space and the area of the two elevators. The elevator shaft area and the office space do not come under the heading of general lighting in Table 220-3(b) but rather will be considered later in the text. The remainder, 6392 sq ft, is the area to be considered now.

NEC Table 220-3(b) lists the unit load for a store as 3 watts per sq. ft. NEC 220-3 (a) states that this value shall be increased by 25 percent where the load on a branch circuit is continuous (load is expected to continue for 3 hours or more) such as store lighting. The unit load in the present case must be 1.25 × 3 watts, or 3.75 watts per sq ft. The minimum provision for lighting load is 6392 × 3.75 watts, or 23970 watts. Panels *D* and *E*, therefore, must have a sufficient number of lighting circuits to furnish this amount of general illumination.

The specifications provide that all lighting circuits shall be run with No. 12 Type THW copper wire, protected by 20-amp circuit-breakers. NEC 210-3 states that branch circuits shall be classified according to the rating of the overcurrent devices. These 20-amp circuits, each basically capable of supplying 2400 watts of power (20 amps × 120 volts), can be loaded to only 80 percent of this value, or 1920 watts, under the limitation imposed by NEC 210-22(c) since the load continues for long periods of time.

Calculations for Panel *E*

The fixture list, fixture specifications, and the blueprint show the load on circuit *1* as seven 200-watt incandescent units, totaling 1400 watts. Continuing to circuit *2*, NEC 210-22(b) requires that the load on a lighting fixture which employs a ballast shall be calculated according to the ampere-rating of the ballast, and not the wattage of the lamps. The current used in the two ballasts which are part of each fixture on the circuit is 3.2 amps or 384 watts. The total wattage for circuit *2*, which supplies four such units, is 1536 watts. Circuit *3* employs two units of this type, and a third whose ballast rating is 1.7 amps, requiring 972 watts for all units. The load for this

group of three circuits is 1400 watts + 1536 watts + 972 watts, or 3908 watts.

Loads on the two groups *4, 5, 6,* and *7, 8, 9* are identical with the above figures so that the total lighting power to be supplied by panel *E* amounts to 3 × 3908 watts, or 11724 watts. Table 220-11 lists a demand factor of 100 percent for lighting loads of this nature.

In addition to the lighting outlets, a number of wall receptacles and floor outlets are supplied from the panel. The specifications require that No. 12 Type THW wire be used for these circuits, too. NEC 220-3(c)(5) designates a load of 180 volt-amperes or 1½ amps (180 divided by 120) for each of such *other outlets*. The eighteen plug receptacles connected to this panel represent a current input of 27 amps or 3240 watts (27 amps × 120 volts).

The load imposed on the feeder is equal to the sum of 11724 watts and 3240 watts, or 14964 watts. At this point, it is well to ascertain if the feeder will be carrying its required proportion of minimum lighting load. Reference to the floor plan shows that circuits from panel *E* provide general illumination for an area which measures 36 ft by 96 ft, a total of 3456 sq ft.

Since the unit load cannot be less than 3.75 watts per sq ft, the amount of general lighting burden allotted to panel *E* cannot be less than 3456 × 3.75 watts, or 12960 watts. This value must be used in the feeder calculations because it is larger than the actual connected load of lighting circuits. It should be noted that the minimum value of lighting wattage does not have to be used but must be present in circuit and feeder capacity. The corrected load on the feeder then is equal to 12960 watts + 3240 watts, or 16200 watts.

With a balanced four-wire, three-phase, 120-208-volt system, the current per conductor may be determined by dividing wattage by 360 (1.73 × 208 volts = 360). The current in the feeder to panel *E* equals 16200 divided by 360 volts, or 45 amps. NEC Table 310-16 lists the nearest size of Type THW copper conductor to carry 45 amps as No. 8, which is rated at 50 amps if

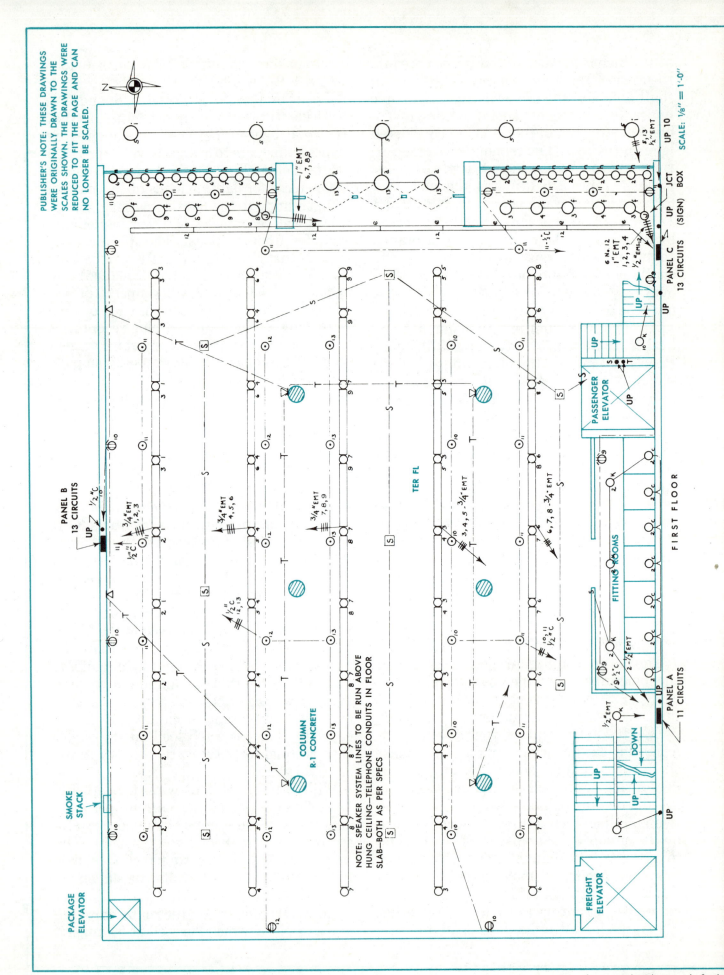

there are not more than three current-carrying conductors in a raceway.

Note 8 of Table 310-16 states that where the number of conductors in a raceway exceeds three, the allowable ampacity of each conductor shall be limited to 80 percent of the values given in the table. Since there are four conductors here, it would seem that the capacity of the No. 8 wire would be reduced to 80 percent of 50 amps, or 40 amps. Reading further, however, Note 10 states that a neutral conductor which carries only the unbalanced current from other conductors shall not be counted in determining ampacities where there are more than three conductors in a raceway.

Such a neutral conductor is included in the feeder under consideration, and the carrying capacity of the No. 8 wire will not be reduced in value. Nevertheless, the rule may be altered somewhat later on because the neutral conductor of a four-wire, three-phase wye system which supplies circuits for electric discharge lamps carries some third harmonic current even when the load is perfectly balanced. Up to this time, however, the Code is not specific on this particular matter.

Another factor is involved here. NEC 215-2 (fine print note) suggests that voltage drop on lighting feeders be limited to 3 percent, but specifications here limit the feeder drop to 1 percent. In the case of a 120-volt feeder (120 volts from phase wire to neutral) this drop amounts to 1.2 volts (.01 × 120 volts). But, since there may be some voltage drop in the neutral conductor due to third harmonic current, the voltage drop in phase conductors should not exceed 1 volt.

American Electricians Handbook and other engineering data (also see 1981 NEC, Chapter 9, Table 8) lists the resistance of No. 6 copper as .41 ohm per 1000 ft. When floor heights and distances on the architectural plans are measured, allowing a sufficient length of conductor for connections at either end of the run, the length of single conductor is found to be 104 ft from panel E to the switchboard in the basement. Voltage drop

equals resistance of the wire times the current. The resistance of the No. 6 conductors equals .104 × .41 ohm. Voltage drop equals .043 × 45 amps, or 1.94 volts, too high a value.

Another common method of calculating volt-drop is by using the formula:

Check by using 3-phase formula:

$$VD = I \times \frac{KL}{CM}$$

$$VD = \frac{KLI\ 1.73}{CM} = \frac{10.7 \times 104 \times 45 \times 1.73}{26240} = 3.3\ V$$

Percent VD = 3.3 V ÷ 208 V = 1.6% VD

Where

CM = Area of conductor in circular-mils
VD = Volt-drop
I = Current
L = Length of conductor in feet
K = Constant (approx 10.7 for copper)

To calculate the volt-drop of a No. 6 copper conductor with a length of 104 feet with a load of 45 amps we find:

$$VD = \frac{KLI}{CM} = \frac{10.7 \times 104 \times 45}{26240} = 1.91\ V\ or\ 1.6\%\ VD$$

The percent of volt-drop in this case would be 1.93 divided by 120, or 1.6 percent.

The resistance of No. 4 conductor is given as .259 ohm per M feet, and the voltage drop in a single conductor equals .104 × .259 ohm × 45 amps, or 1.22 volts. This value is still too great. The resistance of a No. 3 conductor is .205 ohm per M feet, and the drop for a single conductor equals .104 × .205 ohm × 45 amps, or .96 volt, which is acceptable.

$$VD = \frac{KLI}{CM} = \frac{10.7 \times 104 \times 45}{52,620} = .96VD$$

3 Type THW conductors may be installed in a 1¼-inch conduit. Thus, the feeder to panel E will consist of four No. 3 Type THW conductors in a 1¼-inch conduit.

Calculations for Panel D

The loads are: circuit 1—1200 watts, 2—1920 watts, 3—1200 watts, 4—1920 watts, 5—800 watts, 6—1152 watts, 7—1152 watts, 8—1152 watts, 9—720 watts, 10—1080 watts, 11—1080 watts, 12—720 watts, and 13—720 watts. The circuits devoted to general lighting: 1,2,3,4,5, and 6 require a total of 8192 watts.

Measurement shows the floor area supplied with general illumination from this panel to be 2496 sq ft. The minimum wattage required is 2496 × 3.75 watts, or 9360 watts. This value, being somewhat higher than actual load on the circuits, will be used in the feeder calculations.

The office has an area of 440 sq ft, requiring 1650 watts of illumination at the 3.75-watt rate. NEC Table 220-3(b) lists office buildings at a minimum unit load of 3.5 watts per sq ft. For a steadily burning load this minimum would have to be increased to 4.4 watts (1.25 × 3.5 watts). Although this NEC provision does not apply strictly to the one small office space here, it does point to the fact that office buildings and office areas need somewhat more lighting intensity than sales areas. Using the value given by NEC Table 220-3(b) we find this area will require 440 × 4.4 watts, or 1936 watts. The lighting load connected to circuits 7 and 8 amounts to 2200 watts. Because the connected load of 2200 watts × 1.25 = 2750 watts is larger than that required by Table 220-3(b) the higher value of 2750 watts will be used.

Plug receptacle circuits 9, 10, 11, 12, and 13 take 4320 watts. The total load on the feeder is 9360 watts + 2750 watts + 4320 watts, or 16430 watts, the current being 46 amps (16430 watts divided by 360), for which the minimum size of Type THW conductor is No. 8.

The length of single conductor between panel D and the switchboard is approximately 83 feet. The voltage drop for a No. 8 stranded conductor would be .083 × .633 × 46 amps, or 2.42 volts, which is too high. Use of a No. 4 conductor will result in a drop of .083 × .259 ohm × 46 amps, or .99 volt, a desirable value. Table 3A, Chapter 9, shows that four No. 4 Type THW conductors may be installed in a 1¼-inch conduit.

$$\text{(Also VD} = \frac{KLI}{CM} = \frac{10.7 \times 83 \times 46}{41,740} = .99 \text{ VD)}$$

FIRST FLOOR PLAN

Three lighting panels are shown on this floor, panel B in the north wall, panels A and C in the south wall. Panel C is confined, primarily, to supplying outlets in the entrance, canopy, and show window areas, panels A and B to the general store illumination. The wiring plan on this floor is similar to that employed on the floor above.

A single riser is indicated in the north wall, the feeder conduit associated with panel E on the second floor. The riser near the left end of the south wall, marked *up*, is the same one shown on the second floor plan, marked *down*. The riser next to panel A is a feeder conduit to panel D, above. The next two risers, also marked *up*, were mentioned before. The third riser, marked *up 10*, is the other end of the conduit supplying the lighting unit at the head of the stairs on the second floor. The two risers in the east wall of the passenger elevator shaft have been explained.

Calculations for Panel B

The loads are as follows: circuit 1—1800 watts, 2—1920 watts, 3—1556 watts, 4—1800 watts, 5—1920 watts, 6—1556 watts, 7—1800 watts, 8—1920 watts, 9—1556 watts, 10—720 watts, 11—1080 watts, 12—1260 watts, and 13—1080 watts. The circuits devoted to general lighting, 1, 2, 3, 4, 5, 6, 7, 8 and 9, use 15828 watts.

Measurement shows the floor area lighted by these circuits to be 3344 sq ft. The minimum acceptable wattage for this area is 3344 × 3.75 watts, or 12540 watts. Since the connected load is greater than this amount, the higher value will be used in feeder calculations.

Plug receptacle circuits 10, 11, 12, and 13 require 4140 watts. The load on the feeder equals 15828 watts + 4140 watts, or 19968 watts, the current being approximately 56 amps. The minimum size of Type THW conductor is No. 6.

The length of single conductor between panel B and the switchboard is approximately 92 feet. The voltage drop in a No. 6 conductor equals .092 × .410 ohm × 56 amps, or 2.11 volts. A No. 3 conductor reduces voltage drop to .092 × .205 ohm × 56 amps, or 1.06 volts, which is slightly high. A No. 2 conductor will lower the drop still further to .092 × .162 ohm × 56 amps, or .84 volt. Thus, panel B requires four No. 2 Type THW conductors in a 1¼-inch conduit.

$$\text{(Also VD} = \frac{KLI}{CM} = \frac{10.7 \times 92 \times 56}{60,630} = .84 \text{ VD)}$$

Calculations for Panel A

The loads are as follows: circuit *1*—300 watts, *2*—1000 watts, *3*—1800 watts, *4*—1920 watts, *5*—1556 watts, *6*—1800 watts, *7*—1920 watts, *8*—1556 watts, *9*—540 watts, *10*—1260 watts, and *11*—1080 watts. The circuits concerned with general lighting are *1*, *2*, *3*, *4*, *5*, *6*, *7*, and *8* which have a connected load of 11852 watts.

Measurement shows the floor area lighted by these circuits as 3080 sq ft, for which the minimum wattage must be not less than 3080 × 3.75 watts, or 11550 watts. The connected load, being somewhat higher than this value, will be used in feeder calculations.

Plug receptacle circuits *9*, *10*, and *11*, take 2880 watts. The total load on the feeder is 11852 watts + 2880 watts, or 14732 watts, the current being approximately 41 amps and calling for a minimum size of No. 8 Type THW conductor.

The length of single conductor from panel *A* to the switchboard is about 73 feet. Voltage drop in a No. 6 conductor would be .073 × .41 ohm × 41 amps, or 1.23 volts, which is high. A No. 4 conductor has a drop of .073 × .259 ohm × 41 amps, or .78 volt, which is excellent. The feeder to panel *A* will consist of four No. 4 Type THW conductors in a 1¼-inch conduit.

$$(\text{Also VD} = \frac{KLI}{CM} = \frac{10.7 \times 73 \times 41}{41,740} = .78 \text{ VD})$$

Calculations for Panel C

Except for front stair lighting circuit *10* and cove lighting circuit *12*, this panelboard feeds only show window, entrance, and canopy areas. The loads are as follows: circuit *1*—1200 watts, *2*—1000 watts, *3*—1500 watts, *4*—1000 watts, *5*—1500 watts, *6*—1200 watts, *7*—1000 watts, *8*—1500 watts, *9*—1000 watts, *10*—300 watts, *11*—1800 watts, *12*—864 watts, and *13*—900 watts.

The circuits for lighting the show windows, circuits *1*, *2*, *3*, *4*, *5*, *6*, *7*, *8*, and *9*, take 9400 watts. NEC 220-3(c) Ex. 3 provides that a minimum of 200 watts per linear foot be assigned for show window lighting. The two show windows are each 22 feet long, so

that a minimum of 8800 watts must be used in the feeder calculations. Here the greater value of 9400 watts connected load will be employed.

The total load on panel *C* is equal to 14764 watts, the current being approximately 41 amps, the smallest size of Type THW conductor No. 8. Since panel *C* is directly over the switchboard, voltage drop in the feeder conductors will be negligible, and four No. 8 Type THW conductors in 1¾-inch conduit will be satisfactory.

Important Practical Consideration

Since the specifications state that the basement ceiling is unfinished, and since floors are of poured reinforced concrete construction, it will be necessary to set basement lighting outlet boxes and conduit on the first floor deck at the same time that first floor plug receptacle boxes are spotted there. Plans and specifications should always be checked with a view to detecting situations of this kind.

BASEMENT PLAN

The basement lighting panel is included in the main switchboard. A power panel, located in the boiler room, furnishes current to pump motors there and to a motor which operates the package elevator unit. Details connected with the motor installations are not outlined here because the subject of power equipment will be taken up in the next section of this book.

The basement contains a shipping and receiving space, an alteration department, a large stockroom area, employees' toilets, public toilets, and a small office. Both employees' and public toilets have exhaust fans connected directly to respective lighting circuits so they are in operation at all times while light switches are turned on. The specifications furnish details regarding the fans, stating that they consume 130 watts each. Plug receptacle circuit *13*, in the alteration department, has two special outlets described in the specifications.

A pair of risers in the north wall are feeders to panels *B* and *E*. In the south wall, risers at the left are associated with panels

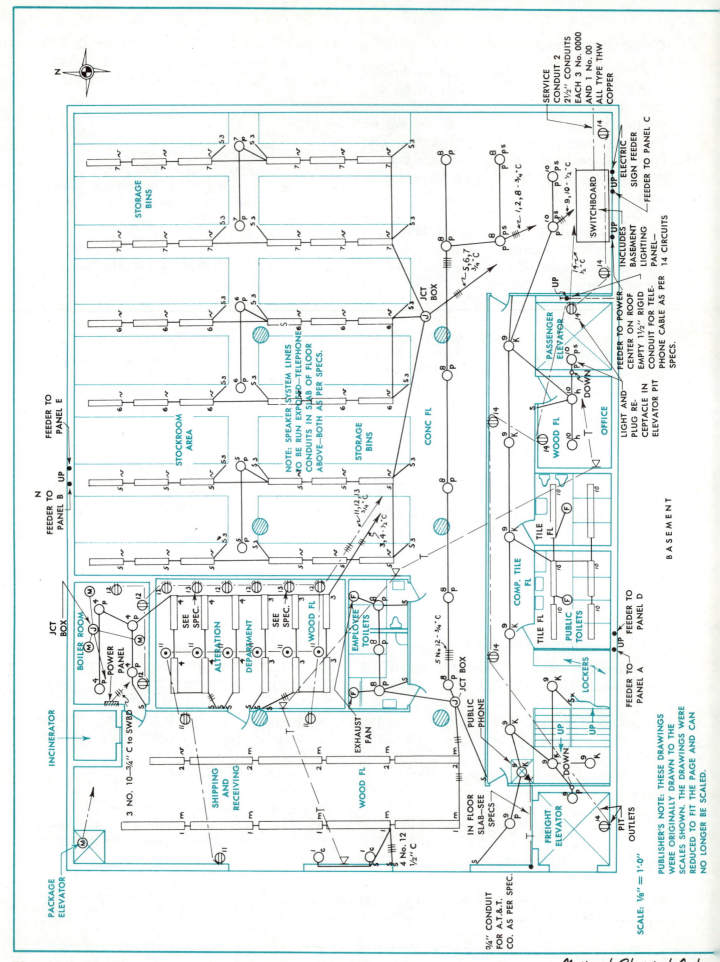

D and *A*. Toward the other end, a conduit rises to the power center on the roof. A feeder conduit to panel *C* is next in line, followed by one to an electric sign on the roof. The *down* marking in the west wall of the freight elevator shaft refers to a conduit leading to the pit light, and the identical marking in the passenger elevator wall shows a run to its pit light.

Calculations for Basement Panel

Loads are as follows: circuit *1*—1200 watts, *2*—870 watts, *3*—1224 watts, *4*—1624 watts, *5*—1220 watts, *6*—1220 watts, *7*—1220 watts, *8*—1360 watts, *9*—1100 watts, *10*—1464 watts, *11*—1620 watts, *12*—1260 watts, *13*—1500 watts, and *14*—1260 watts, the total being 18142 watts. The current is approximately 51 amps so that a No. 6 conductor will be large enough to supply the panel.

Service Calculation

The service load may be obtained by adding together all the feeder requirements, including boiler room power, sign, and roof power. A load of 4800 watts is to be allowed for boiler room power, 8640 watts for the sign, and 36000 watts for roof power. Starting with panel *E* and continuing in the order of calculations, the total load is 16200 watts + 16430 watts + 19968 watts + 14732 watts + 14764 watts + 18142 watts + 8640 watts + 4800 watts + 36000 watts, or 149676 watts. The current is equal to 149676 watts divided by 360, or 416 amps. [Note: in making calculations for balanced 4-wire, 3-phase, 120-208-volt systems, divide total wattage by 360 in order to obtain current per phase wire.]

Table 310-16 shows that a 600 MCM Type THW conductor will carry 420 amps. It is worth noting, however, that two smaller conductors are more easily handled than a single large one. NEC 310-4 permits conductors of No. 0 and larger to be run in parallel, provided they are of the same length, conductor material, circular-mil area, insulation type and are terminated in the same manner. Where parallel conductors are run in

separate raceways or cables, the raceways or cables shall have the same physical characteristics.

Two parallel conductors on each leg of the circuit would each carry one-half the total current, or 208 amps. With all conductors in the same conduit, a total of six current-carrying wires, the rating of each must be reduced to 80 percent of the value indicated in Table 310-16. Reference to the table shows that a size 0000 Type THW conductor has an ampacity of 230 amps where there are not more than three current-carrying conductors in a raceway. For six conductors each must have a normal ampacity equal to 208 amps divided by .8, or 260 amps. A 300 MCM Type THW would be required in this case. But if two sets of 0000 conductors are installed in separate conduits, each conductor's full ampacity will be restored, and there will be considerable saving of copper. This is the practice followed here.

NEC 220-22 states that the neutral feeder *must* be large enough to carry the maximum unbalanced load. The section says, further, that the maximum unbalanced load is that load which can be connected between the neutral and any ungrounded conductor, which means that loads which have no neutral component, such as power loads which are confined to the three phase wires, must be subtracted from total load in order to determine this value of current. Total load in the present instance is 416 amps, and the 3-phase power load 114 amps (boiler-room power plus roof power). The maximum unbalanced load on the neutral is, therefore, equal to 416 amps minus 114 amps or 302 amps.

NEC 220-22 provides, also, that a demand factor of 70 percent may be applied to that portion of the neutral load in excess of 200 amps. At first glance it would appear that the neutral load could be reduced to 200 amps + (.7 × 102 amps), or 272 amps. But a following sentence within NEC 220-22 states that no reduction of neutral capacity shall be allowed for that part of the load which consists of electric discharge lighting. Since electric discharge lighting in the present example accounts for at least 150 amps of load, there is no excess over 200 amps to which a 70 percent factor might be applied. The neutral load, then, must be calculated on the basis of 302 amps.

NEC 300-20 requires that a neutral conductor be run in the same raceway with each set of phase conductors. Either of two neutral wires in this case will carry one-half the unbalanced current (302 amps), or 151 amps. The nearest size listed in the table is No. 00 Type THW, which has an ampacity of 175 amps.

NEC 230-30 permits an uninsulated copper neutral conductor to be installed underground in duct or conduit. It may appear that a larger ampacity might be assigned to an uninsulated conductor, so that a size smaller than No. 00 THW could be used. But Note 5 of Table 310-16 states that where bare conductors are used with insulated conductors their allowable ampacity shall be limited to that permitted for an insulated conductor of the same size. In any case, the present specifications require that an insulated neutral conductor be employed. The service will consist of two sets of three 0000 Type THW conductors and one set of No. 00 Type THW conductors in parallel.

Service Conduit Size

When all conductors in a raceway are of the same size, Tables 3A, 3B, and 3C, Chapter 9 shall be consulted to determine the size of conduit required. Each table includes all sizes of conductors with all of the different types of conductor insulation included in one of the three tables. Table 3A includes such types of insulation as T, TW, and THW. Table 3B includes types such as THWN, THHN, and XHHW, and Table 3C includes Types RHW and RHH. For different sizes of wire in a conduit, it becomes necessary to perform calculations based upon Table 1 and Tables 4 through 8 inclusive, Chapter 9.

Table 4 gives dimensions and percent of area of conduit and tubing. The other tables give dimensions for the various types of conductor coverings including rubber, ther-

moplastic, lead-covered, asbestos-varnished-cambric, cross-linked polyethylene and bare.

In the store service, each conduit has three 0000 Type THW conductors and one No. 00 Type THW conductor. Table 5 gives the cross-sectional area of a 0000 Type THW conductor as .3904 sq in. and a No. 00 Type THW conductor as .2781 sq in. The area of all four conductors equals (3 × .3904 sq in.) + .2781 sq in., or 1.4493 sq in. Table 1 allows 40 percent of conduit cross-sectional area to be utilized when there are three or more conductors of different sizes. Cross-sectional area of the conduit, therefore, must be equal to 1.4493 sq in. divided by .4, or 3.6233 sq in. According to Table 4, the required size of conduit is 2½ inches.

480-Volt Lighting System

A method of lighting which has become increasingly popular in commercial and industrial establishments is the four-wire, three-phase, 277-480-volt electric discharge system. Lighting units are connected between neutral conductor and phase wires, imposing 277 volts on ballast assemblies. Switching may be at a panelboard if circuit breakers are marked SWD (NEC 240-83(d)). Where wall switches are used and the voltage between adjacent switches exceeds 300 volts a permanent barrier is required in outlet boxes between switches (NEC 380-8(b)).

Aluminum Wire and Conduit

In recent years, aluminum wire and conduit have been competing with copper wire and steel conduit. Since aluminum conduit is made in standard trade sizes identical with steel, no Code problem arises. Aluminum conductors, however, vary with respect to copper in both allowable ampacity and ohmic resistance, ampacity for a given size being less than that of copper, resistance higher. The manufacturers' claim that lighter weight of the aluminum conductor more than compensates for these apparent defects will not concern us in the present discussion. We shall calculate sizes of wire and conduit required when Type THW aluminum feeder and service conductors are substituted for Type THW copper conductors in the store building example.

Panel E, on the second floor is 104 conductor-feet from the switchboard and supplies a load of 45 amps per phase wire. Using copper conductors, the feeder consists of four No. 3 conductors in 1¼-inch conduit. Voltage drop in the conductor is limited to approximately 1 volt.

$$(\text{No. 3 copper} - VD = \frac{KLI}{CM} = \frac{10.7 \times 104 \times 45}{52,620} = .95 \text{ VD})$$

Table 310-16 (Column 7) lists the size of aluminum conductor required for 45 amps as a No. 6 THW. The resistance of No. 6 aluminum conductor is .674 ohm per M feet. Voltage drop with this conductor is equal to .104 ft $\times$ .674 ohm $\times$ 45 amps, or 3.15 volts (3.15 volts $\div$ 120 volts equals 2.6 percent voltage drop), too high a value.

$$(\text{also } VD = \frac{KLI}{CM} = \frac{17.4 \times 104 \times 45}{26,240} = 3.15 \text{ VD})$$

The resistance of No. 4 aluminum is .424 ohm per M feet. Voltage drop with this conductor is equal to .104 ft $\times$ .424 ohm $\times$ 45 amps, or 1.98 volts.

$$(\text{also } VD = \frac{KLI}{CM} = \frac{17.4 \times 104 \times 45}{41,740} = 1.98 \text{ VD})$$

The resistance of No. 3 aluminum conductor is .336 ohm per M feet. Using this multiplier, the voltage drop becomes .104 $\times$ .336 ohm $\times$ 45 amps, or 1.57 volts.

$$(\text{also } VD = \frac{KLI}{CM} = \frac{17.4 \times 104 \times 45}{52,620} = 1.57 \text{ VD})$$

The resistance of No. 2 aluminum conductor is .266 ohm per M feet. The voltage drop here is equal to .104 $\times$ .266 ohm $\times$ 45 amps, or 1.2 volts, which is still somewhat high.

$$(\text{also } VD = \frac{KLI}{CM} = \frac{17.4 \times 104 \times 45}{66,360} = 1.2 \text{ VD})$$

The resistance of No. 1 aluminum conductor is .211 ohm per M feet, and the drop is equal to .104 $\times$ .211 ohm $\times$ 45 amps or .99 volt, which is quite acceptable.

$$(\text{also } VD = \frac{KLI}{CM} = \frac{17.4 \times 104 \times 45}{83,690} = .99 \text{ VD})$$

Table 3A, Chapter 9, shows that four No. 1 Type THW conductors may be installed in a 1½-inch conduit. The feeder to panel E, then, consists of four No. 1 Type THW aluminum conductors in a 1½-inch conduit.

In the same way, it may be determined that panel D, originally supplied by four No. 4 copper conductors in 1¼-inch conduit, will require four No. 2 aluminum conductors in 1¼-inch conduit.

Panel B, originally supplied by four No. 2 copper conductors in 1¼-inch conduit, will require four No. 0 aluminum conductors in 2-inch conduit.

Panel A, originally supplied by four No. 4 copper conductors in 1¼-inch conduit, will require four No. 2 aluminum conductors in 1¼-inch conduit.

Panel C, originally supplied by four No. 8 copper conductors in ¾-inch conduit, will require four No. 6 aluminum conductors in a 1-inch conduit.

The basement lighting panel, originally supplied by No. 6 copper conductors, will require No. 4 aluminum conductors.

The service, consisting originally of two sets of 0000 Type THW and one set of No. 00 Type THW copper conductors in 2½-inch conduits, must be replaced by two sets of three 300 MCM Type THW and one set of 000 aluminum conductors in 3-inch conduits.

OTHER SCHEDULES OR DRAWINGS WHICH MAY BE PROVIDED

Lighting fixture schedule, showing number and locations of various types of fixtures.
Panelboard schedule, giving circuit designations.
Panelboard details, showing a separate drawing for each type of panel.
Motor schedule, giving sizes and locations.
Power sketches, showing motor switch assemblies.
One-line diagram of telephone system.
One-line diagram of speaker system.
One-line diagram of fire alarm system (if present).
One-line diagram of low-voltage signal system (if present).

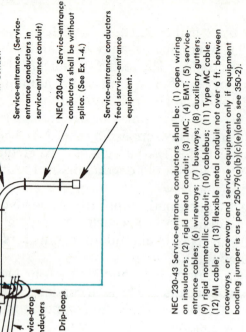

NEC 230-54 (f) Drip loops required to prevent moisture and are used to connect service-entrance wire to service-drop wires below level of service head.

Service Wires in Conduit

Service-entrance. (Service-entrance conductors in service-entrance conduit)

NEC 230-46 Service-entrance conductors shall be without splice. (See Ex 1-4.)

Service-entrance conductors feed service-entrance equipment.

Service Head

Service-drop Conductors

Drip-loops

NEC 230-43 Service-entrance conductors shall be: (1) open wiring on insulators; (2) rigid metal conduit; (3) IMC; (4) EMT; (5) service-entrance cables; (6) wireways; (7) busways; (8) auxiliary gutters; (9) rigid nonmetallic conduit; (10) cablebus; (11) Type MC cable; (12) MI cable; or (13) flexible metal conduit not over 6 ft. between raceways, or raceway and service equipment only if equipment bonding jumper is as per 250-79(a)(b)(c)(e)(also see 350-2).

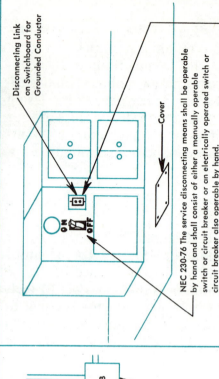

Disconnecting Link on Switchboard for Grounded Conductor

Cover

ON   OFF

NEC 230-76 The service disconnecting means shall be operable by hand and shall consist of either a manually operable switch or circuit breaker or an electrically operated switch or circuit breaker also operable by hand.

NEC 230-75 If the switch or circuit breaker does not interrupt the grounded conductor, other means of disconnection must be provided in the cabinet or switchboard.

NEC 384-3 (c) Switchboard or panelboards used as service equipment shall have means within the service disconnect section for connecting the supply neutral to the metal frame. (Size as per 250-79 (c).)

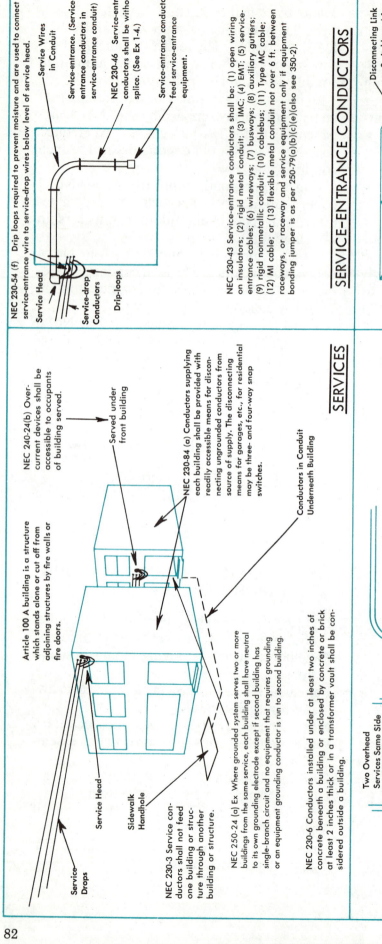

Article 100 A building is a structure which stands alone or cut off from adjoining structures by fire walls or fire doors.

NEC 240-24(b) Over-current devices shall be accessible to occupants of building served.

Served under front building

NEC 230-84 (a) Conductors supplying each building shall be provided with readily accessible means for disconnecting ungrounded conductors from source of supply. The disconnecting means for garages, etc., for residential may be three- and four-way snap switches.

Conductors in Conduit Underneath Building

NEC 250-24 (a) Ex Where grounded system serves two or more buildings from the same service, each building shall have neutral run to its own grounding electrode except if second building has single-branch circuit and no equipment that requires grounding or an equipment grounding conductor is run to second building.

NEC 230-6 Conductors installed under at least two inches of concrete beneath a building or enclosed by concrete or brick at least 2 inches thick or in a transformer vault shall be considered outside a building.

Service Drops

Service Head

Sidewalk Handhole

NEC 230-3 Service conductors shall not feed one building or structure through another building or structure.

B

B's Overcurrent Device Fused at Lower Value Than His Service Switch

Tenant B's Service Switch

Ltg B

Tenant A's Service Switch

Meter

Ltg A

1φ Lighting 115-230 Volts

Two Overhead Services Same Side of Building

Meter

3φ Power 230 Volts

NEC 230-91(c) In a multiple-Occupancy building, each occupant shall have access to his overcurrent devices except as permitted by 240-24(b) EX.

NEC 230-93 If necessary to prevent tampering, the service equipment for a special load may be locked.

NEC 230-92 If service equipment locked or inaccessible, overcurrent devices of lower rating must be installed in an accessible location.

NEC 230-72 If supplied by two or more services of the same side of the building, the service equipment shall be grouped and marked for identification. Service for fire pumps permitted to be in another location. (See Ex)

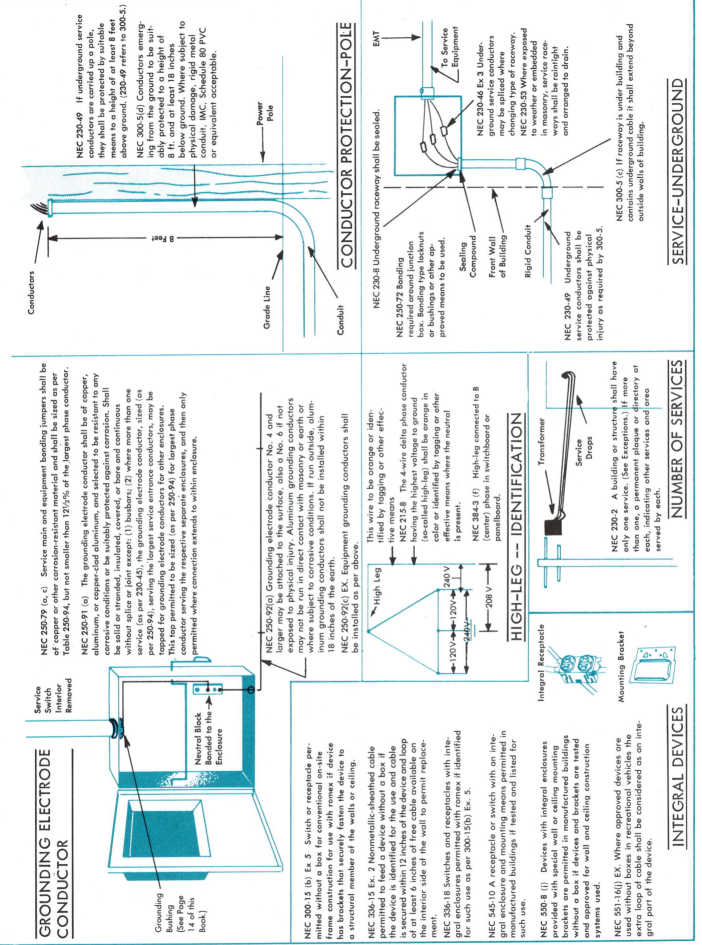

GROUNDING ELECTRODE CONDUCTOR

NEC 230-49 If underground service conductors are carried up a pole, they shall be protected by suitable means to a height of at least 8 feet above ground. (230-49 refers to 300-5.)

NEC 300-5(d) Conductors emerging from the ground to be suitably protected to a height of 8 ft. and at least 18 inches below ground. Where subject to physical damage, rigid metal conduit, IMC, Schedule 80 PVC or equivalent acceptable.

Conductors

Power Pole

8 Feet

Grade Line

Conduit

CONDUCTOR PROTECTION-POLE

EMT

To Service Equipment

NEC 230-8 Underground raceway shall be sealed.

NEC 230-46 Ex 3 Underground service conductors may be spliced where changing type of raceway.
NEC 230-53 Where exposed to weather or embedded in masonry, service raceways shall be raintight and arranged to drain.

NEC 250-72 Bonding required around junction box. Bonding type locknuts or bushings or other approved means to be used.

Sealing Compound

Front Wall of Building

Rigid Conduit

NEC 230-49 Underground service conductors shall be protected against physical injury as required by 300-5.

NEC 300-5 (c) If raceway is under building and contains underground cable it shall extend beyond outside walls of building.

SERVICE-UNDERGROUND

NEC 250-79 (a, c) Service main and equipment bonding jumpers shall be of copper or other corrosion-resistant material and shall be sized as per Table 250-94, but not smaller than 12½% of the largest phase conductor.

NEC 250-91 (a) The grounding electrode conductor shall be of copper, aluminum, or copper-clad aluminum, and selected to be resistant to any corrosive conditions or be suitably protected against corrosion. Shall be solid or stranded, insulated, covered, or bare and continuous without splice or joint except: (1) busbars; (2) where more than one service (as per 230-45), the grounding electrode conductor, sized (as per 250-94), serving the largest service entrance conductors, may be tapped for grounding electrode conductors for other enclosures. This tap permitted to be sized (as per 250-94) for largest phase conductor serving the respective separate enclosures, and then only permitted where connection extends to within enclosure.

NEC 250-92(a) Grounding electrode conductor No. 4 and larger may be attached to the surface, also a No. 6 if not exposed to physical injury. Aluminum grounding conductors may not be run in direct contact with masonry or earth or where subject to corrosive conditions. If run outside, aluminum grounding conductors shall not be installed within 18 inches of the earth.

NEC 250-92(c) EX. Equipment grounding conductors shall be installed as per above.

This wire to be orange or identified by tagging or other effective means

NEC 215-8   The 4-wire delta phase conductor having the highest voltage to ground (so-called high-leg) shall be orange in color or identified by tagging or other effective means where the neutral is present.

NEC 384-3 (f) High-leg connected to B (center) phase in switchboard or panelboard.

High Leg

240 V

120 V
120 V

208 V

240 V

120 V

HIGH-LEG -- IDENTIFICATION

Service Switch Interior Removed

Neutral Block Bonded to the Enclosure

Grounding Bushing
(See Page 14 of this Book.)

NEC 300-15 (b) Ex 5 Switch or receptacle permitted to feed a device without a box if the device is identified for the use with romex if device has brackets that securely fasten the device to a structural member of the walls or ceiling.

NEC 336-15 Ex. 2 Nonmetallic-sheathed cable permitted to feed a device without a box if the device is identified for the use and cable is secured within 12 inches of the device and loop of at least 6 inches of free cable available on the interior side of the wall to permit replacement.

NEC 336-18 Switches and receptacles with integral enclosures permitted with romex if identified for such use as per 300-15(b) Ex. 5.

NEC 545-10 A receptacle or switch with an integral enclosure and mounting means permitted in manufactured buildings if tested and listed for such use.

NEC 550-8 (i) Devices with integral enclosures provided with special wall or ceiling mounting brackets are permitted in manufactured buildings without a box if devices and brackets are tested and approved for wall and ceiling construction systems used.

NEC 551-16(i) EX. Where approved devices are used without boxes in recreational vehicles the extra loop of cable shall be considered as an integral part of the device.

Transformer

Service Drops

NEC 230-2 A building or structure shall have only one service. (See Exceptions.) If more than one, a permanent plaque or directory at each, indicating other services and area served by each.

NUMBER OF SERVICES

Integral Receptacle

Mounting Bracket

INTEGRAL DEVICES

Multioutlet Assembly

Building Finish or Baseboard

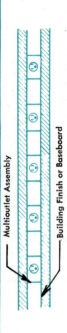

NEC 353-2 Multioutlet assembly may be used in dry locations. It shall not be used: (1) where concealed except that back and sides of metallic assembly may be surrounded by building finish or baseboard, (2) where exposed to severe mechanical injury, (3) where voltage between conductor is 300 volts or more unless the metal is not less than .040 inch thick, (4) where subject to corrosive vapors, (5) in hoistways, nor (6) in any hazardous location.

NEC 353-3 May be used through dry partition where no outlet concealed.

NEC 220-2(c)(5) Ex 1 Each 5 feet or fraction thereof of fixed multioutlet assembly shall be considered as one outlet of 180 VA capacity, except where a number of appliances may be connected simultaneously, then each 1 foot or fraction will be considered one outlet. NOTE: This exception does not apply to dwelling units or guest rooms of motels or hotels.

MULTIOUTLET ASSEMBLY

NEC 720-5 For circuits of less than 50 volts, standard lampholders rated not less than 660 watts are required.

NEC 210-21 (a) Lampholders connected to circuits having a rating greater than 20 amperes shall be of the heavy-duty type and shall be rated not less than 660 watts, if of admedium type, and not less than 750 watts if any other type. (See page 182 in this text for lampholders.)

HEAVY-DUTY LAMPHOLDERS -- RATING

30-Amp Circuit

Heavy-Duty Lampholders (1 in. nominal diameter)

NEC 240-83 (d) Where circuit breakers are used to switch 120-volt and 277-volt fluorescent fixtures, they shall be marked SWD and approved for such switching duty.
NEC 380-11 Circuit breakers used as a switch must conform to 240-81.

Nonadjustable Circuit Breaker

Center of operating handle not over 6½ feet from floor. (380-8 (a))

Shall be clearly marked "On" or "Off". (Up is "On" position.) (240-81)

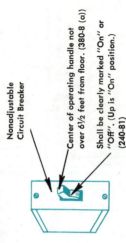

CIRCUIT BREAKERS

NEC 240-3 Ex 1 If allowable ampacity of a conductor does not correspond with a standard rating of overcurrent device, the next larger rating may be used, provided that the current is not greater than 800 amps.

NEC 240-82 An air circuit breaker used for branch circuits shall be of such design that alteration at its operating characteristics will be difficult.

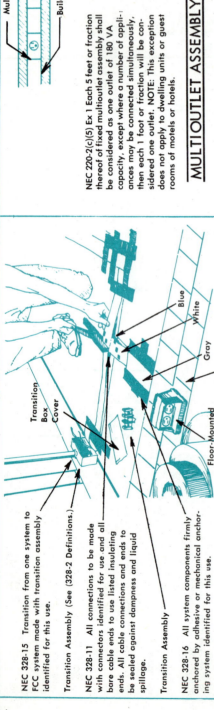

Blue
White
Gray

Transition Box Cover

Floor-Mounted Receptacle

NEC 328-10 Carpet squares not larger than 36 in. by 36 in.

NEC 328-3 System to comply with 210, 220, 240, 250, 300.

NEC 328-1 Type FCC cable is a field-installed wiring system using associated accessories for branch circuits and installed under carpet squares as per Article 328.

FLAT CONDUCTOR CABLE -- TYPE FCC -- ARTICLE 328

NEC 328-15 Transition from one system to FCC system made with transition assembly identified for this use.

Transition Assembly (See (328-2 Definitions.)

NEC 328-11 All connections to be made with connectors identified for use and all bare cable ends to use listed insulating ends. All cable connections and ends to be sealed against dampness and liquid spillage.

Transition Assembly

NEC 328-16 All system components firmly anchored by adhesive or mechanical anchoring system identified for this use.

NEC 328-10 Type FCC cable to be covered with carpet squares that have release-type adhesive.

NEC 328-6 Branch circuits: (a) Voltage not over 300 volts between ungrounded conductors or 150 volts to ground; (b) Not over 20 amps for general-purpose circuit or 30 amps for individual circuit.

NEC 328-5 Not permitted: (1) outdoors or wet locations; (2) corrosive vapor areas; (3) hazardous areas; (4) residential, school and hospital buildings.

Plug Receptacles

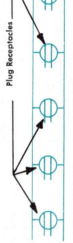

NEC 210-21 (b) Plug receptacles shall have a rating not less than the load served; when connected to circuits having two or more outlets, they shall conform to the following:

15-amp circuit—max. rating 15 amps
20-amp circuit—rating 15 or 20 amps
30-amp circuit—rating 30 amps
40-amp circuit—rating 40 or 50 amps
50-amp circuit—rating 50 amps

Receptacles installed on 15-amp and 20-amp circuits must be of the grounding type. (210-7)

NEC 210-21 (b) (1) A single receptacle installed on an individual branch circuit shall have a rating not less than branch circuit.
NOTE: 15-amp and 20-amp receptacles shall not supply a cord-and-plug connected load where the total load exceeds 80% of the rating of the receptacle. (210-21 (b-2))

RECEPTACLE RATINGS

No. 1 RH-130 Amps

150-Amp Fuses

NEC 240-21 (General Rule) Overcurrent protection required where conductors receive their supply. (See Exceptions.)

NEC 240-3 Ex 1 If allowable ampacity of conductor does not correspond with a standard rating of overcurrent device, the next larger rating may be used, provided that the current is not greater than 800 amps and conductor is not part of multioutlet branch circuit feeding receptacles that are readily accessible.

NEC 210-19 (a) Branch-circuit conductors shall have an ampacity not less than branch circuit and not less than maximum load served.

NEC 240-3 (General Rule) Conductors shall be protected in accordance with their ampacities.

CONDUCTORS -- OVERCURRENT PROTECTION

Conduit

Fitting with Thread Hubs

NEC 370-13(d) Threaded boxes or fittings not over 100 cubic inches in size which contain no devices are considered fastened if two or more conduits are threaded into box on two or more sides and are supported within 3 feet of the box.

NEC 370-13(e) Threaded boxes or fittings with devices not over 100 cubic inches are considered supported if two or more conduits are threaded into box and are firmly secured within 18 inches of the box.

THREADED BOX SUPPORTS

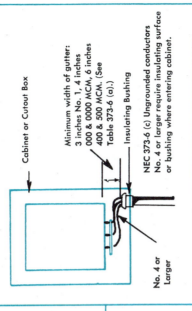

Cabinet or Cutout Box

Minimum width of gutter: 3 inches No. 1, 4 inches 000 & 0000 MCM, 6 inches 400 & 500 MCM. (See Table 373-6 (a).)

Insulating Bushing

NEC 373-6 (c) Ungrounded conductors No. 4 or larger require insulating surface or bushing where entering cabinet.

No. 4 or Larger

DEFLECTION OF CONDUCTORS

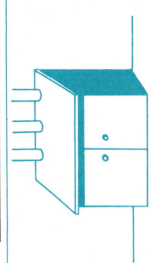

NEC 384-6 A switchboard in a wet location or outside a building shall have a weather-proof enclosure to comply with 373-2.

SWITCHBOARD -- OUTSIDE

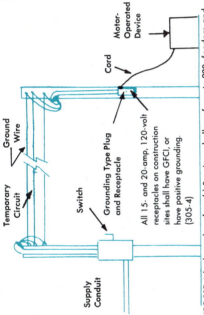

All grounding as per 250.

Ground Wire

Motor-Operated Device

Cord

Temporary Circuit

Switch

Grounding Type Plug and Receptacle

All 15- and 20-amp, 120-volt receptacles on construction sites shall have GFCI, or have positive grounding. (305-4)

Supply Conduit

NEC 305-4 (a, b, c, d, e, f, g, h) Services shall conform to 230; feeders and branch circuits on lighting branch circuits to be of rated ampacity; receptacles to be of grounding type and connected to grounding conductor; receptacles not permitted on lighting branch circuits; suitable disconnecting switches or plug connectors required to disconnect all ungrounded conductors of each temporary circuit; lamps to be protected by suitable fixture or lampholder with guard; box not required for splices or junction connections when using multiconductor cord or cable assemblies (see 110-14(b) and 400-9); flex cords and cables to be protected from damage.

TEMPORARY WIRING

NEC 230-43(13) Flexible metal conduit permitted as part of service raceway if not over 6 ft. long with bonding jumper around flex as per 250-79(a, b, c, e) (also see 350-2).

Greenfield permitted for service conduit if not over 6 ft. and external bond. (230-43(13)).

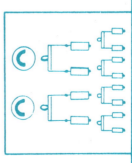

Flexible Metal Conduit (Greenfield)

Flexible metal conduit permitted as grounding means where conduit and fittings are listed for purpose (as per 250-91(b)) but total length shall not exceed 6 feet and not over a 20-amp circuit. Bonding jumper may be installed on outside of conduit as per 250-79(e) and sized as per Table 250-95. (350-5)

NEC 350-3 Minimum size is ½ in. except: (1) underplaster extensions (344-2); (2) motor leads (430-145(b)); (3) 6 ft. to lighting fixtures (410-67(c)); mfg. wiring systems (604-6(a)).

Maximum of 4 quarter bends in one run of conduit. Angle connectors not permitted for concealed work. (350-6)

FOR USE RESTRICTIONS, SEE 350-2.

Flexible conduit shall be secured within 12 inches of box or fitting and not over 4½ feet in remainder of run except: (1) where fished; (2) not over 3 feet at terminals for flexibility; (3) up to 6 feet at fixtures, as per 410-67 (c). (350-4)

FLEXIBLE METAL CONDUIT

Appliance

NEC 240-4 Flexible cords (including tinsel) to be protected at their ampacities per Table 402-5.

Ex. 1. Cord listed with specific appliance or portable lamp is permitted to be connected to a standard branch circuit as follows:
Tinsel No. 18 or larger cord, 20-amp circuit.
No. 16 or larger cord, 30-amp circuit.
Cord rated 20-amp or over for 40- or 50-amp circuit.

Ex. 2. Fixture wire connected to 120-volt or more standard branch circuit permitted to be protected as follows:
No. 18 (up to 50 ft.) or No. 16 (up to 100 ft.), 20 amp circuit.
No. 14 or larger, 20- or 30-amp circuit.
No. 12 or larger, 40- or 50-amp circuit.

EX. 3 Listed extension cordsets with No. 16 or larger conductors permitted on 20-amp branch circuit.

CORDS AND FIXTURE WIRE -- OVERCURRENT PROTECTION

Inductive Load (such as motors or fluorescent fixtures)

Snap Switch

NEC 380-14 (a, b, c) AC general-use snap switches controlling inductive loads shall not exceed rating of load at applied voltage. AC-DC general-use snap switches controlling inductive loads shall have rating of 200% of load at applied voltage. Where rated 20 amps or less and connected to aluminum wire, they shall be marked CO/ALR and be listed for the purpose.

SNAP SWITCHES -- RATING

NEC 384-5 Switchboards with exposed live parts shall be located in permanently dry locations accessible only to qualified persons.

SWITCHBOARDS -- LOCATION

SNAP SWITCHES IN PANELBOARDS

NEC 384-16 (b) Panelboards with snap switches rated 30 amps or less shall have over-current protection not in excess of 200 amps.

Switch

Overcurrent Device Not Over 200 Amperes

Panelboard with Snap Switches

OVERCURRENT PROTECTION OF PANELBOARDS

NEC 384-16 (a) A lighting and appliance panelboard shall be protected by not more than two sets of fuses or circuit breakers having a rating not greater than that of the panelboard.

200-Amp Switch and Overcurrent Devices

Other Loads

100-Amp Switch and Overcurrent Devices

Panelboard Rated at 100 Amps

FLOOR BOXES

Code gives no exception for dwelling units—receptacles in floor require a listed floor box unless enforcing authority judges them free from physical damage, dirt, etc.

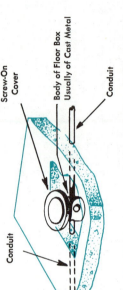

Screw-On Cover

Body of Floor Box Usually of Cast Metal

Conduit

Conduit

NEC 370-17 (b) Receptacles located in floors shall be installed in floor boxes listed specifically for this application. (See Exception for show windows.)

NEC 220-12 Unit load of 200 watts per linear foot required.

SHOW WINDOW WIRING

NEC 600-6 (b) Separate 20-amp circuit to outside outlet for sign required. Not required for entry to inside hall or corridor.

NEC 210-62 At least one receptacle outlet required above show window for each 12 ft or major fraction thereof, measured horizontally.

Advertising Device

NEC 370-17 (b) Ex Receptacles in elevated floors of show windows and similar locations may be installed in standard type receptacle boxes where enforcing authority judges them to be free from physical damage, dirt and moisture.

LIGHTING AND APPLIANCE PANELBOARD

NEC 384-14 A lighting and appliance switchboard is one having more than 10% of its overcurrent devices rated 30 amperes or less for which neutral connections are provided.

NEC 384-15 Not over 42 overcurrent devices of lighting and appliance panelboard, other than the mains, permitted in one cabinet. A two-pole circuit breaker counts as two overcurrent devices, a three-pole breaker as three.

NEC 384-16 (a) Ex 1 Panelboard may be protected by feeder overcurrent device if it does not exceed rating of panelboard.

NEC 384-16 (a) Ex 2 For existing installations individual protection not required in individual residential occupancy.

NEC 384-16 (a) (c) Lighting and appliance switchboards shall be protected individually by not more than two circuit breakers or sets of fuses unless the feeder provides adequate protection. Load on an overcurrent device shall not exceed 80 percent of its rating where the load will continue for three hours or more, unless the device is specifically approved for 100 percent continuous operation.

NEC 384-19 Except as permitted for services, panelboard fuses shall be installed on load side of switches.

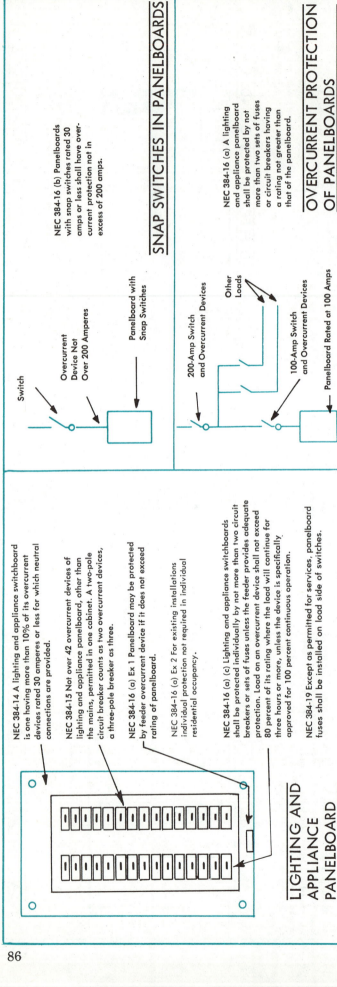

FLEXIBLE CORDS

NEC 400-8 Flexible cords and cables shall NOT be used: (1) as a substitute for fixed wiring; (2) thru holes in walls, floors or ceilings; (3) thru doors, windows or similar openings; (4) attached to building surfaces; (5) concealed by building walls, floors or ceilings.

NEC 400-7 Flexible cords and cables, permitted uses: (1) pendants; (2) fixtures; (3) portable lamps or appliances; (4) elevator cables; (5) cranes and hoists; (6) stationary equipment to facilitate frequent interchange; (7) prevent transmission of vibration or noise; (8) where mechanical connections and fastening means of appliance are designed to permit removal for maintenance and repair; (9) data processing cables as per 645-2; (10) connection of moving parts; (11) temporary wiring as per 305-2 (b) (c).

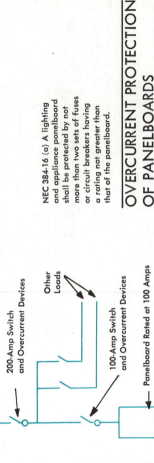

Wall Outlet Cord Concealed

Cord Tacked to Wall and Run Through Doorway

Cord Through Floor

Surface Mounted Receptacle

NEC 400-9 Flexible cords shall be used only in continuous lengths, not spliced or tapped except hard-service flexible cord No. 14 and larger if spliced as per 110-14 (b).

NEC 400-11 Flexible cord used in show windows and show cases shall in general, be S, SO, SJ, SJO, ST, STO, SJT, SJTO, and AFS.

NEC 400-12 In general, flexible cords shall not be smaller than No. 18. (See Tables 400-4, 400-5.)

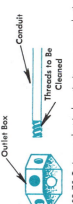

Single Circuit Breaker or Fused Switch

Panel-board

Tap Conductors Not Over 25 Feet in Length

CONDUIT protects wires from physical damage.

Limits Load to Ampacity of Tap Conductor

Tap conductors have at least 1/3 ampacity of supply wires.

Solderless Connectors

Supply Conductors

FEEDER TAP RULES

FEEDER TAP NOT OVER 25 FEET LONG

NEC 240-21 Ex 3 A smaller conductor may be tapped to a larger one without overcurrent protection at that point if the smaller conductor has an ampacity at least 1/3 that of the conductor from which it is supplied, and provided that the tap is suitably protected from physical damage, is not over 25 ft long, and terminates in a single circuit breaker or set of fuses which limits current in the smaller conductor to a suitable value.

FEEDER TAP OVER 25 FEET LONG

NEC 240-21 Ex 10 High bay (over 35 ft high at walls) manufacturing buildings, maintained and serviced by qualified persons, permit feeder taps not over 25 ft horizontally and not over 100 ft total length where the tap conductor: (a) has at least 1/3 ampacity of feeder overcurrent protection; (b) terminates in single overcurrent device that limits load to ampacity of tap conductors; (c) suitably protected from physical damage and in raceways; (d) continuous and contain no splice; (e) No. 6 copper or No. 4 aluminum or larger; (f) does not penetrate walls, floors or ceilings; (g) tap not less than 30 ft from floor.

Expansion Joint

(Appleton Electric Co.)

Equipment Bonding Jumpers

Ground Clamps

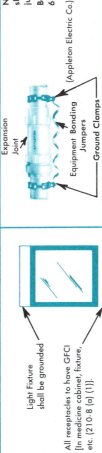

BONDING JUMPERS

NEC 250-77 Expansion joints and telescoping conduits shall be made electrically continuous by bonding jumpers or other means.

Bonding jumper permitted on outside of conduit if not 6 feet long. (250-79 (e))

NEC 250-114(a) The connection shall be made between a grounding conductor and a metallic box, by means of a grounding screw used for no other purpose, or by a listed grounding device.

NEC 250-76 Electrical continuity of metal conduit or cable containing any conductor (other than service entrance) of more than 250 volts to ground shall be assured by bonding jumpers or by other devices approved for the purpose or by: (a) threadless fittings, (b) two locknuts, one inside and one outside the box or cabinet or (c) connectors for EMT, (electrical metal tubing), flexible metal conduit, or cable that seals firmly against box.

Conduit

Threads to Be Cleaned

Outlet Box

GROUNDING-BOXES

NEC 250-114 Where more than one grounding conductor enters a box, the electrical connection shall be such that the removal of a device will not interrupt the grounding continuity.

NEC 250-75 To insure electrical continuity any nonconducting coating shall be removed from conduit or cable armor before fastening to metal enclosure.

NEC 240-2 Equipment shall be protected against overcurrent

Load Switch

CONDUIT protects conductors from physical injury.

Conductors Spliced to Larger Feed Wires

Tap Conductors Not Over 10 Feet in Length

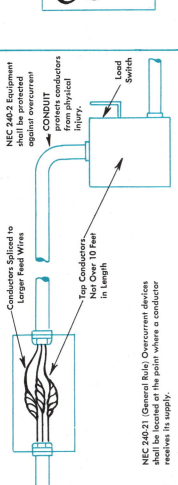

NEC 240-21 (General Rule) Overcurrent devices shall be located at the point where a conductor receives its supply.

NEC 240-21 Ex 2 (Feeder 10 Foot Tap Rule) Overcurrent protection not required where a feeder or transformer secondary conductor receives supply, if tap conductor is not over 10 feet long, ampacity of tap conductor is not less than load served or rating of device supplied, and not less than overcurrent device tap conductor feeds. Also tap conductors not permitted beyond switchboard, panelboard, or control devices they supply, and are protected from damages.

NEC 240-21 Ex 1 If the overcurrent device protecting the larger conductor also protects the smaller, as per Tables 310-16 through 310-19, no further protection is needed.

NEC 240-23 (General Rule) Where a change occurs in the size of the ungrounded conductor, a similar change may be made in the size of the grounded conductor.

FEEDER TO FOOT TAP RULE

Light Fixture shall be grounded

All receptacles to have GFCI [In medicine cabinet, fixture, etc. (210-8 (a) (1)].

Bathroom Sink

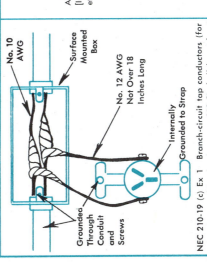

NEC 250-42 (General Rule) Noncurrent-carrying parts of fixed equipment shall be grounded if so located as to be potentially dangerous to persons or capable of creating or of increasing a hazardous condition.

EQUIPMENT GROUNDING--FIXED EQUIPMENT

No. 10 AWG

Surface Mounted Box

No. 12 AWG Not Over 18 Inches Long

Internally Grounded to Strap

Grounded Through Conduit and Screws

NEC 210-19 (c) Ex 1 Branch-circuit tap conductors (for other than cooking loads as per 210-2) shall be at least No. 14 and shall be rated not less than 15 amps when tapped from less than 40-amp circuit and not less than 20 amps when from 40- or 50-amp circuit when loads consist of: (a) lampholder or fixture where tap is not over 18 inches long; (b) fixture has tap conductors as per 410-67; (c) outlet taps not over 18 inches long; (d) industrial, infrared lamp heating appliances; (e) nonheating leads of de-icing and snow-melting cables and mats. For fixture wires and cords see 240-4.

BRANCH-CIRCUIT TAP RULES

NEC 300-19 (b)   Conductors shall be supported by one of the following methods: (b-1) wedges, (b-2) insulating supports, (b-3) deflecting insulators. (See 300-19 (b-3) for method of installation of conductors and supports.)

Intermediate Rigid Metal Conduit, EMT, MI, Type MC Cable, IMC, or ALS Cable

NEC 424-58   Electric duct heaters shall be identified as suitable for the installation.

Ducts or Plenums Used for Environmental Air

NEC 424-65 Electric duct heater controller shall be accessible and disconnect means within sight of the controller.

NEC 300-22 (b)   Where necessary to run wiring system thru air handling ducts or plenums, types MI or MC cables, EMT, flexible metallic tubing, IMC or rigid metal conduit may be used. Physically adjustable equipment and devices in ducts may be wired with flexible metal conduit or liquid-tight flexible metal conduit in lengths not exceeding 4 ft.

NEC 300-22 (a)   No wiring of any kind shall be installed in ducts used for dust, loose stock, flammable vapors, or commercial cooking.

NEC 300-22(c) Wiring methods in hollow spaces above lowered ceilings, that are also used for supply or return plenums for environmental air, are limited to types MI, MC, AC cables, and other factory-assembled multiconductor control or power cable specifically listed for the use. Other cables and conductors to be in EMT, IMC, flexible metallic tubing, rigid metal conduit, flexible metal conduit, busway or wireway with metal covers where accessible. Ex 1 Liquidtight flexible metal conduit in single lengths not over 6 ft permitted. Ex 4 Listed prefabricated cable assemblies of manufactured metallic wiring systems where listed for this use permitted. Ex 5 Stud or joist space in dwelling units where used as air handling permits normal wiring to pass thru if perpendicular to space. See 800-3(b)(3) for communication cables. Also 725-2(b), 760-4(d), and 820-15 Ex.

AIR PLENUMS -- WIRING METHODS

NEC 300-22 (d)   Wiring systems used with data processing systems and located within air handling spaces created by raised floors shall conform to Article 645.

NEC 645-2 (b)   Specifically approved computer cable or flexible cords permitted to connect or interconnect data processing equipment. (See 645-2 (c) for specific rules for wiring under raised floors.)

DATA PROCESSING SYSTEMS

NEC 300-19 (a) Conductors in vertical raceways shall be supported at intervals. (See Table 300-19 (a).) Exception: If total riser is less than 25% of spacing in Table 300-19 (a), no cable support required.

Not exceeding:
100 ft. for No. 18 through No. 8 copper or aluminum
100 ft. for No. 6 through No. 0 copper
200 ft. for No. 6 through No. 0 aluminum
50 ft. for 350,000 CM to 500,000 CM copper
120 ft. for 350,000 CM to 500,000 CM aluminum

(b-2)

(b-3)

Insulating Support

Deflecting Insulators

Wedges   (b-1)

Cable   Deflected Not Less Than 90 Degrees

100 ft

50 ft

CONDUCTORS -- VERTICAL SUPPORTS

NEC 250-75   Where metal raceway serves as grounding conductor, it shall be electrically continuous or bonded to assure ability to safely conduct possible fault currents.

NEC 300-10   Interior metal raceways and enclosures shall be effectively joined throughout the whole installation to provide electrical continuity.

NEC 250-79 (e)   Equipment bonding jumper shall be inside conduit with circuit wires except if not over 6 ft, then may be outside conduit if routed with conduit.

Locknuts and Bushings or Equal

CONTINUITY -- RACEWAYS AND BOXES

NEC 300-3 (b) Ex   Secondary wiring to electric-discharge lamps of 1000 volts or less may occupy the same enclosure as branch-circuit conductors, if insulated for the secondary voltage of the ballast.

NEC 410-31   Branch-circuit conductors within 3 inches of a ballast shall have a temperature rating of at least 90 degrees C, such as RHH, THHN, FEP, FEPB, SA, XHHW, AVA, and THW.

WIRING WITHIN FLUORESCENT FIXTURES

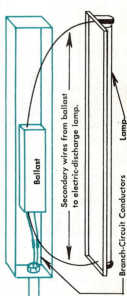

Ballast

Secondary wires from ballast to electric-discharge lamp.

Lamp

Branch-Circuit Conductors

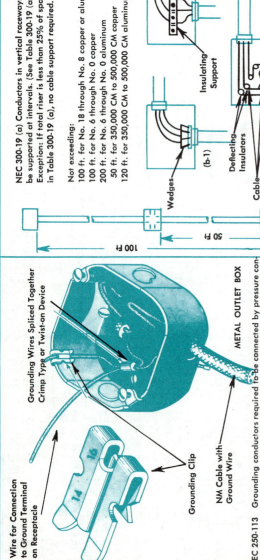

Wire for Connection to Ground Terminal on Receptacle

Grounding Wires Spliced Together Crimp Type or Twist-on Device

METAL OUTLET BOX

Grounding Clip

NM Cable with Ground Wire

NEC 250-113   Grounding conductors required to be connected by pressure connectors, clamps, or other approved means. Solder connections not permitted.
NEC 250-114   Where more than one grounding conductor enters box, they shall be so connected that removal of fixture or receptacle will not interrupt the grounding continuity.

BOXES -- GROUNDING

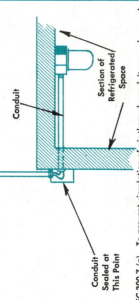

Conduit Sealed at This Point

Conduit

Section of Refrigerated Space

NEC 300-7 (a)   To prevent circulation of air through conduit runs whose ends are exposed to widely different temperatures, conduit shall be sealed off.
NEC 300-7 (b)   Expansion joints shall be provided where necessary to compensate for thermal movement.

RACEWAYS EXPOSED TO DIFFERENT TEMPERATURES

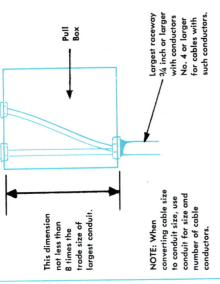

Pull Box

This dimension not less than 8 times the trade size of largest conduit.

Largest raceway ¾ inch or larger with conductors No. 4 or larger for cables with such conductors.

NOTE: When converting cable size to conduit size, use conduit for size and number of cable conductors.

NEC 370-18 (a-1) In straight pulls, length of box shall be not less than 8 times trade diameter of largest raceway, for raceways ¾ inch or larger, with conductors No. 4 or larger.

BOXES -- STRAIGHT PULLS

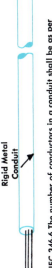

Rigid Metal Conduit

NEC 346-6 The number of conductors in a conduit shall be as per the percentage of fill permitted in Table 1, Chapter 9 of NEC.

NEC Chapter 9, Notes to Table 1:

Note 1. When conductors are the same size, Tables 3A, 3B, and 3C used to determine conduit size. Tables not used where conduit used for physical protection only.

Note 2. Equipment grounding conductors included when calculating conduit fill.

Note 3. Conduit nipples not over 24 inches are permitted and derating required by Note 8 of Tables 310-16 through 310-19 does not apply.

Note 4. The actual dimensions of conductors, such as compact or multiconductor cables, shall be used if not included in Chapter 9.

Note 5. For allowable percentage of conduit fill see Table 1.

NOTE: Table 1 states the percent of conduit fill, Table 4 states cross-sectional area for given percentages of fill, and Table 5 lists dimensions of insulated conductors.

Same rule applies to IMC (345-7); rigid metal conduit (346-6); rigid nonmetallic conduit (347-11); EMT (348-6); flexible metal conduit (350-1 refers to Article 346), etc.

See worked examples in lesson text.

NEC 374-6 Ampacity of bare conductors not to exceed 1000 amps per square inch of cross-sectional area for copper or 700 amps per square inch for aluminum. When 30 or less current-carrying conductors, no derating necessary.

NEC 374-5 Maximum cross-sectional area of all conductors not to exceed 20% of area of gutter. If over 30 current-carrying conductors (excluding signal and motor-control circuits), the derating factors of Note 8 to Tables 310-16 through 310-19 shall apply.

NEC 374-7 Minimum distance between bare conductors mounted on same surface 2 inches. If parts held free in air, 1 inch. Clearance to other metal surfaces, 1 inch.

NEC 374-2 Auxiliary gutter must not extend more than 30 feet beyond the equipment which it supplements, except in elevator work.

NEC 374-8 (b) Taps from bare conductors shall leave the gutter opposite their terminal connections and insulated conductors shall not come in contact with bare conductors of opposite polarity.

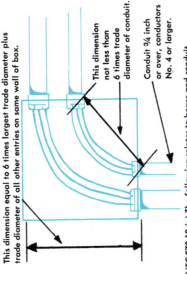

Insulating Support

Connection of Taps to Busbars

CONDUIT

Copper or Aluminum Busbars

Metal Gutter

Bars to Fasten Cover

Auxiliary gutters used to supplement wiring spaces at meter and distribution centers and similar points of wiring systems.

NEC 374-1 Auxiliary gutters may enclose only conductors or busbars. They shall not enclose switches, overcurrent devices, appliances, or similar equipment.

AUXILIARY GUTTERS

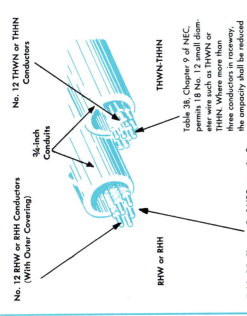

No. 12 RHW or RHH Conductors (With Outer Covering)

No. 12 THWN or THHN Conductors

¾-inch Conduits

THWN-THHN

Table 3B, Chapter 9 of NEC, permits 18 No. 12 small diameter wire such as THWN or THHN. Where more than three conductors in raceway, the ampacity shall be reduced as per Note 8 to Tables 310-16 through 310-19.

RHW or RHH

Table 3C, Chapter 9 of NEC, permits 5 No. 12 conductors with insulation such as RHW or RHH in a ¾-inch conduit.

CONDUCTORS--NUMBER IN CONDUIT

This dimension equal to 6 times largest trade diameter plus trade diameter of all other entries on same wall of box.

This dimension not less than 6 times trade diameter of conduit.

Conduit ¾ inch or over, conductors No. 4 or larger.

NEC 370-18 (a) The following relates to boxes and conduit bodies used as pull or junction boxes, containing raceways ¾-inch or larger, containing wire No. 4 or larger and cables No. 4 or larger.

NEC 370-18 (a-2) For "L" and "U" pulls, distance between raceway entry and opposite side of box at least 6 times trade size of largest conduit plus trade size of all entries on same wall of box. Distance between raceways for same conductor not less than 6 times trade size conduit. For cables the conduit size required for conductors to be used. Exception: Where conduit enters wall of box opposite removable cover, this distance as per Column for one conductor per terminal in Table 373-6 (a).

BOXES -- ANGLE PULLS

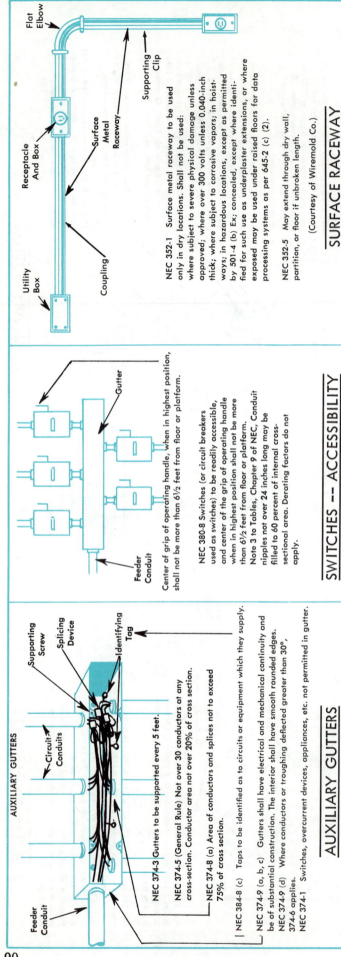

SURFACE RACEWAY

NEC 352-1 Surface metal raceway to be used only in dry locations. Shall not be used: where subject to severe physical damage unless approved; where over 300 volts unless 0.040-inch thick; where subject to corrosive vapors; in hoistways; in hazardous locations, except as permitted by 501-4 (b) Ex; concealed, except where identified for such use as underplaster extensions, or where exposed may be used under raised floors for data processing systems as per 645-2 (c) (2).

NEC 352-5 May extend through dry wall, partition, or floor if unbroken length.

(Courtesy of Wiremold Co.)

SWITCHES -- ACCESSIBILITY

Center of grip of operating handle, when in highest position, shall not be more than 6½ feet from floor or platform.

NEC 380-8 Switches (or circuit breakers used as switches) to be readily accessible, and center of the grip of operating handle when in highest position shall not be more than 6½ feet from floor or platform. Note 3 to Tables, Chapter 9 of NEC, Conduit nipples not over 24 inches long may be filled to 60 percent of internal cross-sectional area. Derating factors do not apply.

AUXILIARY GUTTERS

NEC 374-3 Gutters to be supported every 5 feet.

NEC 374-5 (General Rule) Not over 30 conductors at any cross-section. Conductor area not over 20% of cross section.

NEC 374-8 (a) Area of conductors and splices not to exceed 75% of cross section.

NEC 384-8 (c) Taps to be identified as to circuits or equipment which they supply.

NEC 374-9 (a, b, c) Gutters shall have electrical and mechanical continuity and be of substantial construction. The interior shall have smooth rounded edges.
NEC 374-9 (d) Where conductors or troughing deflected greater than 30°, 374-6 applies.
NEC 374-1 Switches, overcurrent devices, appliances, etc. not permitted in gutter.

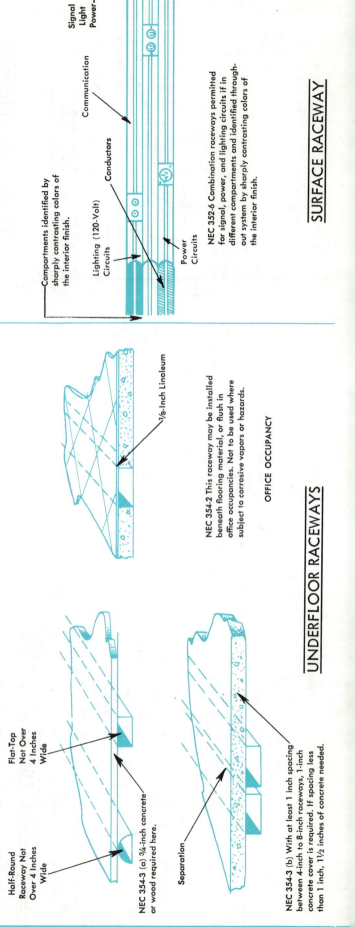

SURFACE RACEWAY

NEC 352-6 Combination raceways permitted for signal, power, and lighting circuits if in different compartments and identified through-out system by sharply contrasting colors of the interior finish.

UNDERFLOOR RACEWAYS

OFFICE OCCUPANCY

NEC 354-2 This raceway may be installed beneath flooring material, or flush in office occupancies. Not to be used where subject to corrosive vapors or hazards.

NEC 354-3 (a) ¾-inch concrete or wood required here.

NEC 354-3 (b) With at least 1 inch spacing between 4-inch to 8-inch raceways, 1-inch concrete cover is required. If spacing less than 1 inch, 1½ inches of concrete cover needed.

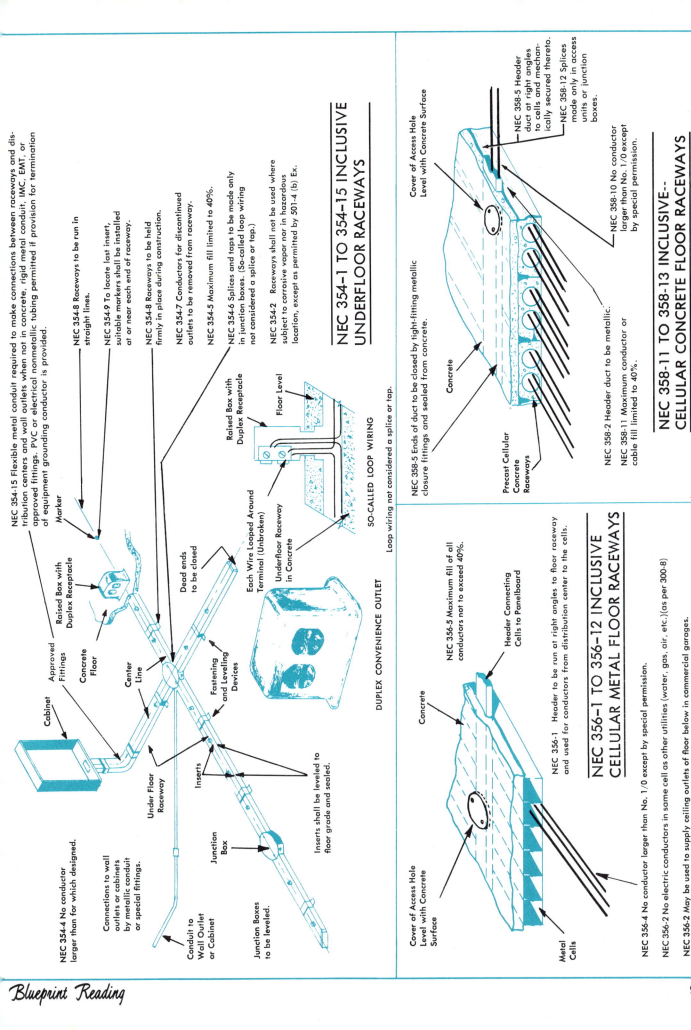

NEC 354-15 Flexible metal conduit required to make connections between raceways and distribution centers and wall outlets when not in concrete, rigid metal conduit, IMC, EMT, or approved fittings. PVC or electrical nonmetallic tubing permitted if provision for termination of equipment grounding conductor is provided.

NEC 354-8 Raceways to be run in straight lines.

NEC 354-9 To locate last insert, suitable markers shall be installed at or near each end of raceway.

NEC 354-8 Raceways to be held firmly in place during construction.

NEC 354-7 Conductors for discontinued outlets to be removed from raceway.

NEC 354-5 Maximum fill limited to 40%.

NEC 354-6 Splices and taps to be made only in junction boxes. (So-called loop wiring not considered a splice or tap.)

NEC 354-2 Raceways shall not be used where subject to corrosive vapor nor in hazardous location, except as permitted by 501-4 (b) Ex.

NEC 354-1 TO 354-15 INCLUSIVE UNDERFLOOR RACEWAYS

Marker

Cabinet

Approved Fittings

Raised Box with Duplex Receptacle

Concrete Floor

Center Line

Under Floor Raceway

Fastening and Leveling Devices

Inserts

Junction Box

Conduit to Wall Outlet or Cabinet

Junction Boxes to be leveled.

Inserts shall be leveled to floor grade and sealed.

Dead ends to be closed

Each Wire Looped Around Terminal (Unbroken)

Underfloor Raceway in Concrete

Raised Box with Duplex Receptacle

Floor Level

SO-CALLED LOOP WIRING

Loop wiring not considered a splice or tap.

DUPLEX CONVENIENCE OUTLET

NEC 354-4 No conductor larger than for which designed.

Connections to wall outlets or cabinets by metallic conduit or special fittings.

Cover of Access Hole Level with Concrete Surface

NEC 358-5 Header duct at right angles to cells and mechanically secured thereto.

NEC 358-12 Splices made only in access units or junction boxes.

NEC 358-10 No conductor larger than No. 1/0 except by special permission.

NEC 358-11 TO 358-13 INCLUSIVE-- CELLULAR CONCRETE FLOOR RACEWAYS

Cover of Access Hole Level with Concrete Surface

NEC 358-5 Ends of duct to be closed by tight-fitting metallic closure fittings and sealed from concrete.

Concrete

Precast Cellular Concrete Raceways

NEC 358-2 Header duct to be metallic.

NEC 358-11 Maximum conductor or cable fill limited to 40%.

NEC 356-5 Maximum fill of all conductors not to exceed 40%.

Header Connecting Cells to Panelboard

NEC 356-1 TO 356-12 INCLUSIVE CELLULAR METAL FLOOR RACEWAYS

NEC 356-1 Header to be run at right angles to floor raceway and used for conductors from distribution center to the cells.

Concrete

Cover of Access Hole Level with Concrete Surface

Metal Cells

NEC 356-4 No conductor larger than No. 1/0 except by special permission.

NEC 356-2 No electric conductors in same cell as other utilities (water, gas, air, etc.)(as per 300-8)

NEC 356-2 May be used to supply ceiling outlets of floor below in commercial garages.

COVE LIGHTING

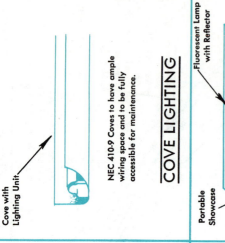

Cove with Lighting Unit

NEC 410-9 Coves to have ample wiring space and to be fully accessible for maintenance.

SHOWCASES -- CORD CONNECTED

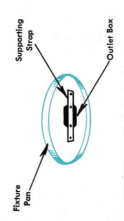

Fluorescent Lamp with Reflector
Portable Showcase
Switch
Remote Ballast
Grounding-Type Plug
Flexible Hard-Service Type Cord

NEC 410-29 Single showcase, not permanently installed, permitted to be cord-and-plug connected to permanent receptacle. Up to six such showcases permitted to be coupled together by flexible cord and locking-type connector, but one showcase of group must connect to permanent receptacle. (See 410-29 (a, b, c, d, e).)

NEC 410-77 (a) Auxiliary equipment not part of light fixture shall be enclosed in accessible permanently installed metal cabinets.

NEC 410-29 (e) In cord-connected showcases the secondary circuit of electric discharge lighting restricted to single case.

FLUORESCENT FIXTURES -- CONNECTION OF

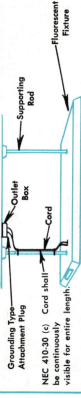

Supporting Rod
Fluorescent Fixture
Outlet Box
Cord
Grounding Type Attachment Plug

NEC 410-30 (c) Cord shall be continuously visible for entire length.

NEC 410-14 Fluorescent fixtures supported independently of outlet box shall be connected by metal raceway, metal-clad cable, or nonmetallic cable except cord connections permitted per 410-30 (b) (c).

NEC 410-30(c) Cord-equipped fixtures may be suspended directly below outlet box if cord terminates in grounding type plug.

NEC 410-74 Fixtures designed for direct current shall be so marked.

NEC 410-78 Autotransformer ballasts which raise voltage to more than 300 volts shall be supplied only by a grounded system.

SCREW-SHELL LAMPHOLDERS

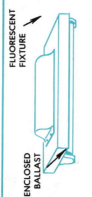

Female Plug
Lampholder

NEC 410-47 Screw-shell lampholders are NOT to be installed for use as plug receptacles. Where supplied by circuit having grounded conductor, then this conductor must connect to screw-shell.

FIXTURE MOUNTING

FLUORESCENT FIXTURE
ENCLOSED BALLAST

NEC 410-76(b) Surface-mounted fixtures containing ballasts must be especially approved for mounting on low-density cellulose fiberboard or spaced not less than 1½ inches therefrom. See (FPN) for definition of combustible low-density cellulose fiberboard sheets, panels, and tiles.

FIXTURES -- USED AS RACEWAY

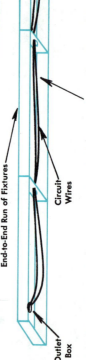

End-to-End Run of Fixtures
Circuit Wires
Outlet Box

NEC 410-73 (e) Fluorescent fixtures installed indoors shall have thermal protection within; ballast (P ballast). Replacement ballasts shall also have integral protection.

NEC 410-31 (General Rule) Fixtures not to be used as raceways for circuit conductors except: (1) where approved as raceway; (2) conductors of a single or multiwire branch circuit supplying the fixtures may be run through an end-to-end assembly of units properly fastened together; (3) in addition to above, one two-wire circuit that supplies at least one fixture in assembly may be run through fixtures. Conductors within 3 inches of ballast shall have 90°C rating. (XHHW, THHN, RHH, AVA, FEP, FEPB) Type THW also permitted within 3 inches of ballast. (See Table 310-13 for special THW rating.)

FIXTURES -- SHOW WINDOWS

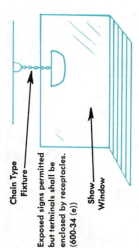

Chain Type Fixture
Show Window

NEC 210-62 At least one overhead receptacle for each 12 ft of show window.

Exposed signs permitted but terminals shall be enclosed by receptacles. (600-34 (e))

Transformers for gas tube signs shall be grounded. (600-37 (e))

NEC 410-7 No externally-wired fixtures other than the chain type permitted in show windows. (See 400-11 for flexible cords.)

FIXTURE SUPPORTS

Supporting Strap
Outlet Box
Fixture Pan

NEC 410-16 (a) (General Rule) Fixture shall be attached to and supported by outlet box.

NEC 410-16 (a) Fixtures weighing more than 50 pounds shall be supported independently of the outlet box.

NEC 410-16 (b) Connections between fixture conductors and circuit conductors shall be accessible for inspection without requiring disconnection of any part of wiring except when connected by attachment plugs and receptacles.

NEC 410-14 (b) When fluorescent fixture is mounted over box, an opening is required in back of fixture so box is accessible.

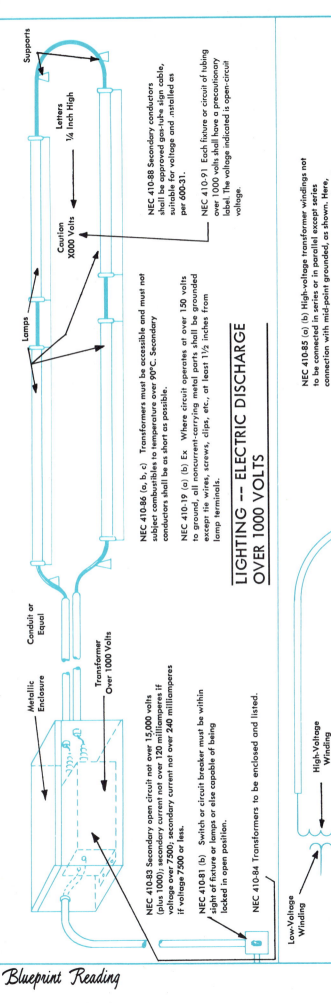

Supports

Letters
1/4 Inch High

Lamps

Caution
X000 Volts

Conduit or
Equal

Metallic
Enclosure

Transformer
Over 1000 Volts

NEC 410-88 Secondary conductors shall be approved gas-tube sign cable, suitable for voltage and .installed as per 600-31.

NEC 410-91 Each fixture or circuit of tubing over 1000 volts shall have a precautionary label. The voltage indicated is open-circuit voltage.

NEC 410-86 (a, b, c) Transformers must be accessible and must not be subject combustibles to temperature over 90°C. Secondary conductors shall be as short as possible.

NEC 410-19 (a) (b) Ex Where circuit operates at over 150 volts to ground, all noncurrent-carrying metal parts shall be grounded except tie wires, screws, clips, etc., at least 1 1/2 inches from lamp terminals.

LIGHTING -- ELECTRIC DISCHARGE OVER 1000 VOLTS

NEC 410-83 Secondary open circuit not over 15,000 volts (plus 1000); secondary current not over 120 milliamperes if voltage over 7500; secondary current not over 240 milliamperes if voltage 7500 or less.

NEC 410-81 (b) Switch or circuit breaker must be within sight of fixture or lamps or else capable of being locked in open position.

NEC 410-84 Transformers to be enclosed and listed.

Low-Voltage Winding

High-Voltage Winding

No. 14 Wire (Insulated)

Transformer

Ground

High-Voltage Winding

Low-Voltage Winding

Also Connected to Enclosure

Neon Tubing

NEC 410-85 (a) (b) High-voltage transformer windings not to be connected in series or in parallel except series connection with mid-point grounded, as shown. Here, grounded ends shall be connected by an insulated conductor not smaller than No. 14 AWG.

NEC 410-88 Secondary conductory wiring methods shall be made with approved gas-tube sign cable suitable for applied voltage and in accordance with 600-31.

SIGN -- ELECTRIC DISCHARGE OVER 1000 VOLTS (NEON)

Combustible Walls

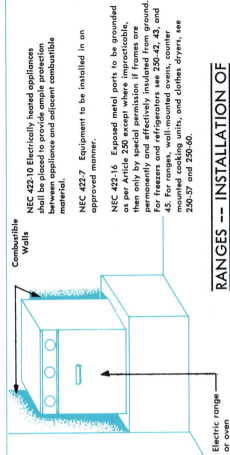

Electric range or oven

NEC 422-10 Electrically heated appliances shall be placed to provide ample protection between appliance and adjacent combustible material.

NEC 422-7 Equipment to be installed in an approved manner.

NEC 422-16 Exposed metal parts to be grounded as per Article 250 except where impracticable, then only by special permission if frames are permanently and effectively insulated from ground. For freezers and refrigerators see 250-42, 43, and 45. For ranges, wall-mounted ovens, counter mounted cooking units, and clothes dryers, see 250-57 and 250-60.

RANGES -- INSTALLATION OF

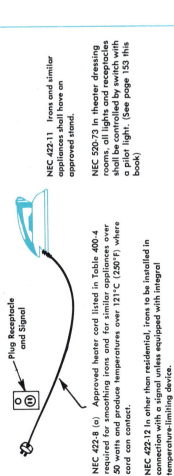

Plug Receptacle and Signal

NEC 422-11 Irons and similar appliances shall have an approved stand.

NEC 520-73 In theater dressing rooms, all lights and receptacles shall be controlled by switch with a pilot light. (See page 153 this book)

SMOOTHING IRONS

NEC 422-8 (a) Approved heater cord listed in Table 400-4 required for smoothing irons and for similar appliances over 50 watts and produce temperatures over 121°C (250°F) where cord can contact.

NEC 422-12 In other than residential, irons to be installed in connection with a signal unless equipped with integral temperature-limiting device.

NEC 422-2 Current-carrying parts to be protected from contact, except where necessarily exposed as in toasters and grills.

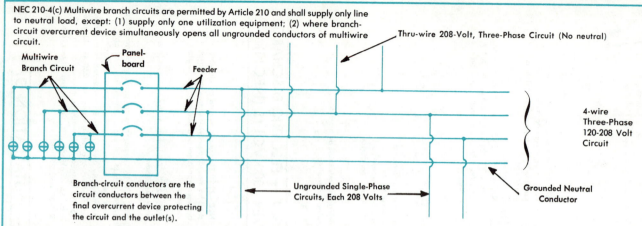

NEC 210-4(c) Multiwire branch circuits are permitted by Article 210 and shall supply only line to neutral load, except: (1) supply only one utilization equipment; (2) where branch-circuit overcurrent device simultaneously opens all ungrounded conductors of multiwire circuit.

Thru-wire 208-Volt, Three-Phase Circuit (No neutral)

Multiwire Branch Circuit

Panel-board

Feeder

4-wire Three-Phase 120-208 Volt Circuit

Branch-circuit conductors are the circuit conductors between the final overcurrent device protecting the circuit and the outlet(s).

Ungrounded Single-Phase Circuits, Each 208 Volts

Grounded Neutral Conductor

NEC 210-4(a) All multiwire branch circuit conductors shall originate in same panelboard.

NEC 210-4(b)In dwelling units where more than one receptacle on the same yoke (split receptacle) is fed by a multiwire branch circuit, means must be provided at the panelboard to simultaneously disconnect all ungrounded conductors.

NEC 210-10, 215-7 Ungrounded conductors for feeders and branch circuits may be tapped from ungrounded wires of circuits which have identified neutrals. Switching devices in these circuits shall have a pole in each conductor. For branch circuits, all poles must open simultaneously where they serve as disconnecting means as required for appliances (see 422-21 (b)); fixed electric space heating (see 424-20); electric de-icing and snow-melting (see 426-21); motor controller (see 430-85); for motor see 430-103.

BRANCH CIRCUIT -- TAPS FROM

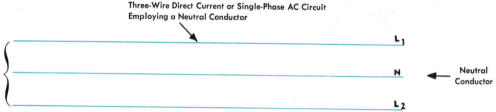

Three-Wire Direct Current or Single-Phase AC Circuit Employing a Neutral Conductor

L_1

N — Neutral Conductor

L_2

NEC 220-22 Feeder neutral sized for maximum unbalance between neutral and phase conductors, except neutral for household ranges, ovens, and dryers permitted to be reduced to 70% of load as calculated per Table 220-18 and Table 220-19. A further demand of 70% permitted for portion of maximum unbalanced load in excess of 200 amperes. Also applies to 4-wire 3-phase and 5-wire 2-phase systems.

NOTE: Exception. This 70% demand factor does not apply for that portion of the feeder load which consists of electric discharge lighting.

NOTE: Where household electric ranges and dryers are supplied by the feeder, this demand factor is additional to that allowed in Table 220-18 and Table 220-19.

Table 220-11 (Footnote) Demand factors in this table do not apply to computed loads on feeders to areas in hospitals, hotels, and motels, where the entire lighting load is used generally at one time.

NEC 220-3(a) Continuous load on branch circuit shall not exceed 80 percent of branch-circuit rating.

Exception: Does not apply where assembly including overcurrent device approved for 100 percent operation.

NEC 215-2   (Fine Print Note) Voltage drop for power, heating, or lighting feeders should not exceed 3 percent. Voltage drop for feeder plus branch circuit should not exceed 5 percent. (See 210-19 (a) Note for branch circuits.)

LOAD CALCULATIONS -- FEEDERS AND BRANCH CIRCUITS

2 Sets of Four-Wire Feeders with Common Neutral

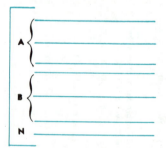

3 Sets of Three-Wire Feeders with Common Neutral

NEC 225-7 (b) On poles or other structures, a common neutral may be used with not more than 8 ungrounded conductors of a multiwire branch circuit. Ampacity of this neutral must be adequate to carry maximum unbalance.

NEC 215-4 (a) A common neutral may be used with two or three sets of 3-wire feeders or two sets of 4-wire or 5-wire feeders.

NEC 220-22 Neutral sized to carry maximum unbalance as determined by Article 220.

COMMON NEUTRAL FOR FEEDERS

TRADE COMPETENCY TEST NO. 3A

INSTRUCTIONS: The following statements are either True or False. Draw a circle around the **T** if the statement is True, or around the **F** if it is False. The questions are to be answered, as nearly as possible, in accordance with the specific provisions of the NEC.

• True or False •

1. In a fuseholder, the screw shell shall be connected to the line side of the circuit.   T   F

2. In general, a building cannot have openings in its fire walls.   T   F

3. Taps not over 18 inches long, for individual lampholders, may be smaller than the branch circuit conductors.   T   F

4. Lighting outlets, installed in other than dwelling occupancies, shall not be connected to 50-amp branch circuits.   T   F

5. The feeder demand factor for store lighting shall be 100% of the connected load.   T   F

6. The neutral feeder conductor shall be the same size as the ungrounded conductors for a resistance load.   T   F

7. In interior wiring a common neutral feeder wire may not be run for two or three sets of 3-wire feeder circuits.   T   F

8. Splices are sometimes permitted in service conductors.   T   F

9. The service overcurrent device is permitted to be located adjacent to the service disconnect means.   T   F

10. Fuses shall never be field-connected in parallel.   T   F

11. The maximum rating of a 250-volt cartridge fuse is 600 amps.   T   F

12. When bonding jumpers are required for grounding interior equipment, they shall be selected from Table 250-94.   T   F

13. No. 18 copper wire may be used for grounding a portable device used on a 20-amp circuit.   T   F

14. Properly made splices are permitted in metal surface raceways.   T   F

15. No conduit shall be run through a duct used for dust removal.   T   F

Continued on Next Page

TEAR OUT HERE

16. Temporary wiring is permitted for Christmas decorative lighting for a period of not over 60 days. T F

17. Surface metal raceway may be extended through dry floors. T F

18. Conduit nipples not over 24 inches long may be filled to 60 percent of their cross-sectional area. T F

19. Conductors in underfloor raceway systems may be spliced or tapped at individual outlets. T F

20. The full cross-sectional area of cellular metal floor raceway may be utilized if Type AC armored cable is employed. T F

21. Header ducts used with cellular concrete floor raceways may be run diagonally across the cells provided they run in straight lines. T F

22. Plastic inserts can be used with metal underfloor raceways. T F

23. The cross-sectional area of conductors in any section of auxiliary gutter shall not exceed 20 percent of its area. T F

24. AC-DC general-use snap switches controlling ordinary inductive loads shall have an ampere rating twice that of the load. T F

25. In a commercial occupancy electric discharge lighting transformers of more than 1000 volts shall not be connected in series. T F

TEAR OUT HERE

Student
Number_____

Instructor's
Name_____

INSTRUCTIONS: In the space provided at the right of each of the following questions, write **a, b, c,** or **d** to indicate which of the alternatives given makes it most nearly correct. The questions are to be answered, as nearly as possible, in accordance with the specific provisions of the NEC.

• Multiple-Choice •

1. One of the approved methods of identifying the high leg of a 4-wire delta connected secondary system is by:

    (a) use of a white wire, (b) use of a green wire,
    (c) use of a red wire, (d) tagging the wire

    1._____

2. Where branch-circuit loads are expected to continue for three hours or more, the minimum unit loads given in Table 220-3(b) shall be increased by:

    (a) 80%, (b) 10%, (c) 25%, (d) 35%

    2._____

3. A unit load for each outlet supplying heavy-duty lampholders that are not used for general illumination in industrial occupancies shall be:

    (a) 600 volt-amperes, (b) 180 volt-amperes, (c) 5 amperes,
    (d) 1½ amperes

    3._____

4. The minimum feeder allowance for show-window lighting expressed in watts-per-linear-foot shall be:

    (a) 100, (b) 200, (c) 300, (d) 500

    4._____

5. 15-amp receptacles on standard 15- and 20-amp branch circuits shall not supply a portable appliance load in excess of:

    (a) 15 amps, (b) 12 amps, (c) 10 amps, (d) 9 amps

    5._____

6. The unit load for a store is listed as:

    (a) 3 watts per sq ft, (b) 3.75 watts per sq ft, (c) 2 watts
    per sq ft, (d) 2.5 watts per sq ft

    6._____

7. Where permissible, the demand factor applied to that portion of unbalanced neutral feeder load in excess of 200 amps is:

    (a) 40%, (b) 80%, (c) 70%, (d) 60%

    7._____

Continued on Next Page

TEAR OUT HERE

8. Underground service conductors carried up a pole must be protected from mechanical injury to a height of at least:

   (a) 12 feet, (b) 8 feet, (c) 15 feet, (d) 9 feet

   8. _____

9. The largest standard cartridge fuse rating is:

   (a) 6,000 amps, (b) 1,200 amps, (c) 1,000 amps, (d) 600 amps

   9. _____

10. A 25-foot tap taken from a 100-amp feeder shall have an ampacity of not less than:

    (a) 15 amps, (b) 20 amps, (c) 25 amps, (d) 33 1/3 amps

    10. _____

11. Metal enclosures for grounding electrode conductors shall be:

    (a) rigid conduit only, (b) not less than ¾ inch in diameter, (c) bonded, (d) electrically continuous

    11. _____

12. The size of a copper grounding electrode conductor utilizing a water pipe electrode for a 120-208-volt service which has three 250 MCM copper conductors and a No. 3/0 neutral is:

    (a) No. 8, (b) No. 2, (c) No. 6, (d) No. 4

    12. _____

13. According to the Code, in a building where water pipe is available, the minimum size of the grounding electrode conductor for an ungrounded system where the service conductors are 3/0 copper shall be:

    (a) No. 4 copper, (b) No. 2 copper, (c) No. 6 copper, (d) No. 4 aluminum

    13. _____

14. No. 0 copper conductors in vertical raceway shall be supported at intervals not exceeding:

    (a) 50 feet, (b) 75 feet, (c) 100 feet, (d) 125 feet

    14. _____

15. Fixed equipment may be grounded by way of:

    (a) copper wire, (b) aluminum conduit, (c) EMT, (d) any one of these

    15. _____

16. The largest size of electrical metallic tubing is:

    (a) 3-inch, (b) 4-inch, (c) 2-inch, (d) 1¼-inch

    16. _____

TEAR OUT HERE

17. The largest conductor permitted in ⅜-inch flexible metal conduit is:

    (a) No. 12, (b) No. 16, (c) No. 14, (d) No. 10

    17. _____

18. Underfloor raceways not over 4 inches wide shall be covered by concrete or wood to a thickness of not less than:

    (a) ¾ inch, (b) 2 inches, (c) 1½ inches, (d) 1 inch

    18. _____

19. Underfloor raceways over 4 inches but less than 8 inches wide and spaced less than 1 inch apart shall be covered with concrete to a depth not less than:

    (a) ¾ inch, (b) 2 inches, (c) 1½ inches, (d) 1 inch

    19. _____

20. The combined cross-sectional area of all conductors or cables in an underfloor raceway shall not exceed:

    (a) 20%, (b) 50%, (c) 30%, (d) 40%

    20. _____

21. The largest size of conductor allowed in a cellular metal floor raceway without special permission is:

    (a) No. 0, (b) 500 MCM, (c) No. 000, (d) 250 MCM

    21. _____

22. Angle-pull dimensional requirements apply to junction boxes only when the size of conductor is equal to or larger than:

    (a) No. 0, (b) No. 4, (c) No. 000, (d) No. 6

    22. _____

23. Splices and wires within any portion of an auxiliary gutter shall not fill the cross-sectional area of the gutter by more than:

    (a) 20%, (b) 50%, (c) 60%, (d) 75%

    23. _____

24. The maximum number of overcurrent devices in a lighting and appliance branch-circuit panelboard shall not exceed:

    (a) 30, (b) 36, (c) 42, (d) 48

    24. _____

25. Panelboards equipped with snap switches rated at 30 amps or less, shall have overcurrent protection not in excess of:

    (a) 150 amps, (b) 300 amps, (c) 100 amps, (d) 200 amps

    25. _____

TEAR OUT HERE

TRADE COMPETENCY TEST NO. 3C

Student
Number_____

Instructor's
Name_____

INSTRUCTIONS: Complete each of the following statements, by writing the missing word or words in the space provided at the right of each question, to make a true statement. The questions are to be answered, as nearly as possible, in accordance with the specific provisions of the NEC.

• Completion •

1. In other than dwellings, each 5 feet or fraction of multioutlet assembly shall normally be considered as a load of _____.

1. _____

2. Total load on a branch circuit shall not exceed its _____ .

2. _____

3. Voltage drop on power feeders should be limited to _____ percent.

3. _____

4. The unit load for a hall in other than single-family dwelling or an apartment is listed as _____ watts per square foot.

4. _____

5. Service conductors of 600 volts or less buried under _____ inches or more of concrete are considered outside the building.

5. _____

6. Where conduit gives physical protection for underground cable, a _____ is required where underground cable leaves conduit.

6. _____

7. If the allowable current-carrying capacity of a conductor does not correspond to the rating of a standard size overcurrent device, the next larger size may be used provided the current does not exceed _____ amps.

7. _____

8. A circuit breaker having an interrupting rating other than _____ amps shall be marked with this rating.

8. _____

9. Circuit breakers used to switch 120 or 277-volt fluorescent fixtures shall be approved for the purpose and marked _____.

9. _____

10. Circuit breakers shall _____ whether in open or closed position.

10. _____

11. Exposed, non-current-carrying metal parts of fixed equipment shall be grounded if located in a wet location and not _____ .

11. _____

12. A No. _____ or larger grounding electrode conductor shall be protected if exposed to severe physical damage.

12. _____

Continued on Next Page

13. Interior raceways installed between areas differing widely in temperatures shall be _____ .

13. _____

14. The _____ of an underfloor raceway shall determine the largest conductor which can be installed.

14. _____

15. Conductors from discontinued outlets in an underfloor raceway installation shall be _____ from the raceway.

15. _____

16. _____ are required at or near the end of each run of underfloor raceway to locate the last insert.

16. _____

17. Dead ends of underfloor raceways shall be _____ .

17. _____

18. Header ducts used with cellular concrete raceways shall be constructed of _____ .

18. _____

19. Threaded boxes not over 100 cubic inches in size that contain devices need no additional support if two or more conduits, which are properly supported within _____ inches of the box, are threaded into the box from the same side.

19. _____

20. Gutters shall be supported throughout their entire length at intervals not exceeding _____ feet.

20. _____

21. The current carried continuously in bare copper bars in auxiliary gutters shall not exceed _____ amps per square inch.

21. _____

22. Three-way and four-way switches shall be wired so that all switching is done in the _____ conductors.

22. _____

23. The wiring between 3-way switches and outlets, where in metal enclosures, shall be run with both _____ in the same enclosure.

23. _____

24. Lampholders of the screw shell type shall be installed for use as _____ only.

24. _____

25. Transformers for electric-discharge lighting systems of more than 1000 volts shall be _____ after installation.

25. _____

National Electrical Code

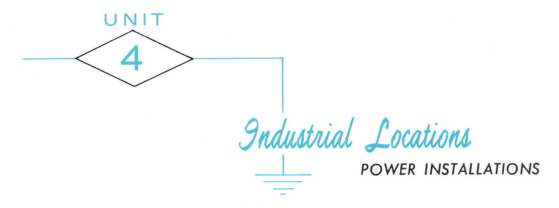

Industrial Locations

POWER INSTALLATIONS

MANUFACTURING PLANT

GENERAL—This blueprint shows only power outlets which are supplied by a 4-wire, 3-phase, 265-460-volt, WYE distribution system. Lighting and receptacle outlets are supplied by a separate 4-wire, 3-phase, 120-208-volt network system and would appear on a separate drawing. Specifications call for Type THW copper conductors throughout the project.

BASIC NEC RULES FOR MOTOR CIRCUITS

To simplify the study of motors we will start by evaluating the section in the NEC that tells us when to use the tables at the end of Article 430 and when to use the motor nameplate rating in determining the ampacities and ratings of the various components of motor circuits. NEC 430-6(a) states that where the current rating of a motor is used to determine the ampacity of conductors or the ampere ratings of switches, controllers, branch-circuit overcurrent devices, etc., the values given in Tables 430-147, 430-148, 430-149, and 430-150, including notes, shall be used instead of the actual current rating that appears on the motor nameplate. If only the full-load current is marked on the nameplate, the horsepower rating shall be determined by the corresponding value given in Tables 430-147, 430-148, 430-149 and 430-150, interpolated if necessary. In the selection of separate motor-running overcurrent protection which will protect motors, motor-control apparatus, and motor branch circuit conductors against excessive heating due to overloads and failure to start, the full-load current rating that appears on the motor nameplate shall be used.

NEC 430-22(a) states that branch-circuit conductors supplying a single motor shall have an ampacity not less than 125 percent of the motor full-load current. NEC 430-32 (a) specifies that continuous-duty motors rated more than 1 hp shall be protected against overload, in general, by a device that will trip at not more than 125 percent of motor full-load current.

NEC 430-52 requires that overcurrent devices protecting branch-circuit conductors shall be capable of carrying motor starting current. Motor starting current varies usually between 150 percent and 300 percent of full-load current, depending on the particular kind of motor. Table 430-152 furnishes information necessary to calculate this current. In rare cases starting current runs somewhat higher than values given there. NEC 430-52 provides for this contingency, permitting larger overcurrent devices which will permit the motor to start.

Application Of These Rules

Consider unit *A1* on the blueprint, a 40-hp, 3-phase, 460-volt, squirrel-cage motor which has no code letters marked on the nameplate and which is started by means of an autotransformer compensator. Table 430-150 gives the full-load current as 52

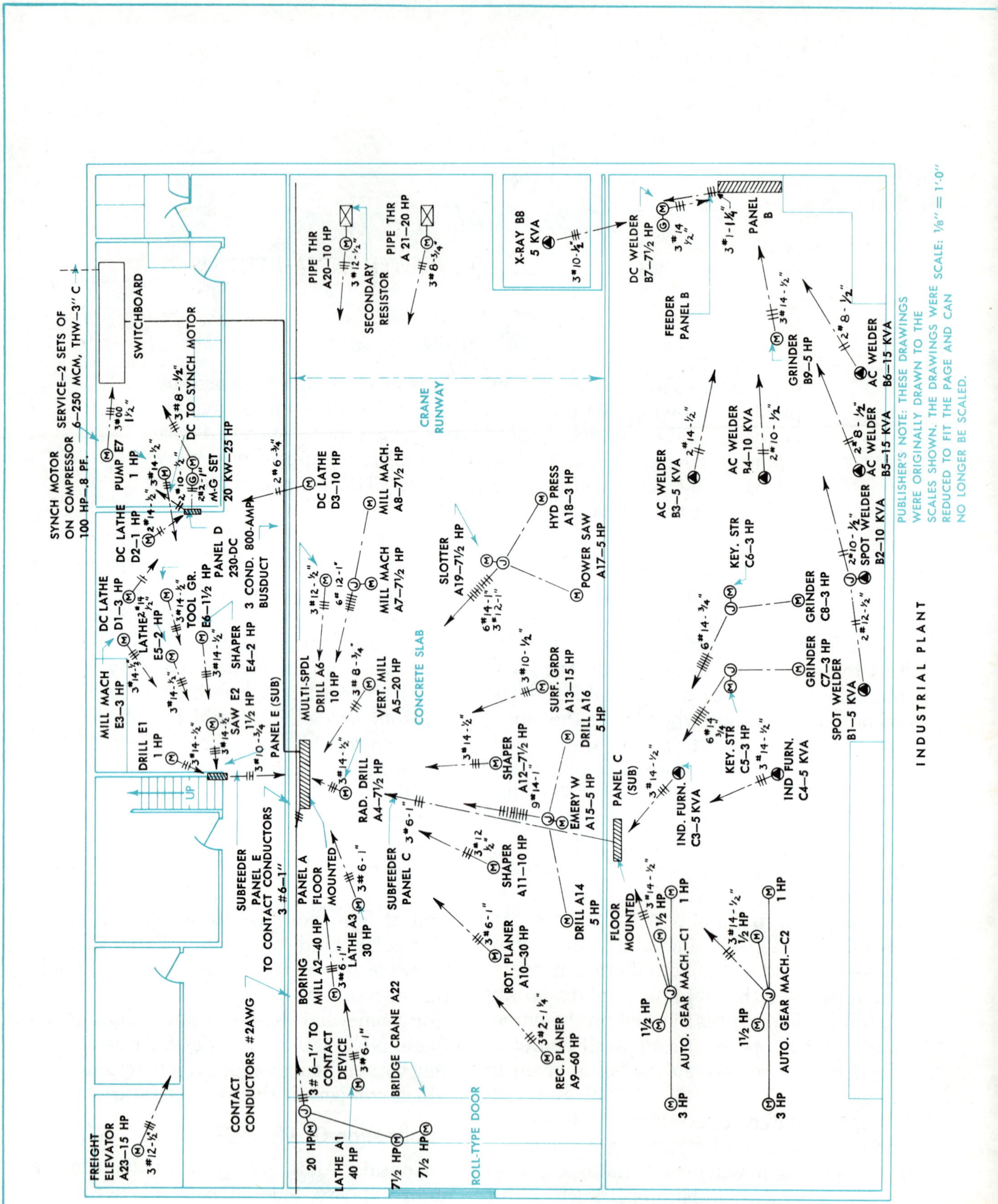

INDUSTRIAL PLANT

SYNCH MOTOR ON COMPRESSOR 100 HP—8 PF.

SERVICE—2 SETS OF 6—250 MCM, THW—3″ C

SWITCHBOARD

PUMP E7 1 HP 3#00 ½″

DC LATHE D2—1 HP

DC LATHE D1—3 HP

LATHE E5—2 HP

TOOL GR. E6—1½ HP

SAW E2 1½ HP SHAPER E4—2 HP

M-G SET 20 KW—25 HP

DC TO SYNCH MOTOR

PANEL D 230-DC 3 COND. 800-AMP BUSDUCT

MILL MACH E3—3 HP

DRILL E1 1 HP

PANEL E (SUB)

SUBFEEDER PANEL E TO CONTACT CONDUCTORS 3#6-1″

CONTACT CONDUCTORS #2AWG

3#6-1″ TO CONTACT DEVICE

FREIGHT ELEVATOR A23—15 HP 3#12-½″

LATHE A1 40 HP

20 HP 7½ HP 7½ HP

BRIDGE CRANE A22

REC. PLANER A9—60 HP

BORING MILL A2—40 HP

LATHE A3 30 HP

PANEL A FLOOR MOUNTED

ROT. PLANER A10—30 HP

SUBFEEDER PANEL C 3#6-1″

RAD. DRILL A4—7½ HP

SHAPER A11—10 HP

DRILL A14 5 HP

MULTI-SPDL DRILL A6 10 HP

VERT. MILL A5—20 HP

SHAPER A12—7½ HP

EMERY W A15—5 HP

SURF. GRDR A13—15 HP DRILL A16 5 HP

PANEL C (SUB)

DC LATHE D3—10 HP

MILL MACH. A8—7½ HP

MILL MACH A7—7½ HP

SLOTTER A19—7½ HP

HYD PRESS A18—3 HP

POWER SAW A17—5 HP

PIPE THR A20—10 HP

SECONDARY RESISTOR

PIPE THR A 21—20 HP

CRANE RUNWAY

CONCRETE SLAB

X-RAY B8 5 KVA

DC WELDER B7—7½ HP

FEEDER PANEL B

PANEL B

GRINDER B9—5 HP

AC WELDER B6—15 KVA

AC WELDER B5—15 KVA

SPOT WELDER B2—10 KVA

AC WELDER B4—10 KVA

AC WELDER B3—5 KVA

KEY. STR C6—3 HP

GRINDER C8—3 HP

GRINDER C7—3 HP

SPOT WELDER B1—5 KVA

KEY. STR C5—3 HP

IND. FURN. C4—5 KVA

IND. FURN. C3—5 KVA

FLOOR MOUNTED

AUTO. GEAR MACH.—C1 1 HP 1½ HP 3 HP

AUTO. GEAR MACH.—C2 1 HP 1½ HP 3 HP

ROLL-TYPE DOOR

UP

National Electrical Code

amps. Under NEC 430-22(a), the branch-circuit conductors must be capable of carrying 1.25 × 52 amps, or 65 amps. Table 310-16 lists the nearest size of Type THW wire as No. 6, and Table 3A of Chapter 9 permits three No. 6 Type THW conductors in a 1-inch conduit.

Branch-circuit protection Table 430-152 —"Maximum Rating or Setting of Motor Branch-Circuit Protective Devices" deals with all types of motors including motors with and without code letters. Column 1 of Table 430-152, "squirrel-cage, autotransformer started" applies here. There are two divisions, one for motors with full-load current not more than 30 amps and the other for those with full-load current greater than 30 amps. The second division pertains to the 40-hp motor. Column 2 shows the branch-circuit nontime-delay fuse rating as 200 percent of full-load current. For a motor current of 52 amps, circuit protection should be on the order of 2 × 52 amps, or 104 amps. The nearest standard fuse or non-adjustable circuit-breaker rating is 110 amperes (NEC 240-6).

Motor running overcurrent protection, as prescribed by NEC 430-32(a), should be not greater than 1.25 × 52 amps, or 65 amps. This protection may be obtained with an adjustable device set at 65 amps or by the nearest fuse or non-adjustable circuit breaker which in this case is 70 amps.

Vertical milling machine A5 is driven by a 20-hp motor, Code letter C, which is started by an autotransformer device. Table 430-150 gives full-load current as 27 amps. NEC 430-22(a) calls for a No. 8 conductor (1.25 × 27 amps = 33.75 amps), and Table 3A, Chapter 9, specifies a ¾-inch conduit. Motor running overcurrent protection as prescribed by NEC 430-32(a) should be not greater than 1.25 × 27 amps, or 33.75 amps. The nearest standard fuse or non-adjustable circuit breaker is rated at 35 amperes (NEC 240-6). Table 430-152 gives 200 percent of full load current for branch circuit protection, 2 × 27 amps = 54 amperes. This would allow a 60-amp nontime delay fuse or time limit circuit breaker for branch-circuit protection.

The 7½-hp, Code F, full-voltage starting

motor on radial drill A4 takes a current of 11 amps, requiring three No. 14 conductors in ½-inch conduit. Motor running protection is 1.25 × 11, or 13.75 amperes, for an adjustable protector or 15 amps for a fuse or non-adjustable circuit breaker (NEC 430-32(a)). Branch-circuit protection would be 300 percent of 11 amps, or 33 amperes, which would permit a 35-amp nontime-delay fuse (Table 430-152).

Pipe-threading machine A21 employs a 20-hp motor of the wound-rotor type. The size of circuit conductors, conduit, and motor running overcurrent protection are exactly the same as in the case of the 20-hp squirrel-cage motor of circuit A5, but the branch-circuit protection will be different. Referring to Table 430-152, we select the wound-rotor motor which allows 150 percent of full-load current, 27 × 1.50 = 40.5 amps, so in this case a 40-amp nontime-delay, dual element fuse or a 40-amp time limit breaker may be used.

The motor nameplate lists a secondary current of 50 amps. NEC 430-23(a) states that for continuous-duty motors the carrying capacity of wires between motor secondary (rotor) and the controller shall not be less than 125 percent of full-load secondary current. This value here is 62.5 amps, the nearest conductor being No. 6. Table 430-23(c) specifies ampacities of conductors between controller and resistors, this value depending upon the type of operation. In most cases asbestos-covered wire is necessary because of heat generated in resistor grids.

Lathe D3 is driven by a 10-hp, 230-volt, direct-current motor. Table 430-147 gives full-load current as 38 amps, requiring two No. 8 conductors in a ¾-inch conduit. Motor running overcurrent protection would be 1.25 × 38 amps, or 47.5 amps and branch-circuit protection would be 1.5 × 38 amps, or 57 amps as per Table 430-152 which would permit a 60-amp dual element (time-delay) fuse.

Elevator Motor

A23 is a 15-hp, high-reactance, 30-60-minute type of motor with full-voltage starting.

NEC Table 430-22 (a-Exception) states that duty on elevator motors shall be classed as intermittent. NEC 430-33 provides that the ampacity of the branch-circuit conductors for intermittent duty motors shall be considered as protected against running overcurrent by a branch-circuit device which does not exceed values specified in Table 430-152. NEC 430-22 (a-Exception) states that the ampacity of the branch-circuit conductors for intermittent duty motors shall be not less than the percentage of the motor nameplate current rating as shown in Table 430-22 (a-Exception) unless the authority having jurisdiction grants special permission for conductors of a smaller size. In this case the ampacity of the branch-circuit conductors for elevator motor *A23*, which has a 30-60-minute rating need be only 90 percent of the motor nameplate full-load current.

The full-load nameplate current rating of elevator motor *A23* is listed as 21 amps. The ampacity of conductors may be rated on the basis of .9 × 21 amps, or 18.9 amps. A No. 12 conductor is needed. Table 430-152 shows that for a high-reactance motor, less than 30 amps full-load current, the branch-circuit protection may be 250 percent of full-load current or 52.5 amps (2.5 × 21 amps) which would permit a 60-amp nontime delay fuse or inverse time circuit breaker.

Feeder Supplying Two or More Motors

NEC 430-24 states that conductors supplying two or more motors shall have an ampacity not less than 125 percent of the full-load current rating of the highest rated motor in the group, plus the sum of full-load current of remaining motors. Thus for two motors, one having a full-load current rating of 20 amps and the other 14 amps, the feeder ampacity must be not less than (1.25 × 20 amps) + 14 amps, or 39 amps. A No. 8 Type THW wire is indicated.

NEC 430-62 provides that overcurrent protection for conductors supplying two or more motors shall be not greater than the branch-circuit overcurrent for the largest rating or setting plus full-load currents of other motors. Assume a 250 percent protec-

tive requirement for the larger motor in the preceding example. The feeder protective device then should be set at not more than (2.5 × 20 amps) + 14 amps, or 64 amps. The nearest standard fuse or circuit breaker that does not exceed this value is 60 amps.

Crane Motors

Bridge crane *A22* employs three motors, a 20-hp and two 7½-hp units, the nameplates of which specify 15-minute ratings. Non-continuous-duty motors are commonly used on this type of service, the ratings being stated according to the nature of the application as 15-minute, 30-minute, or 60-minute.

Table 610-14 (a) sets forth ampacities for conductors supplying this type of motor, permissible currents being larger than those allowed by Table 310-16. Columns of Table 610-14 (a) list values according to types of insulation and ambient operating temperatures, the first of each set of two columns dealing with 60-minute, the second with 30-minute ratings. A footnote states that allowable ampacities of conductors for 15-minute motors shall be equal to the 30-minute ratings increased by 12 percent.

In the present instance the 75 degrees *C* columns are applicable. Since the motors are 15-minute units, sizes of conductors must be determined on the basis of the 30-minute rating increased by 12 percent. Full-load current of the largest motor is 27 amps, and the others 11 amps each. According to the table, a No. 14 Type THW conductor has a 30-minute current-carrying capacity of 26 amps. A 12 percent increase gives this conductor a capacity of 29 amps, making it suitable for use with the large motor.

NEC 610-14 (c) states that a No. 16 conductor may be employed with crane motors, if otherwise suitable. The 30-minute rating of No. 16 is given as 12 amps, and the allowable 12 percent increase raises the 15-minute rating to 13.4 amps, a value satisfactory for the two small motors.

NEC 610-14 (e) states that the ampacity of the power supply conductors on the crane shall be not less than the combined short-time full-load ampere rating of the largest motor or group for any single crane motion, plus 50 percent of short-time full-load ampere rating of the next largest motor or group. Adding the 27 amps of the largest motor to 50 percent of 22 amps (full-load current of other motors), one finds that a No. 8 THW conductor could be used. However, the specifications call for a No. 6 THW conductor here.

NEC 610-14(d) lists sizes of bare contact conductors which, in the present case where the runway is longer than 60 ft, must be No. 2.

Metal Working Machine Tools

NEC 670-3 states that the nameplate shall give necessary data, including full-load current and the ampere rating of the largest motor. NEC 670-4(a) states that the ampacity of supply conductors shall be not less than the full-load current marked on the nameplate, plus 25 percent of the full-load current rating of the highest rated motor as indicated on the nameplate. This requirement, it may be noted, is equivalent to the general rule stated in NEC 430-24.

Automatic gear machines *C1* and *C2* fall within the metal-working-machine-tool classification. The total nameplate rating is 10.2 amps and that of the highest rated motor, 4.8 amps. Under the rules of 670-4(a) the supply conductors must have a carrying capacity of not less than 10.2 amps + (.25 × 4.8 amps), or approximately 11.4 amps. The feeder to panel *C* will consist of three No. 14 conductors in ½-inch conduit. Short-circuit protection for each feeder would be 14.4 amps (3 × 4.8 amps) + 5.4 (2.6 + 1.8 + 1), or 19.8 amps (NEC 430-62). A 20-amp nontime delay fuse is acceptable.

Synchronous Motor

An air compressor in the switchboard room is driven by a 100-hp synchronous motor operating at a leading power factor of 80 percent. Table 430-150 gives full-load current at unity power factor as 101 amps. A footnote states that this figure is to be multiplied by 1.25 when the motor operates at 80 percent power factor, the current here becoming 126 amps. Circuit wires must have an ampacity (NEC 430-22(a)) of at least 1.25 × 126 amps, or 158 amps. Table 310-16 lists No. 00 Type THW copper wire at 175 amps, and Table 3A shows that a 1½-inch conduit is required.

A footnote to Table 430-152 states that for synchronous motors used in applications of this nature, branch-circuit protection need not be greater than 200 percent of full-load current. For a current of 126 amps, the overcurrent device could be set as low as 252 amps. By using an inverse-time breaker, a 250-amp breaker would be needed. It should be noted that two Number 10 conductors in one-half inch conduit, originating at panel D, supply direct-current excitation for the revolving field of the synchronous motor.

X-Ray Unit

The No. 10 conductors specified for this device are much larger than might seem to be required. Their size is dictated by a desire to hold voltage drop to a minimum.

Resistance Welders

NEC 630-31(a-1) states that ampacity of conductors supplying a resistance welder shall be not less than 70 percent of rated primary current for automatically fed welders and 50 percent for manually-operated ones. NEC 630-31(b) requires that rated ampacity of conductors which supply two or more such welders shall be not less than the sum of the above value for the largest welder plus 60 percent of such values for the smaller welders. NEC 630-32(a) and (b) demand that each welder be protected by an overcurrent device set at not more than 300 percent of rated primary current, and that a feeder supplying one or more units shall be protected by a device set at not more than 300 percent of conductor rating.

Unit *B1* here is a 5 kVA spot welder rated at 11 amps, and unit *B2* a 10 kVA spot welder rated at 22 amps, both manually operated. Conductors supplying *B1* may be

No. 14. Those for *B2* may be also No. 14 (.5 × 24 amps). *B1* may be protected by a 30-amp fuse or non-adjustable circuit breaker, *B2* by a 60-amp fuse or non-adjustable circuit breaker. The feeder should not be smaller than (.5 × 22 amps) + (.6 × .5 × 11 amps), or 14.3 amps. In this particular case, however, the question of voltage drop must be considered if the units are to operate efficiently. For this reason, two No. 10 conductors are run in a ½-inch conduit from the junction box to panel *B*. A pair of No. 12 conductors extends from the small welder to the junction box for the same reason.

Transformer Arc Welders

NEC 630-11(b) provides that conductor rating for a group of welders shall be based on the nature of the operation. It is known here that the maximum demand for a similar group of welders used on an identical application is only 50 percent of total capacity of the units. In the present case, the total capacity is equal to 5 kVA + 10 kVA + 15 kVA + 15 kVA, or 45 kVA. One half of this amount is 22.5 kVA, and the current taken when they are balanced across a 460-volt, three-phase supply line is equal to about 30 amps per leg. By using 3 phase formula (kVA = 1.73 × E × I), we find that 1.73 × 460 volts divided into 22,500 VA is equal to 28 amps.

NEC 630-12(a) requires that each welder have overcurrent protection not greater than 200 percent of rated primary current, and section (b) requires that feeder protection be not greater than 200 percent of conductor rating.

Motor-Generator Arc Welder

Unit *B7*, a single-operator, motor-generator, arc welder with a duty cycle of 80 percent comes under the provisions of NEC 630-21(a), which states that conductors supplying such a unit shall have an ampacity equal to only 91 percent of nameplate rating which is 11 amps here. We find that 91 percent of 11 amps is 10 amps which would require a No. 14 conductor, the smallest permissible size.

Feeder Calculations

Load on the sub-feeder from panel *C* to panel *A* totals about 53 amps, and three No. 6 THW conductors are used in a 1-inch conduit. Load on the sub-feeder from panel *E* to panel *A* is approximately 21.6 amps. Specifications call for a No. 10 feeder to this panel. Load imposed on panel *A* by its own units, including 25 percent of the largest motor running current, equals approximately 544 amps. Total load on panel *A* equals approximately 544 amps plus the full-load current of the motors connected to panels *C* and *E*, or 544 + 52 + 22 for a total feeder load to panel *A* of 618 amps. The plan shows an 800-amp busway from the main switchboard to panel *A*.

Feeder conductors for panel *B* must supply the following loads: 14.3 amps, single-phase to resistance welders *B1* and *B2*; 28 amps, three-phase to transformer arc welders *B3*, *B4*, *B5* and *B6*; 10 amps three-phase to motor-generator arc welder *B7*; 11 amps, single-phase to the X-ray unit *B8*; and 7.6 amps, three-phase to the grinder *B9*. When this load is balanced across the three phases, as equally as possible, the most heavily loaded conductor will carry about 70 amps. No. 4 Type THW copper conductors would normally be used, but because of the nature of the equipment, especially the X-ray unit, three No. 1 conductors are specified. NEC Table 3A, Chapter 9, shows that a 1¼-inch conduit is needed.

Service Calculations

Service load is equal to the full-load motor currents of all of the motors plus all other equipment on panels *A* and *B* plus the full-load current of the 25-hp motor-generator, plus that of the 100-hp synchronous motor, plus 25 percent of the full-load current of the synchronous motor, a total of approximately (Panel *A*, 525 full-load amps + Panel *B*, 70 amps + 34 full-load amps for 25-hp MG set + 126 full-load amps for 100-hp synchronous motor + 25 percent of 126 amps, or 158 amps) 525 + 70 + 34 + 158, or 787 amps. This value is larger than that for conductors listed in Table 310-16, so two or more conductors in parallel must be used.

One half of 787 amps equals 393.5 amps, the nearest listed conductor being 600 MCM Type THW. The service may consist of two sets of three 600 MCM Type THW conductors in two 3-inch conduits (as per Table 3A, Chapter 9). For ease of handling, smaller conductors can be used. Table 310-16 gives the ampacity of 250 MCM Type THW as 255 amps. With six conductors in one conduit, the ampacity of each will be reduced to 80 percent of 255 amps, or 204 amps. Two such conductors on each leg will furnish capacity equal to 408 amps. Table 3A, Chapter 9, permits six 250 MCM Type THW conductors in a 3-inch conduit. The service then consists of two sets of six 250 MCM Type THW conductors in 3-inch conduits.

Note: The above service configuration is one of many possible alternatives. Supply conductors could be run in three or more sets of conduit, to avoid penalization with respect to ampacity, or even in the form of busduct if conditions warranted its use.

POWER INSTALLATION FOR RESTAURANT

GENERAL—The power installation is to be confined to the kitchen area, with the exception of a 5-hp exhaust fan on the roof. The service is four-wire delta with three-phase power at 230 volts between phase wires and single-phase power at 115-230 volts between two phase wires and a neutral conductor, as indicated on the small diagram to the right of the drawing.

The panelboard is divided into two sections, *A* and *B*; Section *A* supplies three-phase loads and large single-phase loads which do not require a neutral conductor, while Section *B* supplies any single-phase loads which require a neutral conductor. Type THW copper conductors are used throughout.

It should be noted that the blueprint covers only the power installation; lighting outlets would appear on a separate drawing. This procedure is often used where either the power or the lighting is rather involved, to avoid confusion from crowding both power and lighting onto the same sheet.

Circuit Analysis

Circuits *A-1* and *A-2* are connected to 21-kW three-phase ranges, the nameplate current being 53 amps. These loads in normal operation are likely to continue for long periods of time. NEC 210-22C requires that conductors serving such loads shall carry not more than 80 percent of listed current. The conductors supplying each range, then, must be rated at a minimum ampacity of 53 amps divided by .8, or 67 amps. Table 310-16 rates No. 4, Type THW at 85 amps, and Table 3A, Chapter 9, indicates that a 1-inch conduit is needed for these three No. 4 conductors.

Deep-fat fryers *A-3* and *A-4* are rated at 10 kW single-phase, the current being 44 amps. These loads are similar to those of the ranges in that they will continue for long periods of time so that the 80 percent factor must be employed. The conductors must have a normal ampacity of 44 amps divided by .8, or 55 amps. The tables reveal that two No. 6 conductors in ¾-inch conduit will meet the requirements.

Circuit *A-5* supplies a 5-hp, three-phase blower motor on the roof. A riser leading to this unit may be noted at the left side of the panelboard. NEC 430-6(a) states that, except for hermetic motors (discussed later),

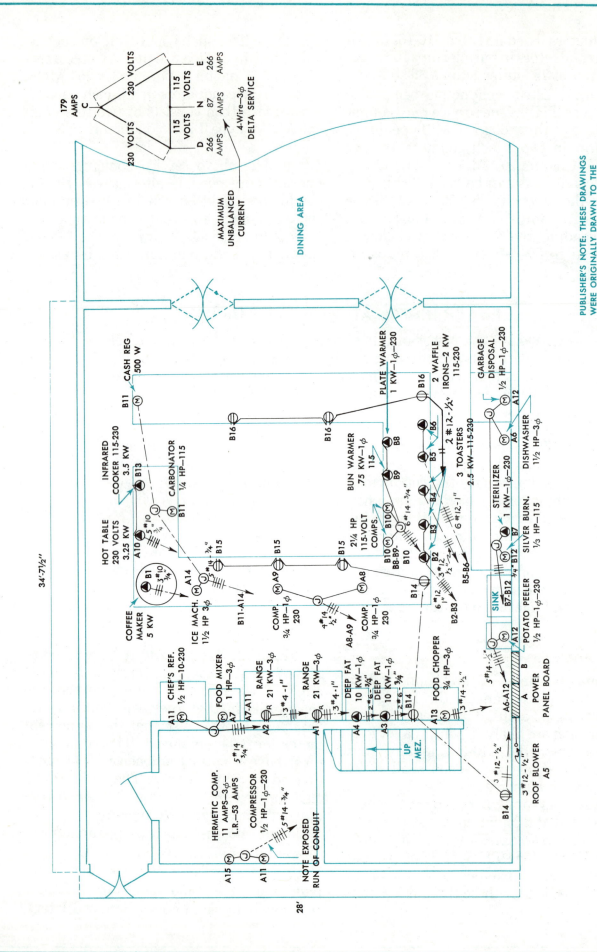

RESTAURANT KITCHEN

PUBLISHER'S NOTE: THESE DRAWINGS WERE ORIGINALLY DRAWN TO THE SCALES SHOWN. THE DRAWINGS WERE REDUCED TO FIT THE PAGE AND CAN NO LONGER BE SCALED.

SCALE: 1/4" = 1'-0"

ampacity of conductors, branch-circuit overcurrent devices, etc., shall be determined from the motor-current tables at the end of Article 430 instead of current ratings on motor nameplates. Table 430-150 lists current of the 5-hp, 3-phase, 230-volt motor as 15.2 amps. NEC 430-22(a) requires that branch-circuit conductors supplying a single motor have an ampacity of 125 percent of 15.2 amps, or 19 amps. No. 12 Type THW is rated at 20 amps, and Table 3A of Chapter 9 shows that three No. 12 conductors may be run in a ½-inch conduit.

NEC 430-52 and Table 430-152, to which it refers, require that the initial rating of a dual-element (time-delay) fuse, used for branch-circuit protection, should not exceed 175 percent of the motor full-load current listed in Table 430-150. The rating of the dual-element fuse for the 5-hp motor should not exceed 1.75 × 15.2 amps, or 26.6 amps. NEC 430-52 permits using the next larger fuse, which in this case would be 30 amps. NEC 430-83 states that a controller shall have a horsepower rating not less than that of the motor, and NEC 430-109 states a similar requirement for the disconnecting means. In the present instance a 30-amp circuit breaker would satisfy the needs of a controller under 430-83 (Exception No. 2) and of a disconnect under 430-110(a). The rating of a manually-operable switch on the roof would have to be 5 hp.

NEC 430-32(a) prescribed that each continuous-duty motor rated at more than 1 hp shall be protected against running overcurrent by a separate overcurrent device rated at not more than 125 percent of motor full-load current, or else contain an integral thermal protector inside the motor. If the blower motor does not have built-in thermal protection, an overcurrent device must be supplied, being included perhaps in a magnetic switch used to start and stop the motor. The rating of the device should be 1.25×15.2 amps, or 19 amps, which is not a standard size. NEC 430-34, however, permits use of the next larger standard size, which is 20 amps in this case.

It is important to note that NEC 430-52 merely prescribes that the *maximum* value of branch-circuit protection shall be limited, in general, to the value given in Table 430-152 (300 percent for nontime-delay fuses). NEC 430-58 states that a circuit breaker may be used for motor branch-circuit protection if it conforms to NEC 430-110(a) which in turn requires it to have an ampacity of 115 percent of the full-load current of motor. In this case the 5-hp motor has a full-load current of 15.2 amps×115% or 17.5 amps. If a standard 20-amp circuit breaker at the panel would carry starting current of this motor without tripping, it would satisfy the need with regard to motor overcurrent protection per NEC 430-58, 52.

Circuit *A-6* supplies a 1½-hp, 3-phase, dishwasher motor which has a built-in thermal device. Table 430-150 gives the current as 5.2 amps, so that No. 14 conductors will be large enough. According to NEC 430-83, Exception No. 1, a controller located at the operating location may be a general-use switch which has an ampere rating at least twice that of motor full-load current.

Three No. 14 conductors in ½-inch conduit extend from the motor location to a junction box where they are joined by two No. 14 conductors from a garbage disposal unit connected to circuit *A-12*. A ½-inch conduit carries the five wires to a second junction box at which point two conductors from a potato peeler motor are tapped to the *A-12* circuit before the five wires continue to the panelboard.

Two motors are connected to circuit *A-12* under provisions of NEC 430-53(a) which states that motors not exceeding 1 hp each and having a full-load current not in excess of 6 amps may be connected to a 20-amp, 120-or-less-volt circuit, or a 15-amp, 600-or-less-volt circuit. Individual running overcurrent protection would be required (see NEC 430-32(b-2)) as these motors are permanently installed. Usually, motors of this type are provided with a thermal protector integral with the motor which would give required protection.

Other motors are combined in this manner on circuits *B-10*, *B-11*, and *A-11*, those on the latter circuit being connected together at the panelboard because they reach there in separate conduits. The ¾-hp compressor motors on circuits *A-8* and *A-9* may be combined on a single circuit but would require running protection, as they are automatically started (NEC 430-32 (c)).

Circuit *A-10* for the hot table and *B-14* for the infrared cooker are taken to the panelboard as five No. 10 conductors in ¾-inch conduit. *A-10* is rated at 3250 watts or 14.1 amps, *B-14* at 3500 watts or 15.2 amps. Since these are continuous loads, NEC 210-22(c) imposes the 80 percent factor so that wires supplying them must have normal ampacities of at least 17.6 amps and 19 amps respectively.

A point of interest here is that when there are from four to six current-carrying conductors supplying a constant load all in the same conduit, they need only be derated once as per NEC 210-22(c) Exception 2. No. 12 conductors would be permissible for *A-10* and *B-14* but in this case a No. 10 Type THW was selected. Table 3A, Chapter 9, permits five No. 10 Type THW conductors in a ¾-inch conduit. Derating considerations also enter into calculations for toasters and waffle irons on circuits *B-2, B-3, B-5,* and *B-6.*

Circuit *A-15*, whose conduit is run exposed, supplies an 11-amp sealed hermetic-type compressor motor. This type receives special consideration in NEC. Section 440-6(a) states that the full-load current marked on the compressor nameplate shall be used to determine ampacity of branch-circuit conductors. NEC 440-4(a) requires that locked-rotor current for such motors shall be marked on the nameplate. Here, locked-rotor current is marked as 53 amps. NEC 440-41(a) provides that controllers for sealed hermetic-type motors shall be selected on the basis of full-load current and locked-rotor current as well, the values given in tables at the end of Article 430 being used for the purpose.

According to Table 430-150, column 3, the nearest current greater than 11 amps, (the nameplate full-load current of the motor) is

15.2 amps, which value applies to a 5-hp unit. In Table 430-151 the nearest three-phase locked-rotor current to the 53-amp value on the motor nameplate is 58 amps applying to a 3-hp motor. NEC 440-41(a) states also that where two different horsepower ratings are obtained from different tables, the higher of the two values shall be used. Here, where the two values are 3 hp and 5 hp, the larger of them will be used, necessitating a 5-hp controller. NEC 440-12(a) imposes a similar demand with respect to the disconnecting means.

Service Calculation

The load on panel *A* is obtained by converting motor loads to watts and adding them to the wattages of the cooking and heating appliances. A value equal to 25 percent of the largest motor rating must be included as per NEC 430-24 as well as a demand factor permitted by NEC Table 220-20 for commercial electric cooking and other kitchen equipment. The total load on panel *A* amounts to 71,032 watts or 179 amps.

MOTOR LOAD ON PANEL A

A5 Roof Fan 5 hp, 3ϕ, 230V, 15.2 amps
(watts = 1.73 EI) = 1.73 × 230 × 15.2 = 6048W

A6 Dishwasher 1½ hp, 3ϕ, 230V, 5.2 amps
(watts = 1.73 EI) = 1.73 × 230 × 5.2 = 2069W

A7 Food Mixer 1 hp, 3ϕ, 230V, 3.6 amps
(watts = 1.73 EI) = 1.73 × 230 × 3.6 = 1432W

A8 Compressor ¾ hp, 3ϕ, 230V, 2.8 amps
(watts = 1.73 EI) = 1.73 × 230 × 2.8 = 1114W

A9 Compressor ¾ hp, 3ϕ, 230V, 2.8 amps
(watts = 1.73 EI) = 1.73 × 230 × 2.8 = 1114W

A11 Compressor ½ hp, 1ϕ, 230V, 4.9 amps
(watts = EI) = 230 × 4.9 = 1127W

A11 Chef's Ref ½ hp, 1ϕ, 230V, 4.9 amps
(watts = EI) = 230 × 4.9 = 1127W

A12 Disposal ½ hp, 1ϕ, 230V, 4.9 amps
(watts = EI) = 230 × 4.9 = 1127W

A12 Potato Peeler ½ hp, 1ϕ, 230V, 4.9 amps
(watts = EI) = 230 × 4.9 = 1127W

A13 Food Chopper ¾ hp, 3ϕ, 230V, 2.8 amps
(w = 1.73 × E × I) = 1.73 × 230 × 2.8 = 1114W

A14 Ice Machine 1½ hp, 3ϕ, 230V, 5.2 amps
(w = 1.73 × E × I) = 1.73 × 230 × 5.2 = 2069W

A15 Hermitic Comp 3ϕ, 230V, 11 amps
(w = 1.73 × E × I) = 1.73 × 230 × 11 = 4377W

Total Motor Watts =	23845W
× 25% of largest motor or .25 × 6048 =	+ 1512W
	25357W

COMMERCIAL ELECTRIC COOKING AND KITCHEN EQUIPMENT ON PANEL A

A1 Range	21.00kW, 3ϕ, 230V
A2 Range	21.00kW, 3ϕ, 230V
A3 Deep Fat Fryer	10.00kW, 3ϕ, 230V
A4 Deep Fat Fryer	10.00kW, 3ϕ, 230V
A10 Hot Table	3.25kW, 3ϕ, 230V
Total	**65.25kW**

Table 220-20 permits a demand factor of 70 percent for 5 units of electric cooking and other kitchen equipment in a commercial kitchen, therefore:

$65,250 \times .7 = 45,675$ watts $+ 25,375$ watts motor load equals 71,032 watts.

$71,032$ W $\div 1.73 \times 230$V $= 178.5$, or 179 amps.

The load on panel *B* found in the same manner is 20,094 watts or 87 amps.

MOTOR LOAD ON PANEL B

B10 Compressor ¼ hp, 1ϕ, 115V,	
5.8A — W = EI = 115 × 5.8 =	667W
B10 Compressor ¼ hp, 1ϕ, 115V,	
5.8A — W = EI = 115 × 5.8 =	667W
B11 Carbonator ¼ hp, 1ϕ, 115V,	
5.8A — W = EI = 115 × 5.8 =	667W
B12 Silver Burn. ⅓ hp, 1ϕ, 115V,	
7.2A — W = EI = 115 × 7.2 =	828W
Total Motor Watts =	**2829W**
+ 25 percent of largest motor = .25 × 828 =	+ 207W
Total Motor watts on panel B =	**3036W**

COMMERCIAL ELECTRIC COOKING AND KITCHEN EQUIPMENT ON PANEL B

B1 Coffee Maker 5kW, 115-230V, 1ϕ	5,000
B2 Toaster 25kW, 115-230V, 1ϕ	2,500
B3 Toaster 25kW, 115-230V, 1ϕ	2,500
B4 Toaster 25kW, 115-230V, 1ϕ	2,500
B5 Waffle Iron 25kW, 115-230V, 1ϕ	2,500
B6 Waffle Iron 25kW, 115-230V, 1ϕ	2,500
B7 Sterilizer 1kW, 230V, 1ϕ	1,000
B8 Plate Warmer 1kW, 230V, 1ϕ	1,000
B9 Bun Warmer 75kW, 115, 1ϕ	750
B13 Infrared Cooker 3.5kW, 115-230V, 1ϕ	3,500
Total connected load	**23,750W**

Table 220-20 permits a demand factor of 65 percent for 6 or more units of electrical cooking and other kitchen equipment in a commercial kitchen, therefore:

$23,750$ watts $\times .65 = 15,438$ watts

Total watts on panel B is:

Electric cooking and kitchen equipment =	15438W
Motor load =	3036W
Nine, 115-volt receptacles at 180 VA each (NEC 220-2(c) (4)) =	1620W
	20094W

$20,094$ watts $\div 230$ volts $= 87.4$ or 87 amps.

These values have been marked on the diagram at the right of the drawing, the load on power leg *C* being 179 amps, that on *D* or *E*, 266 amps and the current in *N* (neutral) being 87 amps. NOTE: This method of adding single-phase current directly to three-phase current in order to determine the total current on a particular supply wire is the one customarily used and gives results sufficiently accurate for practical needs.

Table 310-16 gives the nearest ampacities for Type THW copper conductors as 000 for leg *C*, 300 MCM for *D* and *E*, and No. 3 for *N*. Table 5, Chapter 9, lists cross-sectional areas as .3288 sq. in. for 000, .5581 sq. in. for 300 MCM, and .1263 sq. in. for No. 3.

Total cross-sectional conductor area equals:

1 × .3288 sq. in. =	.3288 sq. in.
2 × .5581 sq. in. =	1.1162 sq. in.
1 × .1263 sq. in. =	.1263 sq. in.
Total =	**1.5713 sq. in.**

For 40 percent fill the cross-sectional area of conduit must be at least 1.5713 sq. in. divided by .4, or 3.9283 sq inches. Table 4 Chapter 9, lists the nearest size of conduit as 2½-inch.

As in the case of the store building, parallel service conduits could be made use of in order to reduce size of conductors. The circuit breaker may be a 400-amp frame, with current settings suited to protect each of the three phase-leg conductors.

Table 250-94 discloses that a No. 2 copper grounding electrode conductor is required.

MINERAL-INSULATED METAL-SHEATHED CABLE

NEC 330-14 Fittings properly identified for use shall be used.

Junction Box

Insulating Sleeves on Conductors

Terminal Fitting

Copper Sheathing

NEC 330-15 Terminating seal required to prevent entrance of moisture.

Conductors

Magnesium-Oxide Insulation

NEC 330-3 Permitted for services, feeders, branch circuits, in wet locations, indoors or outdoors, exposed or concealed. (See 330-3.)

NEC 330-13 Radius of bends at least 5 times cable diameter.

NEC 300-15 (b) Ex 3 Straight-through splice for MI cable is allowed without box provided the splice is accessible and an approved fitting is used.

MAXIMUM VOLTAGE TO LTG. FIXTURE

Electric-Discharge Lamp

22 ft. in tunnels

Medium base lampholders permitted on any premises if not over 120-volts to ground or between conductors.

18 ft over bridges, roads, park lots, etc.

Medium or Mogul Base Lampholder

No Switch in Fixture

Min. ht. 8 ft if screw-shell lampholder.

Over 600-volts between conductors or over 277 volts to ground permitted at 22 ft. over roads, bridges, park lots etc. or 18 ft. in tunnels.

No Switch in Fixture

In other than dwellings, lampholders not exceeding their rating permitted where voltage is 120 to 277 volts to ground and not over 480 volts between conductors.

Location: Industrial establishments, office buildings, schools, stores, and public commercial areas may be less than 8 feet from floor.

Industrial or store location with adequate maintenance and supervision.

NEC 210-6(a) Voltage to ground or between conductors not to exceed 120 volts to medium base lampholders or 125-volt devices. (b) Circuits from 120 to 277 volts to ground and not over 480 volts between conductors permitted to supply: 1. Cord and plug loads not over 1380 VA. 2. In other than dwellings: 1. lampholders not exceeding their rating or mogul base screw-shell lampholders or; (2) auxiliary equip. of discharge lamps or; (3) receptacles rated for circuit voltage. Except: 1. infrared industrial lamps as per 422-15(c); 2. railroads as per 110-19; 3. circuits as per 210-6(c). NEC 210-6(c) Circuits over 600 volts between conductors or over 277 volts to ground permitted to supply: (1) receptacles rated for voltage; (2) permanently wired electric-discharge fixtures when mounted: 1. on poles etc. to light outdoor areas such as roads, bridges, parking lots etc. if at least 22 ft. high; 2. at least 18 ft. in tunnels.

CONDUIT & EQUIPMENT IN CONCRETE & WET LOCATIONS

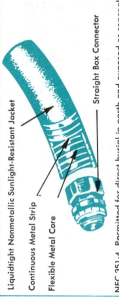

NEC 240-32 Enclosures for overcurrent devices to have at least ¼-inch air space from surface in damp or wet locations and shall be identified for use in such locations.

NEC 300-6 (a) (b) (c) Ferrous raceways, cabinets, etc., must be suitably protected from corrosion. Enameled conduit not permitted outdoors or in wet locations. Unless suitable for existing conditions, conduit, boxes, etc., are not to be installed in concrete, in direct contact with the earth, or where subject to corrosion. In frequently washed areas such as canneries or dairies, the entire wiring system shall have at least ¼-inch air space between it and the wall or supporting surface.

LIQUIDTIGHT FLEXIBLE METAL CONDUIT

Liquidtight Nonmetallic Sunlight-Resistant Jacket

Continuous Metal Strip

Flexible Metal Core

Straight Box Connector

NEC 351-4. Permitted for direct burial in earth and exposed or concealed work. May be used where flexibility required or protection from liquids, vapors, or solids and some hazardous areas (501-4(b), 502-4, 503-3, and 553-7(b)). Not permitted where subject to physical damage or where subjected to heat in excess for which approved.

NEC 351-8 Permitted as fixed raceway if secured every 4½ feet and within 12 inches of box except where – (1) fished, (2) length not over 3 feet at terminals where flexibility is necessary, and (3) not over 6 feet for fixtures.

NEC 351-5 (b) Maximum size shall be 4 inches.

NEC 351-9 Permitted as grounding means where both conduit and fittings approved for purpose, except where – (1) sizes 1¼ inches or smaller and total ground path not over 6 feet may be grounding means if terminated in fittings listed for grounding or (2) flexibility required, a ground wire shall be installed. (See 250-79)

VOLTMETER--AMMETER--OHMMETER

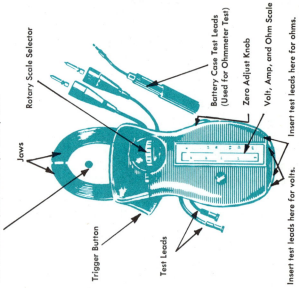

Press trigger button to open jaws. Encircle one conductor with transformer jaws. Read amps on amp scale.

Rotary Scale Selector

Battery Case Test Leads (Used for Ohmmeter Test)

Zero Adjust Knob

Volt, Amp, and Ohm Scale

Jaws

Insert test leads here for ohms.

Trigger Button

Test Leads

Insert test leads here for volts.

UNDERGROUND FEEDER AND BRANCH-CIRCUIT CABLE

NEC 339-3 (a-2) If single conductor used, all wires of same circuit buried in one trench or raceway.

NEC 339-3 (a-3) Minimum depth 24 inches unless specially protected. (See 300-5.)

NEC 339-3 (a-4) May be used for interior wiring.

NEC 300-5 (c) Underground cable run under a building shall be in raceway that extends beyond the outside wall of the building.

NEC 339-1 UF cable in sizes No. 14 copper or No. 12 aluminum thru 4/0 permitted.

NEC 339-3(a)(4) Ampacity determined on 60°C (140°F) basis as per Art. 336.

Tables 310-16 thru 19 Note 8. Bundled conductors (without spacing) required to be derated.

NEC 339-3 (b) Not permitted for: service-entrance cables, commercial garages, theaters, motion picture studios, storage battery rooms, hoistways, hazardous locations, concrete or cement except for nonheating leads as per Article 424, where exposed to the sun, unless approved.

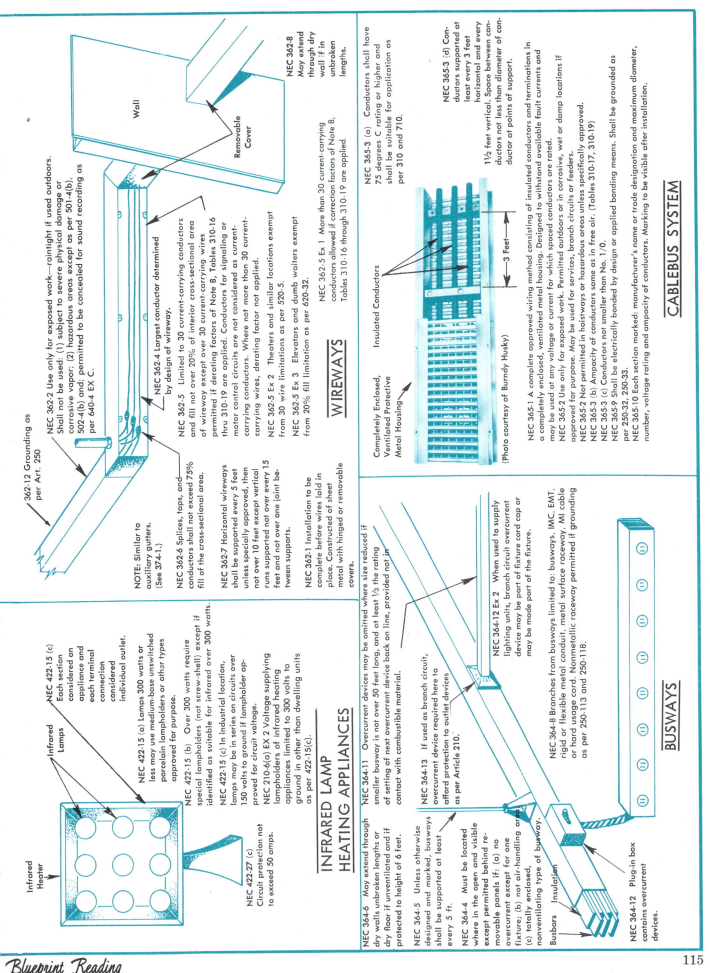

NEC 362-2 Use only for exposed work—raintight if used outdoors. Shall not be used: (1) subject to severe physical damage or corrosive vapor; (2) hazardous areas except as per 501-4(b), 502-4(b) and; permitted to be concealed for sound recording as per 640-4 EX C.

Wall

Removable Cover

NEC 362-8 May extend through dry wall if in unbroken lengths.

362-12 Grounding as per Art. 250

NEC 362-4 Largest conductor determined by design of wireway.

NEC 362-5 Limited to 30 current-carrying conductors and fill not over 20% of interior cross-sectional area of wireway except over 30 current-carrying wires permitted if derating factors of Note 8, Tables 310-16 thru 310-19 are applied. Conductors for signaling or motor control circuits are not considered as current-carrying conductors. Where not more than 30 current-carrying wires, derating factor not applied.

NEC 362-5 Ex 2 Theaters and similar locations exempt from 30 wire limitations as per 520-5.

NEC 362-5 Ex 3 Elevators and dumb waiters exempt from 20% fill limitation as per 620-32.

NEC 362-5 Ex 1 More than 30 current-carrying conductors allowed if correction factors of Note 8, Tables 310-16 through 310-19 are applied.

NOTE: Similar to auxiliary gutters. (See 374-1.)

NEC 362-6 Splices, taps, and conductors shall not exceed 75% fill of the cross-sectional area.

NEC 362-7 Horizontal wireways shall be supported every 5 feet unless specially approved, then not over 10 feet except vertical runs supported not over every 15 feet and not over one joint between supports.

NEC 362-1 Installation to be complete before wires laid in place. Constructed of sheet metal with hinged or removable covers.

WIREWAYS

NEC 365-3 (a) Conductors shall have 75 degrees C rating or higher and shall be suitable for application as per 310 and 710.

NEC 365-3 (d) Conductors supported at least every 3 feet horizontal and every 1½ feet vertical. Space between conductors not less than diameter of conductor at points of support.

3 Feet

Insulated Conductors

Completely Enclosed, Ventilated Protective Metal Housing

(Photo courtesy of Burndy Husky)

CABLEBUS SYSTEM

NEC 365-1 A complete approved wiring method consisting of insulated conductors and terminations in a completely enclosed, ventilated metal housing. Designed to withstand available fault currents and shall be used at any voltage or current for which spaced conductors are rated.
NEC 365-2 Use only for exposed work. Permitted outdoors or in corrosive, wet or damp locations if approved for purpose. May be used for services, branch circuits or feeders.
NEC 365-2 Not permitted in hoistways or hazardous areas unless specifically approved.
NEC 365-3 (b) Ampacity of conductors same as in free air. (Tables 310-17, 310-19)
NEC 365-3 (c) Conductors not smaller than No. 1/0.
NEC 365-9 Shall be electrically bonded by design or applied bonding means. Shall be grounded as per 250-32, 250-33.
NEC 365-10 Each section marked: manufacturer's name or trade designation and maximum diameter, number, voltage rating and ampacity of conductors. Marking to be visible after installation.

INFRARED LAMP HEATING APPLIANCES

Infrared Lamps

NEC 422-15 (c) Each section considered an appliance and each terminal connection considered individual outlet.

NEC 422-15 (a) Lamps 300 watts or less may use medium-base unswitched porcelain lampholders or other types approved for purpose.

NEC 422-15 (b) Over 300 watts require special lampholders (not screw-shell) except if identified as suitable for infrared over 300 watts.

NEC 422-15 (c) In industrial location, lamps may be in series on circuits over 150 volts to ground if lampholder approved for circuit voltage.

NEC 210-6(a) EX 2 Voltage supplying lampholders of infrared heating appliances limited to 300 volts to ground in other than dwelling units as per 422-15(c).

NEC 422-27 (c) Circuit protection not to exceed 50 amps.

Infrared Heater

NEC 364-11 Overcurrent devices may be omitted where size reduced if smaller busway is not over 50 feet long, and at least ⅓ the rating of setting of next overcurrent device back on line, provided not in contact with combustible material.

NEC 364-13 If used as branch circuit, overcurrent device required here to afford protection to outlet devices as per Article 210.

NEC 364-12 Ex 2 When used to supply lighting units, branch circuit overcurrent device may be part of fixture cord cap or may be made part of the fixture.

NEC 364-8 Branches from busways limited to: busways, IMC, EMT, rigid or flexible metal conduit, metal surface raceway, MI cable or hard usage cord. Nonmetallic raceway permitted if grounding as per 250-113 and 250-118.

NEC 364-6 May extend through dry walls unbroken lengths or dry floor if unventilated and if protected to height of 6 feet.

NEC 364-5 Unless otherwise designed and marked, busways shall be supported at least every 5 ft.

NEC 364-4 Must be located where in the open and visible except permitted behind removable panels if: (a) no overcurrent except for one fixture; (b) not air-handling area; (c) totally enclosed, nonventilating type of busway.

Busbars   Insulation

NEC 364-12 Plug-in box contains overcurrent devices.

BUSWAYS

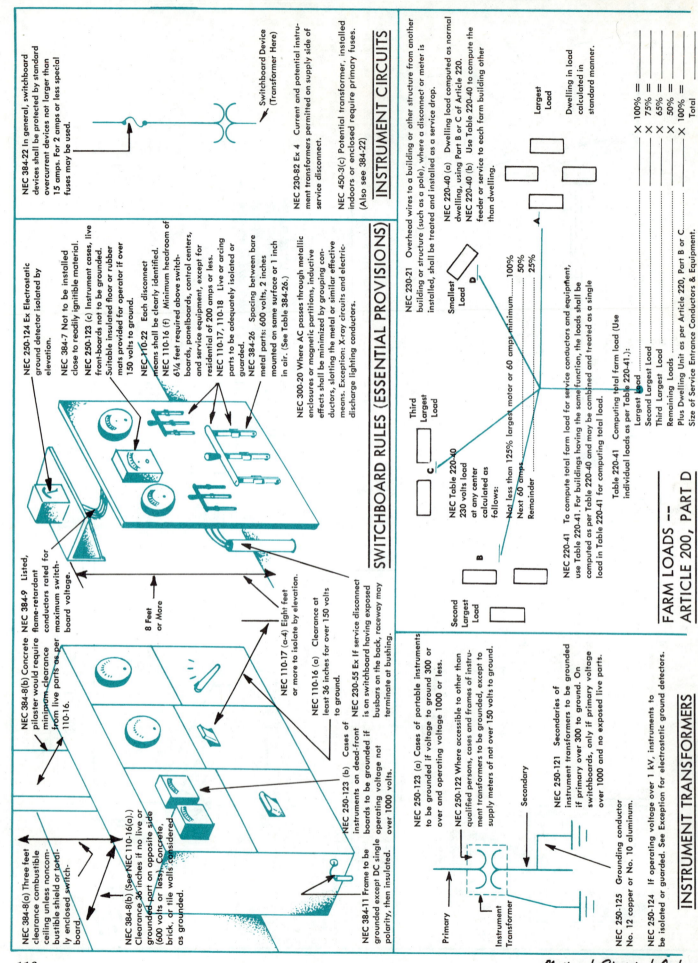

INSTRUMENT CIRCUITS

NEC 384-22 In general, switchboard devices shall be protected by standard overcurrent devices not larger than 15 amps. For 2 amps or less special fuses may be used.

Switchboard Device (Transformer Here)

NEC 230-82 Ex 4 Current and potential instrument transformers permitted on supply side of service disconnect.

NEC 450-3(c) Potential transformer, installed indoors or enclosed require primary fuses. (Also see 384-22)

NEC 230-21 Overhead wires to a building or other structure from another building or structure (such as a pole), where a disconnect or meter is installed, shall be treated and installed as a service drop.

NEC 220-40 (a) Dwelling load computed as normal dwelling, using Part B or C of Article 220.
NEC 220-40 (b) Use Table 220-40 to compute the feeder or service to each farm building other than dwelling.

Largest Load

Dwelling in load calculated in standard manner.

X 100% =
X 75% =
X 65% =
X 50% =
X 100% =
Total

Smallest Load
D

NEC Table 220-40 230 volts load at any center calculated as follows:

Not less than 125% largest motor or 60 amps minimum......100%
Next 60 amps.....50%
Remainder.....25%

Third Largest Load
C

Second Largest Load
B

NEC 220-41 To compute total farm load for service conductors and equipment use Table 220-41. For buildings having the same function, the loads shall be computed as per Table 220-40 and may be combined and treated as a single load in Table 220-41 for computing total load.

Table 220-41 Computing total farm load (Use individual loads as per Table 220-41.):
Largest Load
Second Largest Load
Third Largest Load
Remaining Loads
Plus Dwelling Unit as per Article 220, Part B or C.
Size of Service Entrance Conductors & Equipment.

FARM LOADS -- ARTICLE 200, PART D

SWITCHBOARD RULES (ESSENTIAL PROVISIONS)

NEC 250-124 Ex Electrostatic ground detector isolated by elevation.

NEC 384-7 Not to be installed close to readily ignitible material.

NEC 250-123 (c) Instrument cases, live front-boards not to be grounded. Suitable insulated floor or rubber mats provided for operator if over 150 volts to ground.

NEC 110-22 Each disconnect means shall be clearly identified.

NEC 110-16 (f) Minimum headroom of 6¼ feet required above switchboards, panelboards, control centers, and service equipment, except for residential of 200 amps or less.

NEC 110-17, 110-18 Live or arcing parts to be adequately isolated or guarded.

NEC 384-26 Spacing between bare metal parts: 600 volts, 2 inches mounted on same surface or 1 inch in air. (See Table 384-26.)

NEC 300-20 Where AC passes through metallic enclosures or magnetic partitions, inductive effects shall be minimized by grouping conductors, slotting the metal or similar effective means. Exception: X-ray circuits and electric-discharge lighting conductors.

NEC 384-8(a) Three feet clearance combustible ceiling unless noncombustible shield or totally enclosed switchboard.

NEC 384-8(b) Concrete pilaster would require minimum clearance from live parts as per 110-16.

NEC 384-9 Listed, flame-retardant conductors rated for maximum switchboard voltage.

NEC 384-8(b) (See NEC 110-16(a).) Clearance 36 inches if no live or grounded part on opposite side (600 volts or less). Concrete, brick, or tile walls considered as grounded.

8 Feet or More

NEC 110-17 (a-4) Eight feet or more to isolate by elevation.

NEC 110-16 (a) Clearance at least 36 inches for over 150 volts to ground.

NEC 230-55 Ex If service disconnect is on dead-front having exposed busbars on the back, raceway may terminate at bushing.

NEC 250-123 (b) Cases of instruments on dead-front boards to be grounded if operating voltage not over 1000 volts.

NEC 384-11 Frame to be grounded except DC single polarity, then insulated.

INSTRUMENT TRANSFORMERS

NEC 250-123 (a) Cases of portable instruments to be grounded if voltage to ground 300 or over and operating voltage 1000 or less.

NEC 250-122 Where accessible to other than qualified persons, cases and frames of instrument transformers to be grounded, except to supply meters of not over 150 volts to ground.

NEC 250-121 Secondaries of instrument transformers to be grounded if primary over 300 to ground. On switchboards, only if primary voltage over 1000 and no exposed live parts.

Secondary

Primary

Instrument Transformer

NEC 250-125 Grounding conductor No. 12 copper or No. 10 aluminum.

NEC 250-124 If operating voltage over 1 kV, instruments to be isolated or guarded. See Exception for electrostatic ground detectors.

MOTORS -- RUNNING PROTECTION

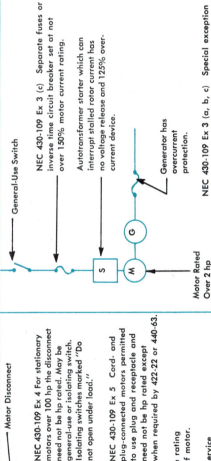

Branch-Circuit Disconnect and Overcurrent Devices

NEC 430-32 (b) (1) Motors 1 hp or less, not permanently installed, manually started within sight of controller, need no running protection if branch circuit is 120 volts or less and 20 amps or less.

NEC 430-81 (c) Attachment plug and receptacle may be controller for portable motor of 1/3 hp or less.

NEC 430-32 (c-3) Automatically-started motors, 1 hp or less, if part of approved non-overloading assemblies, viz., oil burners, need no running overcurrent protection.

NEC 430-32 (c-4) Motors with high impedance windings, viz., clock motors, need no running overcurrent protection. Does not include most split-phase motors.

NEC 430-81 (b) Stationary motors 1/8 hp or less normally left running may be controlled by branch-circuit overcurrent device provided they will not be damaged by overload or failure to start.

MOTOR -- MAXIMUM RUNNING OVERLOAD PROTECTION

NEC 430-34 Where overcurrent value does not correspond to standard size, next larger size may be used—not to exceed 140% full-load current for motors with service factor not less than 1.15 or motors not over 40°C. All others 130%.

NEC 430-35(a) Running overcurrent device may be shunted at starting of manual-start motor if no hazard is introduced and branch circuit device of not over 400% is operative in the circuit during starting period. (b) Shunting not permitted if automatically started. (See Exception)

NEC 430-113 Where equipment is supplied from more than one source, each disconnect shall be adjacent to the equipment. (See Ex 1, 2.)

MOTORS -- GENERAL

Motor Disconnect Switch

Controller

Usually Combined

NEC 430-22 Ampacity of branch-circuit conductors not less than 125% of motor current rating. (See Table 430-22 (a) Ex for Duty-Cycle motors.)

NEC 430-32 (a-1) Running overcurrent protection of 125% of motor current for motors with service factor not less than 1.15 or not over 40 degrees C, 115% of motor current for others.

NEC 440-52 (a-1, 2, 3) Hermetic-type motor overload and failure to start protection.

NEC 430-40 Thermal cutouts and overload relays for motor-running protection not capable of opening short circuits shall be protected as per 430-52 unless approved for group installation and marked with maximum size of required protection.

NEC 430-33 Application to be deemed continuous duty unless positive information to the contrary.

Branch-Circuit Disconnect

NEC 430-52 Motor branch-circuit overcurrent device must be able to carry starting current (150% to 300% as per Table 430-152, absolute maximum 400% with nontime-delay fuses, 225% with time-delay fuses).

MOTOR AND CONTROLLER DISCONNECT MEANS

NEC 430-102(b) Must be in sight of controller location. For over 600 volts see Exception.

NEC 430-106 Service switch suffices where one motor.

NEC 430-107 One or more to be readily accessible and in sight of controller.

NEC 430-108 Every disconnect means in motor circuit shall comply with 430-109 & 430-110.

NEC 430-109 Hp-rated switch or circuit breaker, except if 2 hp or less and 300 volts or less, may be general-use twice motor current. Over 100 hp, a general-use or isolating switch marked "Do Not Operate Under Load".

NEC 430-110 (a) Ampacity at least 115% motor current.

NEC 430-111 Suitable switch or circuit breaker may serve as both disconnect and controller.

NEC 430-83 Controller shall have hp rating not less than motor, with few exceptions. (430-83 (Ex 1, 2, 3))

NEC 430-84 Controller need not open all conductors.

NEC 430-82 (b) If autotransformer, must be arranged so cannot remain in starting position.

NEC 430-102(b) If controller not in sight of motor, it must be lockable, or else a disconnect switch within sight of the motor is required.

MOTOR GENERATOR -- DISCONNECT MEANS

General-Use Switch

Generator has overcurrent protection.

Motor Rated Over 2 hp and Including 100 HP

NEC 430-109 Ex 3 (c) Separate fuses or inverse time circuit breaker set at not over 150% motor current rating.

Autotransformer starter which can interrupt stalled rotor current has no voltage release and 125% over-current device.

NEC 430-109 Ex 3 (a, b, c) Special exception allowing general-use switch for motor which drives a generator that has overload protection.

MOTOR DISCONNECT -- TYPE

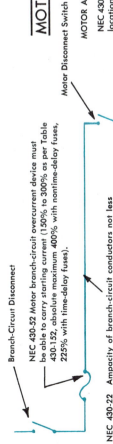

Motor Disconnect

NEC 430-109 Ex 4 For stationary motors over 100 hp the disconnect need not be hp rated. May be general-use or isolating switch. Isolating switches marked "Do not open under load."

NEC 430-109 Ex 5 Cord- and plug-connected motors permitted to use plug and receptacle and need not be hp rated except when required by 422-22 or 440-63.

NEC 430-109 Ex 2 A general-use AC only snap switch may be used to disconnect an AC motor 2 hp or less and 300 volts or less whose full-load current does not exceed 80% of switch amp rating. (See definition of AC general-use snap switch in Article 100.)

NEC 430-109 Ex 2 For stationary motor 2 hp or less, 300 volts or less, disconnecting means may be general-use switch having ampere rating not less than twice full-load current of motor.

NEC 430-106 Where single motor, service switch may serve as motor disconnect if properly rated and within sight of controller.

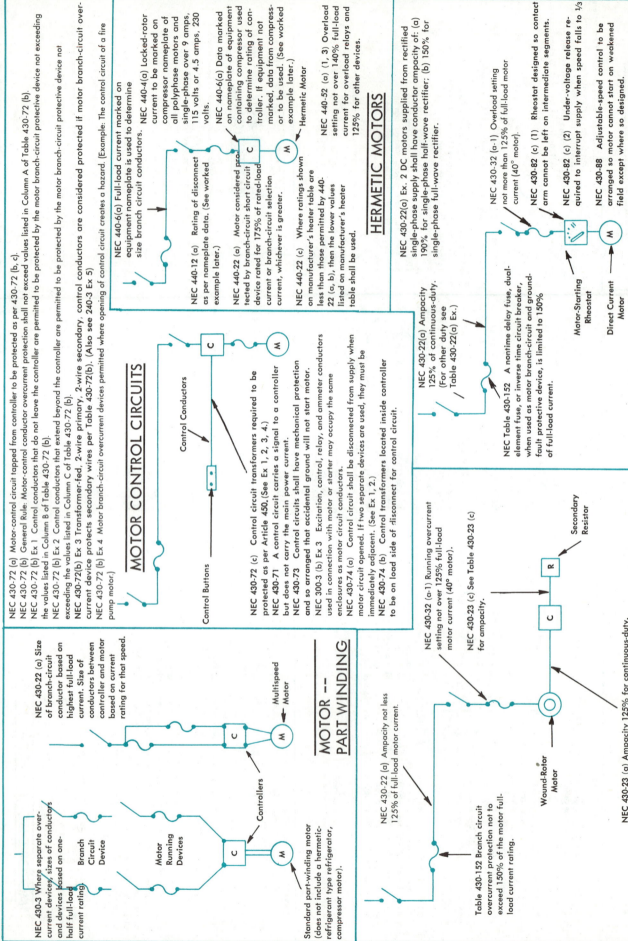

MOTOR CONTROL CIRCUITS

NEC 430-72 (a) Motor-control circuit tapped from controller to be protected as per 430-72 (b, c).
NEC 430-72 (b) General Rule: Motor-control conductor overcurrent protection shall not exceed values listed in Column A of Table 430-72 (b).
NEC 430-72 (b) Ex 1 Control conductors that do not leave the controller are permitted to be protected by the motor branch-circuit protective device not exceeding the values listed in Column B of Table 430-72 (b).
NEC 430-72 (b) Ex 2 Control conductors that extend beyond the controller are permitted to be protected by the motor branch-circuit protective device not exceeding the values listed in Column C of Table 430-72 (b).
NEC 430-72(b) Ex 3 Transformer-fed, 2-wire primary, 2-wire secondary, control conductors are considered protected if motor branch-circuit over-current device protects secondary wires per Table 430-72(b). (Also see 240-3 Ex 5)
NEC 430-72 (b) Ex 4 Motor branch-circuit overcurrent devices permitted where opening of control circuit creates a hazard. (Example: The control circuit of a fire pump motor.)

Control Conductors

Control Buttons

NEC 430-72 (c) Control circuit transformers required to be protected as per Article 450. (See Ex 1, 2, 3, 4.)
NEC 430-71 A control circuit carries a signal to a controller but does not carry the main power current.
NEC 430-73 Control circuits shall have mechanical protection and so arranged that accidental ground will not start motor.
NEC 300-3 (b) Ex 3 Excitation, control, relay, and ammeter conductors used in connection with motor or starter may occupy the same enclosures as motor circuit conductors.
NEC 430-74 (a) Control circuit shall be disconnected from supply when motor circuit opened. If two separate devices are used, they must be immediately adjacent. (See Ex 1, 2.)
NEC 430-74 (b) Control transformers located inside controller to be on load side of disconnect for control circuit.

HERMETIC MOTORS

NEC 440-6(a) Full-load current marked on equipment nameplate is used to determine size branch circuit conductors.
NEC 440-4(a) Locked-rotor current to be marked on compressor nameplate of all polyphase motors and single-phase over 9 amps, 115 volts or 4.5 amps, 230 volts.
NEC 440-6(a) Data marked on nameplate of equipment containing compressor used to determine rating of controller. If equipment not marked, data from compressor or to be used. (See worked example later.)

Hermetic Motor

NEC 440-52 (a) (1, 3) Overload setting not over 140% full-load current for overload relays and 125% for other devices.

NEC 440-12 (a) Rating of disconnect as per nameplate data. (See worked example later.)
NEC 440-22 (a) Motor considered protected by branch-circuit short circuit device rated for 175% of rated-load current or branch-circuit selection current, whichever is greater.
NEC 440-22 (c) Where ratings shown on manufacturer's heater table are less than those permitted by 440-22 (a, b), then the lower values listed on manufacturer's heater table shall be used.

NEC Table 430-152 A nontime delay fuse, dual-element fuse, or inverse time circuit breaker, when used as motor branch-circuit and ground-fault protective device, is limited to 150% of full-load current.

NEC 430-22(a) Ampacity 125% of continuous-duty. (For other duty see Table 430-22(a) Ex.)

DIRECT CURRENT MOTORS

NEC 430-22(a) Ex. 2 DC motors supplied from rectified single-phase supply shall have conductor ampacity of: (a) 190% for single-phase half-wave rectifier; (b) 150% for single-phase full-wave rectifier.

NEC 430-32 (a-1) Overload setting not more than 125% of full-load motor current (40° motor).
NEC 430-82 (c) (1) Rheostat designed so contact arm cannot be left on intermediate segments.
NEC 430-82 (c) (2) Under-voltage release required to interrupt supply when speed falls to 1/3.
NEC 430-88 Adjustable-speed control to be arranged so motor cannot start on weakened field except where so designed.
NEC 430-89 Speed limiting device sometimes required. (See a, b, c.)

Motor-Starting Rheostat

Direct Current Motor

MOTOR -- PART WINDING

NEC 430-3 Where separate over-current devices, sizes of conductors and devices based on one-half full-load current rating.

NEC 430-22 (a) Size of branch-circuit conductor based on highest full-load current. Size of conductors between controller and motor based on current rating for that speed.

Multispeed Motor

Controllers

Branch Circuit Device

Motor Running Devices

Standard part-winding motor (does not include a hermetic-refrigerant type refrigerator, compressor motor).

WOUND-ROTOR MOTOR

Secondary Resistor

NEC 430-32 (a-1) Running overcurrent setting not over 125% full-load motor current (40° motor).

NEC 430-23 (c) See Table 430-23 (c) for ampacity.

NEC 430-22 (a) Ampacity not less than 125% of full-load motor current.

Table 430-152 Branch circuit overcurrent protection not to exceed 150% of the motor full-load current rating.

Wound-Rotor Motor

NEC 430-23 (a) Ampacity 125% for continuous-duty. (For other duty see Table 430-22 (a) Ex.)

NEC 430-32 (d) Secondary circuit requires no overcurrent devices.

TWO OR MORE MOTORS ON ONE BRANCH CIRCUIT

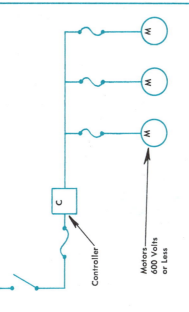

NEC 430-24 Minimum feeder current rating: 125% largest motor current plus sum of others.

Branch Circuit

NEC 430-62(a) Maximum feeder overcurrent largest motor branch setting plus full load of others. (Use Table 430-152 or 440-22(a) for hermetic.)

NEC 430-53 (c) Two or more motors of any size may be connected to one branch circuit if each has running overcurrent protection and all of the five requirements of NEC 430-53 (c) are complied with.

Motors 600 Volts or Less

NEC 430-53 (d) Ampacity of taps not less than ⅓ that of branch circuit conductors, not over 25 feet long and protected from physical injury.

NEC 430-112 Ex (a) (c) Individual disconnects required unless group is part of single machine or if group is in a single room in sight of disconnect.

NEC 430-42 May be a general-purpose branch circuit.

NEC 430-53 (a) Maximum setting overcurrent device: 20 amps, 120 volts or less; 15 amps, 600 volts or less. (See illustrations.)

600 Volts or Less

Motors Not Over 1 HP, Not Over 6 Amps Each

NEC 430-53 (a) Two or more motors not over 1 hp permitted on same 120-volt, 20-amp branch circuit or 600-volt, 15-amp branch circuit if: (1) not over 6 amps, (2) controller max. branch circuit O.C. device is not exceeded; (3) each overload is per 430-32. (b) Where all motors have overload protection and smallest motor protected as per 430-52 and where determined that the branch-circuit device will not open under most severe normal service conditions that might be encountered.

NEC 430-112 (b) Individual disconnects not required where provisions of NEC 430-53 (a) are complied with. (See above illustrations.)

MOTOR CONTROLLERS

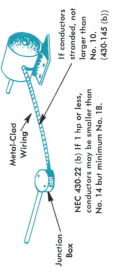

Controller

Motors 600 Volts or Less

NEC 430-87 (General Rule) Each motor of 600 volts or less to have its own controller.
Exceptions (a) (b) (c): Single controller permissible if motors are part of same machine, or if group within provisions of NEC 430-53 (a) or if group in single room in sight of controller.

MISCELLANEOUS MOTOR RULES

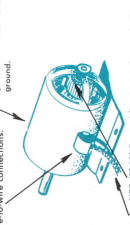

NEC 430-145 (a) Where motor terminals of fixed motors are in metal raceways or metal enclosed cable, junction boxes for motor terminals required. Metal raceways and Type AC cable connected as per Article 250. (FPN See 430-12(e)) for grounding requirements.)

NEC 430-145 (b) Junction box not more than 6 ft from motor under conditions specified in Code.

NEC 430-145 (c) Instrument transformer secondaries and exposed metal parts of instruments or relays used in connection with motors shall be grounded as per 250-121 thru 250-125.

Metal-Clad Wiring

If conductors stranded, not larger than No. 10. (430-145 (b))

Junction Box

NEC 430-22 (b) If 1 hp or less, conductors may be smaller than No. 14 but minimum No. 18.

MOTOR FEEDER CALCULATIONS

NEC 430-24 Minimum feeder ampacity 125 percent full-load current largest motor plus full-load currents of others. Where motors for types of service listed in 430-22 Exception, similar procedure followed, taking into consideration values obtained from Table 430-22 Exception. Where circuitry prevents all motors from operating at same time, conductor size is determined on basis of maximum load at any given time.

NEC 430-62 (a) Maximum feeder protection determined by largest branch-circuit protection. ((FPN) See Example No. 8, Chapter 9—NEC.)

NEC 430-6 (a) For general motor applications, except for torque motors, Tables 430-147, 148, 149, 150, including notes, shall be used to determine conductor ampacity, ampere rating of switches, short-circuit and ground-fault protection, etc. Motor nameplate FLA used to determine overload protection. Where motor marked in amps but not hp, the hp rating is assumed to be value as per Tables 430-147, 148, 149, 150, interpolated if necessary.

NEC 430-12(a) Terminal housings to be metal except non-metallic permitted in other than hazardous area. (b) Table 430-12(b) gives size for wire-to-wire connections.

NEC 430-142 Frames of stationary motors to be grounded if: (1) metal enclosed wiring, (2) wet location, (3) hazardous location, or (4) over 150 volts to ground. If frame not grounded it shall be insulated from ground.

NEC 430-132 Exposed live parts to be guarded if over 50 volts.

NEC 430-14 (a) Motor shall be located where adequate ventilation is provided; also so maintenance can be accomplished.

NEC 430-16 To be protected from accumulation of dust.

NOTE: Motor grounded by metallic raceway.

NEC 430-43 An overload which can automatically restart a motor shall be approved for the motor and if motor restarts could not result in injury.

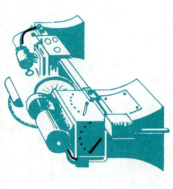

NEC 670-3 The nameplate of a metal working machine tool must show full-load current, ampere rating of the largest motor, and short-circuit interrupting capacity of the machine overcurrent device if one is furnished.

NEC 670-4(b) A machine tool is considered a single unit and shall have a disconnect which may be fed by branch circuits protected by breakers or fuses.

METAL WORKING MACHINE TOOLS

CRANES AND HOISTS

NEC 610-14 (d) Bare conductors No. 6 span 0-30 feet; No. 4 span 30-50 feet; No. 2 over 60 feet.

NEC 610-21 (f) Grounded track may serve as a conductor if supply by way of insulating transformer on each of not over 300 volts and other conductors insulated.

NEC 610-14 (a) and Table 610-14 (a) Provides increased ampacity for conductors to crane motors.

NEC 610-15 Where more than one motor a common-return conductor is permitted.

NEC 430-112 Ex (a) Individual disconnects not required.

NEC 610-42 (a) Each motor to have branch-circuit protection. Ex 1 Two motors operating as a unit may have single overload protector.

NEC 610-61 Grounding of metal parts required.

NEC 610-55 Limit switch or other device required on hoist unit.

NEC 610-41 Overcurrent device required for main contact conductors.

NEC 610-32 A lockable motor disconnect required in leads from runway contact conductors or other power supply on all cranes and monorail hoists. (See exceptions.) Where disconnect not readily accessible from operating station, then means required at operating station to open power circuit of crane or monorail hoist.

NEC 610-31 Disconnecting means operable from ground, and in sight of crane and runway contact conductors.

NEC 610-33 Minimum rating of disconnecting means; 50% of combined short-time ampere ratings of all motors and 75% of sum of short-time ampere ratings of motors for single crane operation.

NEC 610-14 (e-2) Ampacity of supply conductors shall be not less than the sum of short-time full-load current for largest motor, or group, for a single crane motion, plus 50% of that for the next larger motor or group.

Track

Collector

Ground

MOTOR-RUNNING OVERLOAD PROTECTION

Table 430-37 Three motor-running protective devices shall be used to protect three-phase motors unless protected by other approved means. NEC 430-36 Where fuses are used for motor-running protection in a 3-wire, 3-phase AC system with one wire grounded, a fuse shall be used in the grounded conductor.
NEC 430-38 The number of overcurrent devices other than fuses is governed by Table 430-37. (See above illustration.)

Supply — From Table 430-37

4-Wire 3φ
N
M

3-Wire 3φ
M

1φ or DC
M

Overcurrent Devices Other Than Fuses

GENERATORS

NEC 445-4 (a) Constant potential DC generators (other than exciters) shall have overcurrent protection.

NEC 445-5 Ampacity of conductors from generator to first overcurrent device not less than 115% of nameplate amps, except:
(1) Not less than 100% where design and operation prevent overloading; (2) Where factory-wired unit has overcurrent device that is an integral part of generator set.

NEC 445-1 Generators and associated wiring and equipment shall comply with NEC 230, 250, 700, 701, and 702.

NEC 445-4 (b) Two-wire generator may have overcurrent device in one lead; (c) If generator 65 volts or less, driven by motor, motor overcurrent device satisfies requirements if setting limits DC output to 150% of normal; (d) Control devices must be arranged to disconnect balancer sets if they are part of system; (e) Three-wire generators must have overcurrent devices in each armature lead, devices so interlocked as to disconnect generator and equalizers in case of trouble.

NEC 445-3 Nameplate shall show kVA or kW, amps, volts, frequency, power factor, maker's name, rpm, ambient temperature and other.

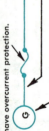
G

MOTORS OVER 600 VOLTS

NEC 430-125 (c-1 (b)) Fault current protection required. May be provided by suitable type and rated fuses that cannot be serviced while energized.

NEC 430-125 (c-1 (a)) Circuit breaker permitted for fault current protection. (May sense fault current by integral or external sensing elements.)

NEC 430-127 Disconnect for controller shall be able to be locked in off position.

NEC 430-122 Controllers marked with maker's name, voltage, current, hp rating and other necessary data (see 430-8) and control voltage.

Motor Over 600 Volts

C
M

NEC 430-124 Ampacity of conductors not less than current at which motor overload overcurrent device is selected to trip.

NEC 430-125 (a) Motor shall have overload protection and motor, motor conductors and control apparatus shall have fault current protection. This protection shall be coordinated and automatically operated except vital motor may use supervised annunciator or alarm.

NEC 430-125 (b-1) Each motor to be protected against overloads or failure to start by thermal protector integral with motor or external sensing devices.

NEC 430-125 (b-2) Secondary circuits of wound-rotor motors are considered to be protected by motor overcurrent protection.

NEC 430-125 (b-3) Operation of overload device shall open all ungrounded conductors at same time.

NEC 430-125 (b-4) Overload sensing devices shall not automatically restart motor unless restarting is no hazard.

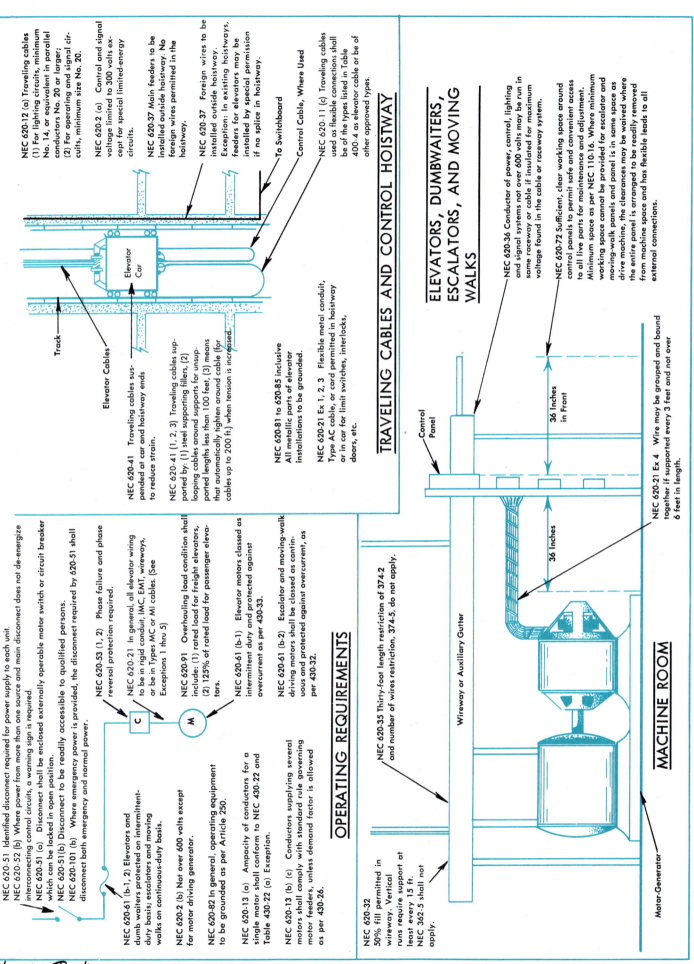

NEC 620-12 (a) Traveling cables (1) For lighting circuits, minimum No. 14, or equivalent in parallel conductors No. 20 or larger; (2) For operating and signal circuits, minimum size No. 20.

NEC 620-2 (a) Control and signal voltage limited to 300 volts except for special limited-energy circuits.

NEC 620-37 Main feeders to be installed outside hoistway. No foreign wires permitted in the hoistway.

NEC 620-37 Foreign wires to be installed outside hoistway. Exception: In existing hoistways, feeders for elevators may be installed by special permission if no splice in hoistway.

To Switchboard

Control Cable, Where Used

NEC 620-11 (c) Traveling cables used as flexible connections shall be of the types listed in Table 400-4 as elevator cable or be of other approved types.

Track

Elevator Car

Elevator Cables

NEC 620-41 Traveling cables suspended at car and hoistway ends to reduce strain.

NEC 620-41 (1, 2, 3) Traveling cables supported by: (1) steel supporting fillers, (2) looping cables around supports for unsupported lengths less than 100 feet, (3) means that automatically tighten around cable (for cables up to 200 ft.) when tension is increased.

NEC 620-81 to 620-85 inclusive
All metallic parts of elevator installations to be grounded.

NEC 620-21 Ex 1, 2, 3 Flexible metal conduit, Type AC cable, or cord permitted in hoistway or in car for limit switches, interlocks, doors, etc.

TRAVELING CABLES AND CONTROL HOISTWAY

ELEVATORS, DUMBWAITERS, ESCALATORS, AND MOVING WALKS

NEC 620-36 Conductor of power, control, lighting and signal systems not over 600 volts may be run in same raceway or cable if insulated for maximum voltage found in the cable or raceway system.

NEC 620-72 Sufficient, clear working space around control panels to permit safe and convenient access to all live parts for maintenance and adjustment. Minimum space as per NEC 110-16. Where minimum working space cannot be provided for escalator and moving-walk panels and panel is in same space as drive machine, the clearances may be waived where the entire panel is arranged to be readily removed from machine space and has flexible leads to all external connections.

Control Panel

36 Inches in Front

36 Inches

36 Inches

NEC 620-21 Ex 4   Wire may be grouped and bound together if supported every 3 feet and not over 6 feet in length.

MACHINE ROOM

Motor-Generator

Wireway or Auxiliary Gutter

NEC 620-35 Thirty-foot length restriction of 374-2 and number of wires restriction, 374-5, do not apply.

NEC 620-32
50% fill permitted in wireway. Vertical runs require support at least every 15 ft. NEC 362-5 shall not apply.

OPERATING REQUIREMENTS

NEC 620-51 Identified disconnect required for power supply to each unit.
NEC 620-52 (b) Where power from more than one source and main disconnect does not de-energize interconnecting control circuits, a warning sign is required.
NEC 620-51 (a) Disconnect shall be enclosed externally operable motor switch or circuit breaker which can be locked in open position.
NEC 620-51 (b) Disconnect to be readily accessible to qualified persons.
NEC 620-101 (b) Where emergency power is provided, the disconnect required by 620-51 shall disconnect both emergency and normal power.

NEC 620-53 (1, 2) Phase failure and phase reversal protection required.

NEC 620-21 In general, all elevator wiring to be in rigid conduit, IMC, EMT, wireways, or be in Types MC or MI cables. (See Exceptions 1 thru 5)

NEC 620-91 Overhauling load condition shall include: (1) rated load for freight elevators, (2) 125% of rated load for passenger elevators.

NEC 620-61 (b-1) Elevator motors classed as intermittent duty and protected against overcurrent as per 430-33.

NEC 620-61 (b-2) Escalator and moving-walk driving motors shall be classed as continuous and protected against overcurrent, as per 430-32.

NEC 620-61 (b-1, 2) Elevators and dumb waiters protected on intermittent-duty basis; escalators and moving walks on continuous-duty basis.

NEC 620-2 (b) Not over 600 volts except for motor driving generator.

NEC 620-82 In general, operating equipment to be grounded as per Article 250.

NEC 620-13 (a) Ampacity of conductors for a single motor shall conform to NEC 430-22 and Table 430-22 (a) Exception.

NEC 620-13 (b) (c) Conductors supplying several motors shall comply with standard rule governing motor feeders, unless demand factor is allowed as per 430-26.

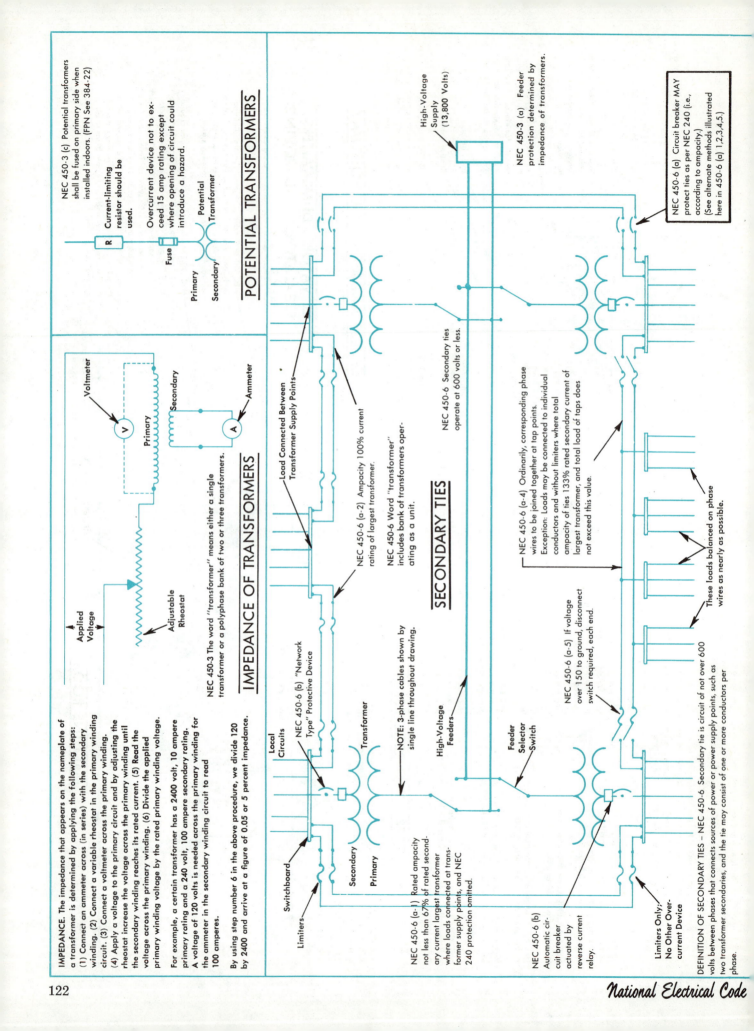

IMPEDANCE. The impedance that appears on the nameplate of a transformer is determined by applying the following steps: (1) Connect an ammeter across (in series) with the secondary winding. (2) Connect a variable rheostat in the primary winding circuit. (3) Connect a voltmeter across the primary winding. (4) Apply a voltage to the primary circuit and by adjusting the rheostat increase the voltage across the primary winding until the secondary winding reaches its rated current. (5) Read the voltage across the primary winding. (6) Divide the applied primary winding voltage by the rated primary winding voltage.

For example, a certain transformer has a 2400 volt, 10 ampere primary rating and a 240 volt, 100 ampere secondary rating. A voltage of 120 volts is needed across the primary winding for the ammeter in the secondary winding circuit to read 100 amperes.

By using step number 6 in the above procedure, we divide 120 by 2400 and arrive at a figure of 0.05 or 5 percent impedance.

NEC 450-3 (c) Potential transformers shall be fused on primary side when installed indoors. (FPN See 384-22)

Current-limiting resistor should be used.

Overcurrent device not to exceed 15 amp rating except where opening of circuit could introduce a hazard.

R

Fuse

Primary

Secondary

Potential Transformer

POTENTIAL TRANSFORMERS

Voltmeter

V

Primary

Secondary

A

Ammeter

Applied Voltage

Adjustable Rheostat

NEC 450-3 The word "transformer" means either a single transformer or a polyphase bank of two or three transformers.

IMPEDANCE OF TRANSFORMERS

High-Voltage Supply (13,800 Volts)

NEC 450-3 (a) Feeder protection determined by impedance of transformers.

NEC 450-6 (a) Circuit breaker MAY protect ties as per NEC 240 (i.e., according to ampacity.) (See alternate methods illustrated here in 450-6 (a) 1,2,3,4,5)

Load Connected Between Transformer Supply Points

NEC 450-6 (a-2) Ampacity 100% current rating of largest transformer.

NEC 450-6 Word "transformer" includes bank of transformers operating as a unit.

NEC 450-6 Secondary ties operate at 600 volts or less.

SECONDARY TIES

NEC 450-6 (a-4) Ordinarily, corresponding phase wires to be joined together at tap points. Exception: Loads may be connected to individual conductors and without limiters where total ampacity of ties 133% rated secondary current of largest transformer, and total load of taps does not exceed this value.

NEC 450-6 (a-5) If voltage over 150 to ground, disconnect switch required, each end.

These loads balanced on phase wires as nearly as possible.

Local Circuits

NEC 450-6 (b) "Network Type" Protective Device

Transformer

NOTE: 3-phase cables shown by single line throughout drawing.

High-Voltage Feeders

Feeder Selector Switch

Switchboard

Limiters

Secondary

Primary

NEC 450-6 (a-1) Rated ampacity not less than 67% of rated secondary current largest transformer where loads connected at transformer supply points, and NEC 240 protection omitted.

NEC 450-6 (b) Automatic circuit breaker actuated by reverse current relay.

Limiters Only: No Other Overcurrent Device

DEFINITION OF SECONDARY TIES – NEC 450-6 Secondary tie is circuit of not over 600 volts between phases that connects sources of power or power supply points, such as two transformer secondaries, and the tie may consist of one or more conductors per phase.

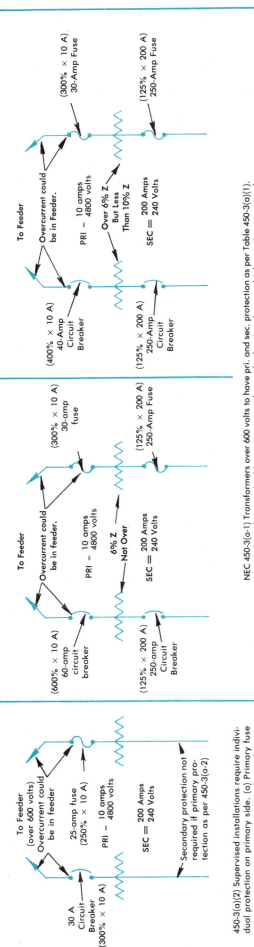

450-3(a)(2) Supervised installations require individual protection on primary side. (a) Primary fuse continuous rating not over 250% of transformer and for circuit breakers rating not over 300%. EX. 1 Next higher fuse or circuit breaker rating permitted if not corresponding to next standard size. EX. 2 Circuit protection permitted to protect transformer.

NEC 450-3(a-1) Transformers over 600 volts to have pri. and sec. protection as per Table 450-3(a)(1). EX. 1 Where fuse or ckt. bkr. rating exceeds standard rating the next higher rating permitted. Table 430-3(a-1) permits pri. protection of 600% for ckt. brkrs. and 300% for fuses if not over 6% Z and 400% for ckt. bkrs. and 300% for fuses if over 6% but not over 10% Z. See Table 450-3(a-1) for sec. protection. (See 450-3(a-1) EX. 2

NEC 450-3 (a) (1) (2) OVERCURRENT PROTECTION OF TRANSFORMERS (OVER 600 VOLTS)

NEC 450-3 (b-2) Individual transformer primary overload protection not required if transformer has coordinated thermal overload protection by manufacturer, arranged to open primary circuit, and if primary overcurrent device is not over 6 times rated primary current for transformers with not more than 6% impedance and not more than 4 times rated primary current for transformers having not more than 6 but not more than 10% impedance except: where secondary current is 9 amps or more and 125% is not standard fuse or circuit breaker, then next larger device may be used. Where secondary is less than 9 amps the overcurrent device may be rated or set at not more than 167% rated primary current.

NEC 450-3 and 450-6 indicate that when the word "transformer" is used it shall mean a transformer or polyphase bank of 2 or 3 single-phase transformers operating as a unit. (See NEC 384-16.)

NEC 450-3 (b) (1) (2) OVERCURRENT PROTECTION OF TRANSFORMERS (600 VOLTS OR LESS)

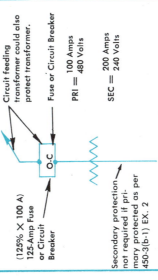

NEC 450-3 (b-2) Individual primary overcurrent not required if primary feeder protected at 250% of primary rated current and secondary protected not over 125% of rated secondary current.

NEC 450-3 (b-1) Each transformer of 600 volts or less is required to have individual primary overcurrent protection. This overcurrent device is to be rated or set at not over 125% of rated primary current except: (1) if primary current is 9 amps or more and 125% is not standard fuse or circuit breaker, then next larger device may be used. Where primary is less than 2 amps, overcurrent device may be rated or set at not more than 300% of rated primary amps, (2) primary circuit protection may also protect transformer if of proper size; (3) if secondary is protected as per 450-3 (b-2).

All conductors feeding in primary or out of secondary shall be protected at their rated ampacities as per 240-3. This protection shall be provided where the conductors receive their supply as per 240-21.

Where secondary conductors feed a lighting and branch-circuit panelboard, each panelboard shall be individually protected on supply side at ampacity not exceeding rating of panelboard. (See NEC 384-16.)

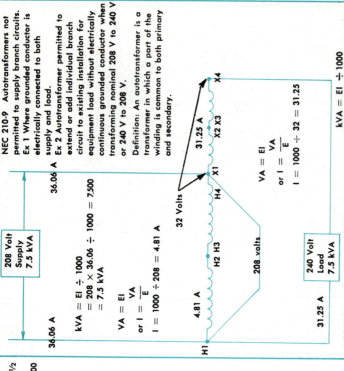

NEC 210-9 Autotransformers not permitted to supply branch circuits. Ex 1 Where grounded conductor is electrically connected to both supply and load.

Ex 2 Autotransformer permitted to extend or add individual branch circuit to existing installation for equipment load without electrically continuous grounded conductor when transforming nominal 208 V to 240 V or 240 V to 208 V.

Definition: An autotransformer is a transformer in which a part of the winding is common to both primary and secondary.

208 Volt Supply 7.5 kVA

36.06 A

$$kVA = EI \div 1000$$
$$= 208 \times 36.06 \div 1000 = 7,500$$
$$= 7.5 \ kVA$$

$$VA = EI$$
$$or \ I = \frac{VA}{E}$$
$$I = 1000 \div 208 = 4.81 \ A$$

31.25 A

X4 X2 X3 X1

32 Volts H4

208 volts H2 H3

4.81 A H1

31.25 A

240 Volt Load 7.5 kVA

$$kVA = EI \div 1000$$
$$= 240 \times 31.25 \div 1000$$
$$= 7.5 \ kVA$$

$$VA = EI$$
$$or \ I = \frac{VA}{E}$$
$$I = 1000 \div 32 = 31.25$$

AUTOTRANSFORMER

NEC 210-9 Ex 2 1 kVA, 208/32 volt, 2-winding, single-phase transformer connected as an auto-transformer to boost the voltage from 208 V to 240 V. This 1 kVA transformer feeds a 7.5 kVA load.

OIL-INSULATED TRANSFORMERS--INSTALLED INDOORS AND OUTDOORS

NEC 450-27 Oil-insulated transformers installed outdoors shall be installed so combustible material, combustible buildings, parts of buildings, door and window openings, fire escapes are safeguarded from transformer oil fires when transformer is installed, attached to, or adjacent to a building or combustible material, or on roofs. (See National Electrical Safety Code ANSI C2-1981)

NEC 450-26 Ex 3 Electric furnace transformers totalling not more than 75 kVA may be installed in fire-resistive location if provisions are made to prevent spread of oil fire.

NEC 450-26 Ex 2 Vault not required if not over 600 volts, not over 10 kVA in combustible area, or 75 kVA in fire-resistant area.

NEC 450-26 Ex 4 Vault not required if in suitable detached building accessible only to qualified persons.

NEC 450-26 (General Rule) Approved vault required for oil-insulated transformer indoors. Ex 1 If not over 112½ kVA, thickness of vault wall need be only 4 inches. (See Exceptions 2, 3, 4, 5.)

Oil-Insulated Transformer

NEC 450-21 (a) Dry-type transformers permitted indoors if rated 112½ kVA or less and 12 in. separation from combustible material unless separated by a fire-resistant, heat-insulating barrier, or unless rated 600 volts or less and totally enclosed except for ventilating openings.

12 inches from combustible surface or suitable barrier.

Dry Transformer, Totally Enclosed Except Vent's

NEC 450-21(b) Ex 1, 2 If over 112½ kVA then required to be in transformer room of fire-resistant construction. Ex 1 Transformer room not required if transformer has a 150°C or higher insulation system and separated from combustibles by at least 12 ft vertically and 6 ft horizontally unless (Ex 2) totally enclosed and ventilated-type construction.

NEC 450-21 (c) Dry-type transformers over 35,000 volts require vault as per Article 450 Part C.

NEC 110-31 (a) Where over 600 volts, openings in dry transformer cases to be guarded if necessary for protection of persons.

TRANSFORMERS -- DRY-TYPE INSTALLED INDOORS

NEC 450-25 Where askarel-insulated transformers are installed in poorly ventilated area, means shall be provided to absorb any gas, or a chimney or flue connected to carry such gases outside buildings.

Askarel-Insulated Transformer Installed Indoors

Pressure-Relief Vent

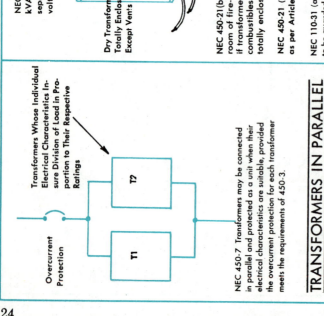

NEC 450-24 Fluid-insulated transformers insulated with a dielectric fluid identified as nonflammable permitted indoors and outdoors. When over 35,000 volts and indoors, a vault is required. (In this section a nonflammable dielectric fluid does not have a flash or fire point and not flammable in air.)

NEC 450-25 Askarel-insulated transformers 25 kVA or over must have pressure-relief vent. If more than 35,000 volts, they shall be installed in an approved vault.

ASKAREL OR NONFLAMMABLE TRANSFORMERS INSTALLED INDOORS

Overcurrent Protection

T2

T1

Transformers Whose Individual Electrical Characteristics Insure Division of Load in Proportion to Their Respective Ratings

NEC 450-7 Transformers may be connected in parallel and protected as a unit when their electrical characteristics are suitable, provided the overcurrent protection for each transformer meets the requirements of 450-3.

TRANSFORMERS IN PARALLEL

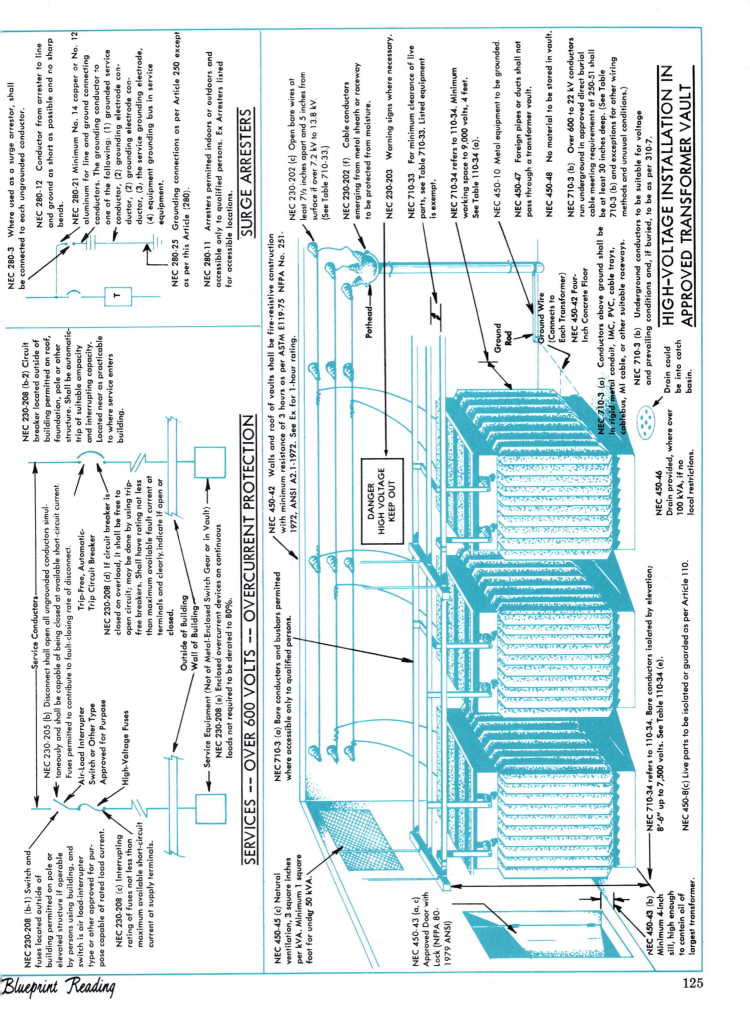

NEC 280-3 Where used as a surge arrestor, shall be connected to each ungrounded conductor.

NEC 280-12 Conductor from arrester to line and ground as short as possible and no sharp bends.

NEC 280-21 Minimum No. 14 copper or No. 12 aluminum for line and ground connecting conductors. The grounding conductor to one of the following: (1) grounded service conductor, (2) grounding electrode conductor, (3) the service grounding electrode, (4) equipment grounding bus in service equipment.

NEC 280-25 Grounding connections as per Article 250 except as per this Article (280).

NEC 280-11 Arresters permitted indoors or outdoors and accessible only to qualified persons. Ex Arresters listed for accessible locations.

SURGE ARRESTERS

NEC 230-202 (c) Open bare wires at least 7½ inches apart and 5 inches from surface if over 7.2 kV to 13.8 kV. (See Table 710-33.)

NEC 230-202 (f) Cable conductors emerging from metal sheath or raceway to be protected from moisture.

NEC 230-203 Warning signs where necessary.

NEC 710-33 For minimum clearance of live parts, see Table 710-33. Listed equipment is exempt.

NEC 710-34 refers to 110-34. Minimum working space to 9,000 volts, 4 feet. See Table 110-34 (a).

NEC 450-10 Metal equipment to be grounded.

NEC 450-47 Foreign pipes or ducts shall not pass through a transformer vault.

NEC 450-48 No material to be stored in vault.

NEC 710-3 (b) Over 600 to 22 kV conductors run underground in approved direct burial cable meeting requirements of 250-51 shall be at least 30 inches deep. (See Table 710-3 (b) and exceptions for other wiring methods and unusual conditions.)

NEC 710-3 (b) Underground conductors to be suitable for voltage and prevailing conditions and, if buried, to be as per 310-7.

HIGH-VOLTAGE INSTALLATION IN APPROVED TRANSFORMER VAULT

NEC 230-208 (b-2) Circuit breaker located outside of building permitted on roof, foundation, pole or other structure. Shall be automatic-trip of suitable ampacity and interrupting capacity. Located near as practicable to where service enters building.

NEC 230-208 (d) If circuit breaker is closed on overload, it shall be free to open circuit; may be done by using trip-free breakers. Shall have rating not less than maximum available fault current at terminals and clearly indicate if open or closed.

Service Conductors

NEC 230-205 (b) Disconnect shall open all ungrounded conductors simultaneously and shall be capable of being closed at available short-circuit current. Shall be automatic-trip of suitable ampacity and interrupting capacity. Located to contribute to fault-closing rate of disconnect.

Fuses permitted to contribute to fault-closing rate of disconnect.

Trip-Free, Automatic Trip Circuit Breaker

Air-Load Interrupter Switch or Other Type Approved for Purpose

High-Voltage Fuses

Outside of Building Wall of Building

Service Equipment (Not of Metal-Enclosed Switch Gear or in Vault)
NEC 230-208 (e) Enclosed overcurrent devices on continuous loads not required to be derated to 80%.

SERVICES -- OVER 600 VOLTS -- OVERCURRENT PROTECTION

NEC 230-208 (b-1) Switch and fuses located outside of building permitted on pole or elevated structure if operable by persons using building, and switch is air load-interrupter type or other approved for purpose capable of rated load current.

NEC 230-208 (c) Interrupting rating of fuses not less than maximum available short-circuit current at supply terminals.

NEC 450-42 Walls and roof of vaults shall be fire-resistive construction with minimum resistance of 3 hours as per ASTM E119-75 NFPA No. 251-1972, ANSI A2.1-1972. See Ex for 1-hour rating.

DANGER HIGH VOLTAGE KEEP OUT

NEC 710-3 (a) Bare conductors and busbars permitted where accessible only to qualified persons.

NEC 710-34 refers to 110-34. Bare conductors isolated by elevation; 8'-6" up to 7,500 volts. See Table 110-34 (e).

NEC 450-8(c) Live parts to be isolated or guarded as per Article 110.

NEC 450-45 (c) Natural ventilation, 3 square inches per kVA. Minimum 1 square foot for under 50 kVA.

NEC 450-43 (a, c) Approved Door with Lock (NFPA 80-1979 ANSI)

NEC 450-43 (b) Minimum 4-inch sill, high enough to contain oil of largest transformer.

Pothead

Ground Rod

Ground Wire (Connects to Each Transformer)
NEC 450-42 Four-Inch Concrete Floor

NEC 710-3 (a) Conductors above ground shall be in rigid metal conduit, IMC, PVC, cable trays, cablebus, MI cable, or other suitable raceways.

NEC 450-46 Drain provided, where over 100 kVA, if no local restrictions.

Drain could be into catch basin.

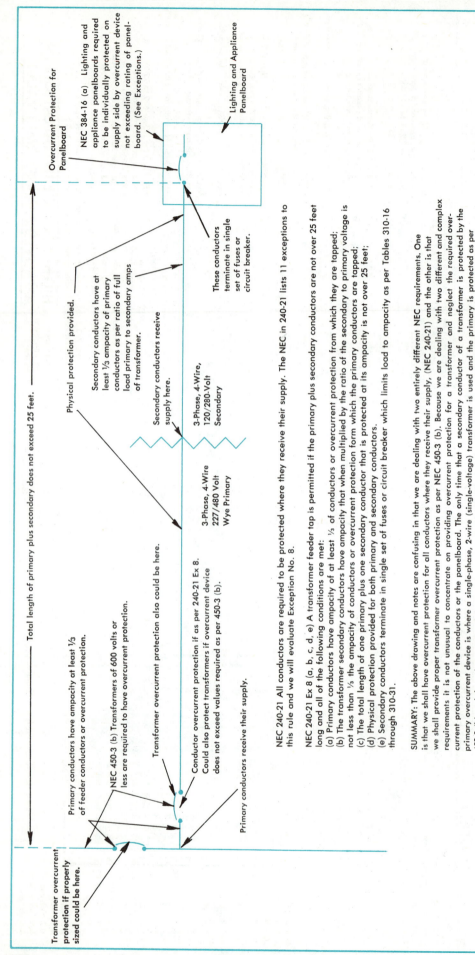

Total length of primary plus secondary does not exceed 25 feet.

Overcurrent Protection for Panelboard

NEC 384-16 (a) Lighting and appliance panelboards required to be individually protected on supply side by overcurrent device not exceeding rating of panelboard. (See Exceptions.)

Lighting and Appliance Panelboard

Physical protection provided.

Secondary conductors have at least ⅓ ampacity of primary conductors as per ratio of full load primary to secondary amps of transformer.

These conductors terminate in single set of fuses or circuit breaker.

Secondary conductors receive supply here.

3-Phase, 4-Wire, 120/280-Volt Secondary

3-Phase, 4-Wire 227/480 Volt Wye Primary

Primary conductors have ampacity at least ⅓ of feeder conductors or overcurrent protection.

Transformer overcurrent protection also could be here.

NEC 450-3 (b) Transformers of 600 volts or less are required to have overcurrent protection.

Conductor overcurrent protection if as per 240-21 Ex 8. Could also protect transformers if overcurrent device does not exceed values required as per 450-3 (b).

Primary conductors receive their supply.

Transformer overcurrent protection if properly sized could be here.

NEC 240-21 All conductors are required to be protected where they receive their supply. The NEC in 240-21 lists 11 exceptions to this rule and we will evaluate Exception No. 8.

NEC 240-21 Ex 8 (a, b, c, d, e) A transformer feeder tap is permitted if the primary plus secondary conductors are not over 25 feet long and all of the following conditions are met:
(a) Primary conductors have ampacity of at least ⅓ of conductors or overcurrent protection from which they are tapped;
(b) The transformer secondary conductors have ampacity that when multiplied by the ratio of the secondary to primary voltage is not less than ⅓ the ampacity of conductors or overcurrent protection form which the primary conductors are tapped;
(c) The total length of one primary plus one secondary conductor that is protected at its ampacity is not over 25 feet;
(d) Physical protection provided for both primary and secondary conductors.
(e) Secondary conductors terminate in single set of fuses or circuit breaker which limits load to ampacity as per Tables 310-16 through 310-31.

SUMMARY: The above drawing and notes are confusing in that we are dealing with two entirely different NEC requirements. One is that we shall have overcurrent protection for all conductors where they receive their supply, (NEC 240-21) and the other is that we shall provide proper transformer overcurrent protection as per NEC 450-3 (b). Because we are dealing with two different and complex requirements it is not unusual to concentrate on providing overcurrent protection for a transformer and neglect the required overcurrent protection of the conductors or the panelboard. The only time that a secondary conductor of a transformer is protected by the primary overcurrent device is where a single-phase, 2-wire (single-voltage) transformer is used and the primary is protected as per 450-3 (b) and the voltage ratio of the transformer protects the secondary conductors to their ampacity. (See 240-3 Ex 5.)

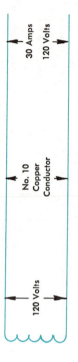

Single-phase 240/120-Volt, 2-Wire (Single Voltage) Secondary

30 Amps
120 Volts

No. 10 Copper Conductor

120 Volts

15-Amp primary fuse protects 30-amp secondary conductor.

No. 14 Copper Conductor

15-Amp overcurrent protection conductor.

15 Amps 240 Volts

NEC 240-3 Ex 5 The secondary conductors of a single-phase, 2-wire (single-voltage) secondary are considered protected if the primary overcurrent conforms to 450-3 and is not more than the current determined by multiplying the secondary conductor amps by the secondary-to-primary transformer voltage ratio.

TRANSFORMER FEEDER TAP 25 FOOT RULE

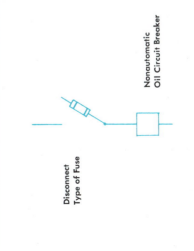

Metal-Clad Draw-Out Switchgear

Oil Switch or Air or Oil Circuit Breaker

Circuit Breaker

NEC 230-204 (a)   Where service disconnect is oil switch or air or oil circuit breaker, then air-break isolating switch required on line side of disconnect and all service equipment.

NEC 230-204 (c) (d) Air-break isolators shall be accessible to qualified persons only and arranged for grounding on load side.

NEC 230-208 (e) Enclosed overcurrent devices operating at over 600 volts are not restricted to 80 percent of rating for continuous loads.

SERVICES OVER 600 VOLTS -- ISOLATING SWITCHES

NEC 710-12 refers to 110-31 (b) "Outdoor Installations."

110-31 (b-1) Shall comply with Article 225 if open to unqualified persons.

110-31 (b-2) Outdoor installations with exposed live parts shall be accessible to qualified persons only and shall also comply with 110-34, 710-32 and 710-33.

NEC 710-13 refers to 110-31 (c) which permits metal-enclosed equipment to be installed outdoors and accessible to the public, if bolts and nuts are difficult to remove. If bottom of enclosure is less than 8 feet from floor it shall be locked.

NEC 710-34 refers to 110-34 that gives the working space required for live parts.

SERVICES OVER 600 VOLTS -- DISCONNECT MEANS

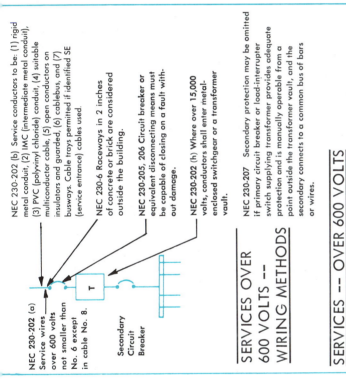

NEC 230-202 (b) Service conductors to be: (1) rigid metal conduit, (2) IMC (intermediate metal conduit), (3) PVC (polyvinyl chloride) conduit, (4) suitable multiconductor cable, (5) open conductors on insulators and guarded, (6) cablebus, and (7) busways. Cable trays permitted if identified SE (service entrance) cables used.

NEC 230-6 Raceways in 2 inches of concrete or brick are considered outside the building.

NEC 230-205, 206 Circuit breaker or equivalent disconnecting means must be capable of closing on a fault without damage.

NEC 230-202 (h) Where over 15,000 volts, conductors shall enter metal-enclosed switchgear or a transformer vault.

NEC 230-207   Secondary protection may be omitted if primary circuit breaker or load-interrupter switch supplying transformer provides adequate protection and is manually operable from a point outside the transformer vault, and the secondary connects to a common bus of bars or wires.

SERVICES OVER 600 VOLTS -- WIRING METHODS

NEC 230-202 (a) Service wires over 600 volts not smaller than No. 6 except in cable No. 8.

Secondary Circuit Breaker

SERVICES -- OVER 600 VOLTS

Nonautomatic Oil Switch Oil Fuse Cutout or Air Interrupter Switch

Switch Capable of Interrupting No-Load Transformer Current

Circuit Breaker Interlocked with Switch

NEC 710-21 (c-2) Sign: "Do Not Open Under Load"

Automatic-Trip Circuit Breaker Having Proper Number of Over-current Devices

NEC 230-208 (a-3)   Line switch interlocked with secondary switch or circuit breaker to prevent line switch from operating when secondary disconnect is closed.

NEC 230-208 (a-2)   Disconnecting means permitted to be automatic trip circuit breaker with proper interrupting and current-carrying capacity.

NEC 230-208 (a-1) Disconnecting means shall be: oil fuse cutout, nonautomatic oil switch, air load interrupter switch or other approved for purpose with proper fuses. Switch interrupting rating shall be equal or more than continuous fuse rating.

NEC 230-208 (e) Enclosed overcurrent devices for service of over 600 volts need not be restricted to 80 percent on continuous loads.

NEC 710-22 Isolating switches are not required where equipment may be safely de-energized through metal-clad switchgear units and removable truck panels.

SERVICES OVER 600 VOLTS -- DISCONNECT MEANS

Primary Conductors

Secondary Circuit Breaker

Load

NEC 230-201 (a) (b) Secondary conductors regarded as service wires where step-down transformers are: (1) outdoors, (2) in separate buildings, (3) in vault as per 450 located inside building, or (4) in locked room or locked enclosure located inside building and accessible only to qualified persons, (5) inside building and in metal-enclosed gear. The primary conductors to be considered service conductors in all other cases. See Exceptions where both primary and secondary are over 600 volts.

SECONDARY CONDUCTORS AS SERVICE CONDUCTORS

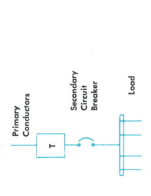

Disconnect Type of Fuse

Nonautomatic Oil Circuit Breaker

NEC 230-204 (b)   Fuses operable as disconnect means permitted as isolating switch where: (1) oil disconnect is non-automatic, and (2) the fuses disconnect the oil switch and all service equipment.

SERVICES OVER 600 VOLTS -- FUSED ISOLATING SWITCH

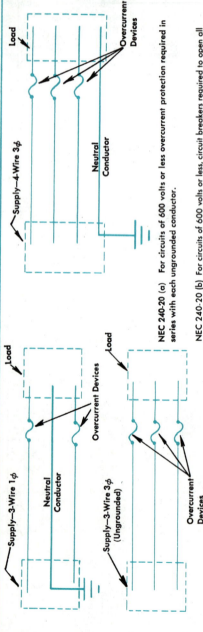

Flame-Retardant Braid

NEC 710-4 Where braid-covered insulated conductors are used for open runs for circuits over 600 volts, then braid shall be flame retardant.

NEC 710-4 Braid shall be stripped back a safe distance. Shall be 1 inch for each kilovolt to ground in circuit.

OVER 600 VOLTS - BRAID-COVERED INSULATED CONDUCTORS, OPEN RUNS

Stress-Cone — Unshielded Cable or Termination

Shielded Cable

NEC 710-6 Metallic shielding and all conducting or semi-conducting static shielding parts of shielded cable to be stripped back at terminations a safe distance determined by circuit voltage. All metallic shielding components to be grounded.

OVER 600 VOLTS -- GROUNDING OR SHIELDING TAPE

Motor

M

Capacitor

Motor Overload Protection

Controller

NEC 460-9 Where capacitor is installed on load side of motor overload device, the rating or setting of motor overload device shall be determined per 430-32 using a lower value of current than that listed on motor nameplate. NOTE: The improved power factor will permit the motor to run at full load with a lower line current than that listed on motor nameplate. When capacitor is installed on line side of motor overload device, then the FLA on nameplate used to determine motor overload protection.

NEC 460-8 (c-1) Ex No separate disconnect required for capacitor.

NEC 460-8(b-1) Ex. Overcurrent device not required in each conductor if connected on load side of motor overload.

NEC 460-8 (a) Rating of conductors at least $\frac{1}{3}$ that of motor circuit conductors and 135% of capacitor current rating.

CAPACITORS -- MOTOR CIRCUITS

NEC 460-6 (a) (b) Discharge resistor connected permanently or automatic to drain charge to 50 volts in 1 minute.

NEC 460-28 (a) For over 600 volts discharge means required to reduce to 50 volts or less in 5 minutes or less.

NEC 460-8 (a) Ampacity of conductors at least 135% of capacitor current rating.

NEC 460-8 (b-1, 2) Overcurrent protection as low as practicable.

NEC 460-8 (c) 1, 2, 4) Disconnecting means shall open all ungrounded conductors simultaneously. Its continuous ampacity must not be less than 135% of rated capacity current.

NEC 460-10 Capacitor cases to be grounded as per 250, except if on a structure operating at other than ground potential.

NEC 460-2 (b) Live parts to be enclosed or guarded except when enclosed and accessible only to qualified personnel.

CAPACITORS -- 600 VOLTS OR LESS

Load

Overcurrent Devices

Supply-4-Wire 3φ

Neutral Conductor

NEC 240-20 (a) For circuits of 600 volts or less overcurrent protection required in series with each ungrounded conductor.

NEC 240-20 (b) For circuits of 600 volts or less, circuit breakers required to open all conductors of circuit, except single-pole breakers permitted for 3-wire DC or single-phase circuits, and lighting and appliance branch circuits connected to 4-wire 3-phase systems or 5-wire 2-phase systems, if circuits supplied from system having a grounded neutral, and voltage does not exceed that permitted by 210-6.

NEC 710-20 For circuits of over 600 volts, overcurrent protection required for each ungrounded conductor.

NUMBER OF OVERCURRENT DEVICES TO PROTECT CIRCUIT CONDUCTORS: 600 VOLTS OR LESS -- NEC 240-20; OVER 600 VOLTS -- NEC 710-20

Load

Overcurrent Devices

Supply—3-Wire 1φ

Neutral Conductor

Supply—3-Wire 3φ (Ungrounded)

Overcurrent Devices

NEC 710-20 (b) For circuits of over 600 volts, where fuses are used, a fuse is required in each ungrounded conductor.

NEC 710-20 (a) (General Rule) On circuits over 600 volts, where circuit breakers are used for AC 3-phase circuits, they shall have at least 3 overcurrent relays operated from 3 current transformers. (See Ex 1, 2.)

Three-Phase Line, Over 600 Volts

NEC 460-24 (a) Group-operated switch rated to carry continuously at least 135% of rated capacitor bank current, rated to interrupt load of bank, rated for maximum in rush currents and rated to carry capacitor fault currents.

NEC 460-25 (a) (b) Overcurrent protection required to detect and interrupt dangerous fault current. Three-phase device permitted to protect 3-phase bank.

NEC 460-2 (a) Vaults or outdoor enclosures as per NEC 710 required for capacitors containing over 3 gallons of flammable liquid.

Three-Phase Delta Connected Capacitor Bank (Over 600 Volts)

3-Phase Overcurrent Device

NEC 460-28 (a) (b) Discharge means required to reduce to 50 volts or less in 5 minutes or less. To be permanently connected or have automatic connection after capacitor bank disconnected from source.

Three-phase Capacitor Bank

NEC 460-27 If grounded, the capacitor enclosure and neutrals shall comply with 250, except if on a structure operating at other than ground potential.

NAMEPLATE—NEC 460-26 Lists maker's name, rated voltage, kVar or amps, frequency, phases, and type and amount of liquid.

CAPACITORS -- OVER 600 VOLTS

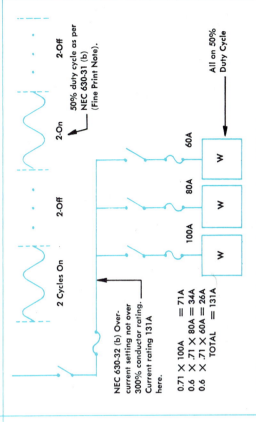

Welders

W	W	W	W	W	W
20A	20A	30A	40A	50A	60A

NEC 630-12 (b) Overcurrent device set at not more than 200% of conductor rating.

Feeder 134 Amps

.71 × 60 amps = 43 amps
.71 × 50 amps = 36 amps
.71 × 40 amps = 28 amps
.71 × 30 amps = 21 amps
.71 × 20 amps = 14 amps
.71 × 20 amps = 14 amps

TOTAL 134 amps

1 × 43 amps = 43 amps
1 × 36 amps = 36 amps
.85 × 28 amps = 24 amps
.7 × 21 amps = 15 amps
.6 × 14 amps = 8 amps
.6 × 14 amps = 8 amps

TOTAL 134 amps

Based on 50% duty-cycle; multiplying factor .71 as per NEC 630-11 (a).

NEC 630-11 (b) Conductor rating for a group of welders shall be determined in each case by the nature of use. It is suggested that adequate conductor size will be obtained if the following percentages of current values determined as per NEC 630-11 (a) are applied: 100 percent for the two largest welders, plus 85 percent for the third largest, plus 70 percent for the fourth largest, plus 60 percent for all others.

GROUP OF TRANSFORMER ARC WELDERS

2-Off   2-On   2-Off

50% duty cycle as per NEC 630-31 (b) (Fine Print Note).

2 Cycles On

W	W	W
100A	80A	60A

All on 50% Duty Cycle

NEC 630-32 (b) Overcurrent setting not over 300% conductor rating. Current rating 131A here.

0.71 × 100A = 71A
0.6 × .71 × 80A = 34A
0.6 × .71 × 60A = 26A
TOTAL = 131A

NEC 630-31 (a-1, 2) (b) Conductor rating based on Table 630-31 (a-2) value obtained for largest welder plus 60% of that for the others.

NEC 630-33 Rating of disconnect not less than that of supply conductors.

GROUP OF RESISTANCE WELDERS

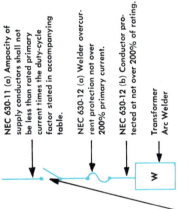

NEC 630-11 (a) Ampacity of supply conductors shall not be less than rated primary current times the duty-cycle factor stated in accompanying table.

NEC 630-12 (a) Welder overcurrent protection not over 200% primary current.

NEC 630-12 (b) Conductor protected at not over 200% of rating.

Transformer Arc Welder

NEC 630-13 Disconnect switch or circuit breaker shall have ampere rating sufficient to accommodate required overcurrent protection, as per 630-12.

TRANSFORMER ARC WELDER

NEC 630-33 Rating of disconnect not less than that of supply conductors as required by 630-31.

NEC 630-32 (a) Maximum rating of overcurrent device 300% rated primary current.

NEC 630-31 (a-1) Minimum rating of supply conductors, varying duty 70% rated primary current automatic; 50% manual.

NEC 630-31 (a-2) Rating of supply conductors for specific cooperation as per duty-cycle table: Multiplier 0.71 for 50% duty, 0.45 for 20% duty.

RESISTANCE WELDER

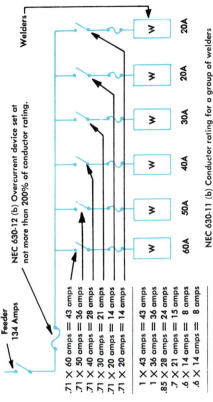

Resistor or Reactor

Not Exposed to Physical Injury

Where 1-foot clearance, may be mounted on slab or panel.

Non-Absorptive Noncombustible Material

R

NEC 470-2 Resistors and reactors are not to be installed where exposed to physical injury.

NEC 470-3 A space of 12 inches is required between resistors and reactors and any combustible material. If space less than 12 inches, a thermal barrier is required. (See above for alternate method where 12 inches cannot be obtained.)

NEC 470-4 Insulated conductors shall be rated for 90°C when used for connections between resistance elements and controllers, except, for motor starting service, other insulations permitted.

RESISTORS AND REACTORS (600 VOLTS AND UNDER)

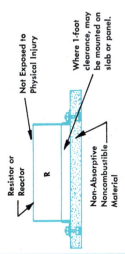

Disconnect required. Must be large enough to accommodate required overcurrent protection.

Not over 200% of primary amps of welder and not over 200% of conductor amps.

NEC 630-21 (a) Supply conductor rating for M-G arc welder having 50% duty cycle is .75 of rated primary amps as per welder nameplate.

NEC 630-22 (a) Overcurrent protection of welder not over 200% of rated primary current. (Protection may also protect supply conductors.)

NEC 630-22 (b) Welder supply conductors to be protected at not over 200% of conductor rating.

NEC 630-23 Each welder shall have disconnect. Rating not less than required to accommodate overcurrent protection required by 630-22.

MOTOR-GENERATOR WELDER

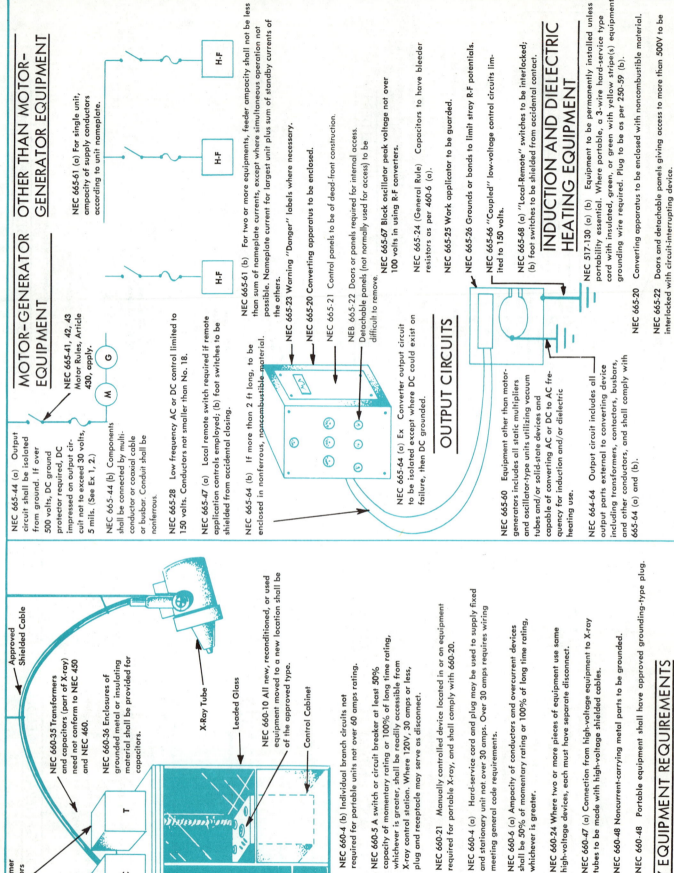

MOTOR-GENERATOR EQUIPMENT

OTHER THAN MOTOR-GENERATOR EQUIPMENT

H-F

H-F

H-F

NEC 665-61 (a) For single unit, ampacity of supply conductors according to unit nameplate.

NEC 665-41, 42, 43 Motor Rules, Article 430, apply.

G

M

NEC 665-44 (a) Output circuit shall be isolated from ground. If over 500 volts, DC ground protector required, DC impressed on output circuit not to exceed 30 volts, 5 mils. (See Ex 1, 2.)

NEC 665-44 (b) Components shall be connected by multiconductor or coaxial cable or busbar. Conduit shall be nonferrous.

NEC 665-28 Low frequency AC or DC control limited to 150 volts. Conductors not smaller than No. 18.

NEC 665-47 (a) Local remote switch required if remote application controls employed; (b) foot switches to be shielded from accidental closing.

NEC 665-64 (b) If more than 2 ft long, to be enclosed in nonferrous, noncombustible material.

NEC 665-61 (b) For two or more equipments, feeder ampacity shall not be less than sum of nameplate currents, except where simultaneous operation not possible. Nameplate current for largest unit plus sum of standby currents of the others.

NEC 665-23 Warning "Danger" labels where necessary.

NEC 665-20 Converting apparatus to be enclosed.

NEC 665-21 Control panels to be of dead-front construction.

NEB 665-22 Doors or panels required for internal access. Detachable panels (not normally used for access) to be difficult to remove.

NEC 665-67 Block oscillator peak voltage not over 100 volts in using R-F converters.

NEC 665-24 (General Rule)   Capacitors to have bleeder resistors as per 460-6 (a).

NEC 665-25 Work applicator to be guarded.

NEC 665-26 Grounds or bonds to limit stray R-F potentials.

NEC 665-66 "Coupled" low-voltage control circuits limited to 150 volts.

NEC 665-68 (a) "Local-Remote" switches to be interlocked; (b) foot switches to be shielded from accidental contact.

INDUCTION AND DIELECTRIC HEATING EQUIPMENT

NEC 517-130 (b)   Equipment to be permanently installed unless portability essential. Where portable, a 3-wire hard-service type cord with insulated, green, or green with yellow stripe(s) equipment grounding wire required. Plug to be as per 250-59 (b).

NEC 665-20   Converting apparatus to be enclosed with noncombustible material.

NEC 665-22   Doors and detachable panels giving access to more than 500V to be interlocked with circuit-interrupting device.

OUTPUT CIRCUITS

NEC 665-64 (a) Ex   Converter output circuit to be isolated except where DC could exist on failure, then DC grounded.

NEC 665-50   Equipment other than motor-generators includes all static multipliers and oscillator-type units utilizing vacuum tubes and/or solid-state devices and capable of converting AC or DC to AC frequency for induction and/or dielectric heating use.

NEC 664-64   Output circuit includes all output parts external to converting device including transformers, contactors, busbars, and other conductors, and shall comply with 665-64 (a) and (b).

Transformer
Capacitors

Approved
Shielded Cable

NEC 660-35 Transformers and capacitors (part of X-ray) need not conform to NEC 450 and NEC 460.

NEC 660-36 Enclosures of grounded metal or insulating material shall be provided for capacitors.

X-Ray Tube

Leaded Glass

Control Cabinet

T

C

C

NEC 660-10 All new, reconditioned, or used equipment moved to a new location shall be of the approved type.

NEC 660-4 (b) Individual branch circuits not required for portable units not over 60 amps rating.

NEC 660-5 A switch or circuit breaker at least 50% capacity of momentary rating or 100% of long time rating, whichever is greater, shall be readily accessible from X-ray control station. Where 120V, 30 amps or less, plug and receptacle may serve as disconnect.

NEC 660-21   Manually controlled device located in or on equipment required for portable X-ray, and shall comply with 660-20.

NEC 660-4 (a)   Hard-service cord and plug may be used to supply fixed and stationary unit not over 30 amps. Over 30 amps requires wiring meeting general code requirements.

NEC 660-6 (a)   Ampacity of conductors and overcurrent devices shall be 50% of momentary rating or 100% of long time rating, whichever is greater.

NEC 660-24   Where two or more pieces of equipment use same high-voltage devices, each must have separate disconnect.

NEC 660-47 (a)   Connection from high-voltage equipment to X-ray tubes to be made with high-voltage shielded cables.

NEC 660-48   Noncurrent-carrying metal parts to be grounded.

NEC 660-48   Portable equipment shall have approved grounding-type plug.

X-RAY EQUIPMENT REQUIREMENTS

NEC 551-3 When requirements for travel trailers (Article 551) differ from other articles of NEC, the requirements of 551 shall prevail.

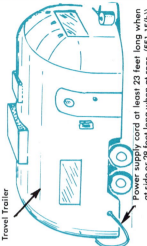

Travel Trailer

Power supply cord at least 23 feet long when at side or 28 feet long when at rear. (551-15(b))

NEC 551-2 Definition of RECREATIONAL VEHICLE: A vehicular type unit which is self-propelled or is mounted on or pulled by another vehicle and is primarily designed as temporary living quarters for camping, travel, or recreational use. Normally identified as Travel Trailer, Camping Trailer, Truck Trailer, Truck Camper, and Motor Home.

NEC 551-2 Travel trailer is: mounted on wheels, has a living area less than 220 square feet (excluding wardrobes, closets, cabinets, kitchen units, fixtures, etc.), is of such size and weight so that a special highway use permit is not required, and is designed as temporary living quarters while camping or traveling.

NEC 551-15(e) Power-supply assembly located with 15 feet of rear, left (driver or road) side, or at rear, left within 18 inches of outside wall.

NEC 551-9(c) Each 120-volt, single phase, 15- or 20-amp receptacle outlet shall have GFCI where installed in following locations: (1) adjacent to bathroom lavatory (receptacle to be at least 24 inches from floor); (2) adjacent to any lavatory; (3) area used for toilet, toilet and/or shower, or toilet and tub-shower enclosure; (4) on exterior of vehicle except as per 551-8(e).

NEC 551-5(a) Low-voltage circuits installed by manufacturer subject to NEC except if used for braking. Lights subject to Federal and State regulations but not lower than code.
NEC 551-5(b) Low-voltage wiring shall be: (1) copper except metal chassis or frame permitted as return path if connections are accessible and are copper and mechanically secure; (2) Type HDT, SGT, SGR, SLX or insulated as per Table 310-13 or equal. Conductors No. 6 thru 18 shall be listed; (3) single wires to be stranded; (4) wires to be marked on surface every 4 ft. and be listed.
NEC 551-21 All receptacle outlets shall be grounding type and installed as per 210-7 and 210-21.
NEC 551-10 RV with 120 V system for lights, receptacles, and appliances: (a) 15 Amp supply if one 15 Amp circuit. (b) 20 Amp supply if one 20 Amp circuit, (c) 30 Amp supply if two or more 15 Amp or 20 Amp circuits. [See 210-23(a) for permitted loads; (See 551-13(c) for disconnect and O.C. protection.) 551-10(d) covers 40 or 50 amp supply.
NEC 551-23(a) Grounding conductor in power supply shall be connected to grounding bus or approved grounding means in distribution panelboard.
NEC 551-24(a) All exposed metal parts, enclosures, frames, light fixture canopies, etc., shall be bonded to grounding terminals or enclosure of distribution panelboard.
NEC 551-8(a) Equipment and material indicated for connection to system rated 120 volts, 2-wire with ground, or 120-240 volts, 3-wire with ground, shall be approved and installed as per NEC 551, Part A.

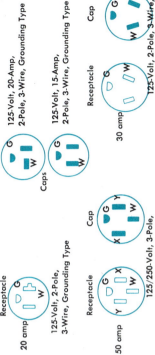

125-Volt, 20-Amp, 2-Pole, 3-Wire, Grounding Type

125-Volt, 15-Amp, 2-Pole, 3-Wire, Grounding Type

30 amp — 125-Volt, 2-Pole, 3-Wire, Grounding Type

125/250-Volt, 3-Pole, 4-Wire, Grounding Type

20 amp — 125-Volt, 2-Pole, 3-Wire, Grounding Type

50 amp — 125/250-Volt, 3-Pole, 4-Wire, Grounding Type

The receptacle and attachment-plug-cap configurations shown here are used for recreational vehicle supply cords and recreational vehicle parks and are required by NEC 551-15(c). (Details can be found in ANSI C73.13-1972.)

MOBILE HOMES

Service Drops

Mobile Home Service Equipment

Ground Rod

Power-supply cord—not less than 20 ft or more than 36½ ft long. (NEC 550-5(d))

NEC 550-2 Definition of MOBILE HOME: A factory-assembled structure provided with service and readily moveable as a unit on its own running gear and designed to be used as a dwelling unit without permanent foundation.

NEC 550-8(b)(e) At least one bathroom and one outdoor receptacle required. Outdoor receptacle permitted in compartment of home if accessible from outside. All 15- and 20-amp, 120-volt, outdoor and bathroom receptacles to be on GFCI.

Attachment-plug-cap configurations used for mobile home supply cords and mobile home parks. (See Figure 550-5(f))

NEC 550-5(i) Where load exceeds 50 amps, or permanent feeder used, then supply shall be (1) mast weatherhead installed, as per 230, and contain 4 continuous, insulated, color-coded feeder conductors, or (2) raceway from home disconnect to under home with provisions to attach suitable junction box or fitting on underside of house.

NEC 550-4(c) Rules of Article 550 apply when 120/240-volt, 3 wire used for service. (Also see 550-4(a))

NEC 550-6(a) Disconnecting means required inside home; neutral strip to be insulated.

NEC 550-7(a) Outside dimensions to be used in calculating minimum lighting load (3 VA per square foot).

NEC 550-10 (a thru i) Nonmetallic-sheathed cable to be protected if exposed less than 15 inches from floor; if rigid conduit, double-locknuts required. Slack to be left in range cable. Under-chassis wiring is to be in rigid conduit or IMC. EMT or PVC if in protected area.

NEC 550-13 See 550-13(a) for calculating lighting and small appliance load. See 550-13(b) for total power supply load.

NEC 550-13(c) This section permits use of 220-30 and Table 220-30 for calculating lighting and load in mobile homes.

NEC 550-5 (a through i) Service equipment shall be located outside but adjacent to home. Power supply to home shall be one approved 50-amp mobile home power-supply cord, or a permanently installed circuit, except if home has oil or gas-fired heating and cooking, then 40-amp cord acceptable. Cord to be secured to distribution panelboard with suitable cord clamp and have 4 wires, one to be the grounding conductor and to be green or green with yellow stripe(s). Cord 21 to 36½ feet and have a molded attachment plug cap, 125/250-volt, 50-amp, 3-pole, 4-wire grounding type, as per NEC Figure 550-5(f). Cord shall enter home in exterior wall, floor, or roof and have suitable protection where entering.

NEC 550-8 (a, c, d) Plug receptacles shall be of grounding type. Receptacles shall be installed at countertops, refrigerator, gas range, vanities, and underneath wall-mounted cabinets.

NEC 550-11(a, b, c) No equipment shall be grounded to the neutral wire inside the mobile home. All exposed metal parts shall be bonded to grounding terminal or panel enclosure. Bond required (No. 8 copper or equal not exposed to physical damage) between panel and accessible terminal on chassis. Panel ground connected from service ground. Metallic (gas, water, etc.) pipes may be grounded by connecting to terminal on chassis.

NEC 550-12 Wiring must be tested for ground after installation.

The receptacle and attachment-plug-cap configurations shown here are used for mobile home supply cords and mobile home parks and are required by 550-5(c). (Details can be found in ANSI C73.17-1972.)

125/250-Volt, 50-Amp, 3-Pole, 4-Wire, Grounding Type

NEC 680-6 (b) (2) Lighting outlets and fixtures, installed between 5 and 10 feet horizontally from walls of pool, to be protected by GFCI unless at least 5 feet above maximum water level and rigidly mounted.

NEC 680-6 (b) (3) Where cord-connected lighting fixtures are installed within 16 feet of nearest water surface, they shall comply with 680-7.

NEC 680, Sections 4, 11, 22 (a-5), 25 (c), and 26 give definitions and list specific requirements for electrically operated pool covers.

NEC 680-20 (a-1) Underwater fixture, during normal use (not relamping) shall be free from shock hazard by design. If operating at over 15 volts, GFCI required to eliminate shock hazard during relamping. If supplied by transformer, as per 680-5 (a), and circuit at not over 15 volts, GFCI not required if fixture is of proper design and properly installed. Only approved underwater fixtures and GFCI permitted.

NEC 680-2 All equipment shall be approved for the purpose installed to comply with this Article.

NEC 680-6 (b-1) Lighting outlets and fixtures not to be installed over pool or over area 5 feet horizontally from inside of pool, unless at least 12 feet above maximum water level. Except: Existing located less than 5 feet horizontally from walls of pool and at least 5 feet above maximum water level and rigidly mounted.

NEC 680-20 (a)(3) Pool lights in walls so top of lens at least 18 inches below normal water level. Ex lights approved for purpose need be only 4 inches below normal water level.

NEC 680-20 (a) (4) Where safe operation of fixture is dependent upon submersion, then inherent fixture over-heating protection required.

NEC 680-21 (a-2, 3) Junction boxes to underwater pool-light forming shell shall be of copper, brass, suitable plastic, or other approved material, and provide continuity by means of copper, brass, or other approved corrosion-resist-ant metal that is integral with box.

Lighting Panel

NEC 680-21 (d) Junction boxes shall have one more grounding terminal than conduit entries.

Deck Junction Box

NEC 680-5 (a) Transformers shall be of two-winding type, with metallic barrier between primary and secondary windings.

Lighting Fixtures

NEC 680-20 (b-3) Noncurrent parts of wet-niche fixtures shall be grounded to forming shell by a positive locking device and requires tool to remove fixture from forming she

NEC 680-20(b-1, 2) Wet-niche underwater fixtures shall have approved forming shell with threaded conduit entries. Rigid metal conduit or IMC of brass or similar metal or PVC shall extend from forming shell to suitable junction box or enclosure located as per 680-21. Where PVC used, a No. 8 insulated, copper wire required in PVC with provisions to terminate in junction box or transformer or GFCI enclosure. Termination of No. 8 in forming shell to be protected by covering or encapsulating with listed potting compound. Where flexible cord is used, the end of cord jacket and wire terminals in forming shell to be similarly protected.

NEC 680-22(a) Bonding. The following parts shall be bonded: All metal parts of electric pool water handling equipment and other metallic elements within or attached to pool structure, including reinforcing rods, coping stones, deck, piping systems, ladders, rails, diving board platforms, drains, skimmers, refill pipes, forming shells and all other metal piping or fixed metal parts within 5 feet of inside walls of pool (unless separated from pool by permanent barrier). Except: (1) tie wires acceptable as bond re-rod; (2) common bonding grid may be re-rod; (3) isolated metal parts, no more than 4 inches in any dimension and do not penetrate into pool structure more than 1 inch, need not be bonded.

NEC 680-22 (b) The parts required to be bonded by 680-22 (a) shall be connected to a common bonding gr with not less than No. 8 solid copper wire, insulated, covered or bare, and connected by pressure connectors or clamps of brass, copper, or copper alloy. The common bonding grid is permitted to be: (1) reinforcing steel with usual tie wires; (2) wall of bolted or welded metal pool; (3) No. 8 or larger solid copper wire, insulated, covered or bare.

NEC 680-22 (c) Pool heaters over 50 amps to be grounded and bonded, as per specific instructions, as to which parts to be grounded and bonded.

NEC 680-6 (a) (1) Ex (2) (3) Receptacles on property required to be at least 10 feet from inside walls of pool. Where pool installed for existing dwelling, at lease one receptacle required between 10 and 20 feet of pool. All receptacles between 10 and 20 feet of pool to be protected by GFCI. A single, locking and grounding type receptacle for a permanent water-pump motor as per 680-7 permitted between 5 and 10 ft. from pool. All 120-volt receptacles within 20 feet of pool to have GFCI. (See 210-8(a)(3).)

NEC 680-25(f) Other electrical equipment to be grounded as per Art. 250 and wiring as per Chapter 3.

NEC 680-20 (b-1) Metal parts of fixtures or forming shell shall be of brass or other corrosion-resistant metal.

NEC 680-25 (b-3) Flexible cords to wet-niche lighting fixtures shall contain a grounding conductor not smaller than No. 16.

NEC 680-20 (a-2) No underwater lighting fixture may operate at a voltage greater than 150V between conductors.

NEC 680-8 Proper clearances shall be maintained between open overhead wiring and swim pools, diving structures, observation stands, towers, and platforms. (See EX. 1, 2)

NEC 680-21 (b-4) Enclosures for electrical equipment that extends directly to underwater pool-light forming shell shall be not less than 4 feet from pool, unless separated by barrier and not less than 8 inches above ground level, pool deck, or maximum pool water level, whichever is greatest.

NEC 680-25 (b) (1) Conduit contains insulated copper equipment grounding conductor, sized as per Table 250-95; minimum of No. 12.

NEC 680-25 (d) Where equipment ground wire connected to sub-panel, then equipment ground to continue to service ground.

Rigid Metal Conduit, IMC or PVC
Junction Box
To Grounding Terminal of Service Equipment
Supply Cord with Equipment Ground Wire
Underwater Fixture
Diving Board
No. 8 Solid Copper Bond
Metal Fence, Etc.
Forming Shell
No. 8 Solid Copper Bond
No. 8 Solid Copper Bond
5'
Drain
Skimmer
Equip. Ground Cond.
To Service
Reinforcing Rods
Sub-Panel
Re-Rods
Pump
Filter
Steel Tie Wires
Heater
No. 8 Solid Copper Bond
No. 8 Solid Copper Bond
POOL
Fill Spout

NEC 680-22 (b) Common Bonding Grid may consist of: (1) Structural reinforcing steel rods that are bonded together by usual tie wires; (2) Wall of bolted or welded metal pool; (3) Solid copper wire, insulated, covered or bare, and not smaller than No. 8.

NEC 680-24 Electrical equipment to be grounded. Wet-and-dry-nitche fixtures; equipment within 5 feet of pool; all associated recirculating equipment; junctions boxes; transformer enclosures; GFCI; panels not part of service equipment.

BONDING & GROUNDING PERMANENT SWIMMING POOLS

SWIMMING POOLS--ARTICLE 680

CLEARANCE	VOLTS		
Service Drop or Overhead Wiring	0-750 V	750-15 kV	15-50 kV
A = To Water	18'	25'	27'
B = To Diving Platform	14'	16'	18'
C = Horizontal from Inside Pool Walls	Limit Extends from Outer Edge of Above But Not Less Than 10 Feet		

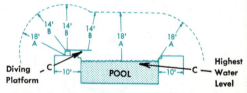

Overhead conductor clearances at 0-750 volts to ground.

NEC 680-8 Following parts of swim or wading pools not permitted under existing service drops or other overhead wiring, and new overhead wiring not permitted to be installed above the following: (1) swim or wading pools, including area extending 10 feet horizontally from walls of pool; (2) diving structure; (3) towers, platforms, or observation stands. Ex 1, 2 Utility power lines to have clearances as per above table. Utility communications and CATV to have 10 ft minimum clearance from above Table areas.

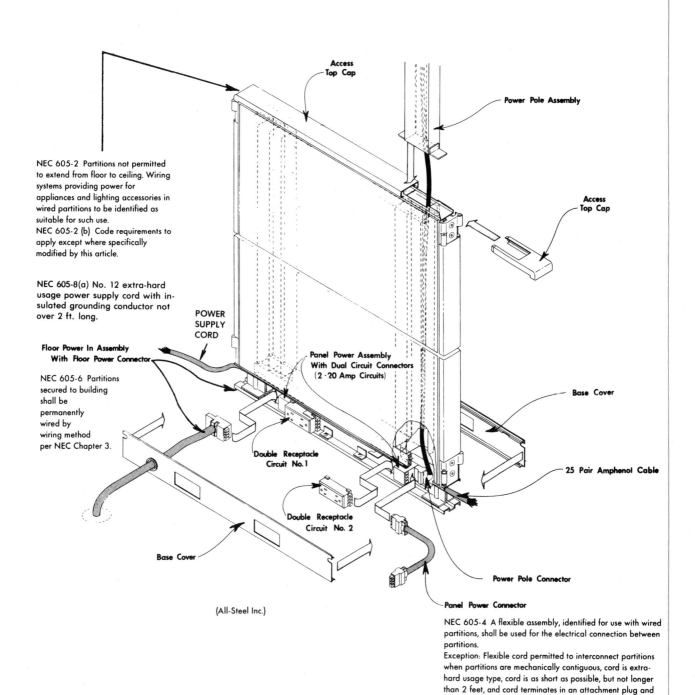

Access Top Cap

Power Pole Assembly

Access Top Cap

NEC 605-2 Partitions not permitted to extend from floor to ceiling. Wiring systems providing power for appliances and lighting accessories in wired partitions to be identified as suitable for such use.
NEC 605-2 (b) Code requirements to apply except where specifically modified by this article.

NEC 605-8(a) No. 12 extra-hard usage power supply cord with insulated grounding conductor not over 2 ft. long.

POWER SUPPLY CORD

Floor Power In Assembly With Floor Power Connector

NEC 605-6 Partitions secured to building shall be permanently wired by wiring method per NEC Chapter 3.

Panel Power Assembly With Dual Circuit Connectors (2 - 20 Amp Circuits)

Base Cover

Double Receptacle Circuit No.1

25 Pair Amphenol Cable

Double Receptacle Circuit No. 2

Base Cover

Power Pole Connector

Panel Power Connector

(All-Steel Inc.)

NEC 605-4 A flexible assembly, identified for use with wired partitions, shall be used for the electrical connection between partitions.
Exception: Flexible cord permitted to interconnect partitions when partitions are mechanically contiguous, cord is extra-hard usage type, cord is as short as possible, but not longer than 2 feet, and cord terminates in an attachment plug and cord-connector with strain relief.

NEC 605-1 This article deals with lighting accessories, electrical equipment, and wiring systems that are contained within or installed on relocatable wired partitions.

NEC 605-3 Wireways in partitions containing conductors and connections shall be identified as suitable for the condition of use. (FPN prohibits use of flexible cords).

NEC 605-5 Lighting equipment shall be identified for use with wired partitions and shall: (a) have adequate support; (b) where cord-and-plug-connected shall not exceed 9 ft., not smaller than No. 18, contain an equipment grounding conductor and shall be of the hard usage type; (c) lighting accessories shall not contain receptacle outlets.

NEC 605-8 Free standing individual partitions that are not over 30 feet long and are mechanically contiguous are permitted to be cord and plug connected providing: (a) cord is at least No. 12 AWG with insulated grounding conductor of extra-hard usage type and not over 2 feet long, (b) permanent building wiring connection permitted by cord if receptacle is on separate circuit and located not more than 12 inches from partition (c) not more than 13, 15 Amps, 125 Volt receptacles permitted in individual or group of interconnected partitions, (d) multiwire circuits not permitted.

OFFICE FURNISHINGS (LIGHTING ACCESSORIES AND WIRED PARTITIONS) — ARTICLE 605

NEC 770-1 Article 770 covers the installation of optical fiber cables along with electrical conductors. It does not cover construction of cables or installation in situations other than those covered in this article.

NEC 770-5 Optical Fibers and Electrical Conductors.

NEC 770-5(a) Optical fibers permitted in hybrid cables with electric light, power, or Class I circuits of 600 volts or less only where functions of all conductors are associated. Non-conductive cable conductors permitted in same raceway with light, power, or Class I circuits of 600 volts or less. Conductive and hybrid cables not permitted in same raceway with light, power, or Class I circuits. (See Exceptions)

NEC 300-17 Shall apply to conductors in raceways.

NEC 770-4 (b) Optical fibers permitted in same cable and Conductive and Nonconductive cables permitted in same raceway or enclosure with any of the following: (1) Class 2 & 3 circuits per Article 725. (2) Fire protection systems per Article 760. (3) Communication circuits as per Art. 800. (4) Community antenna TV and radio distribution systems as per Art. 820.

NEC 770-5(c) Grounding of conductive members of optical fiber cables as per Art. 250.

NEC 770-6(a) Optical fiber cables in buildings to be listed as resistant to the spread of fire.

NEC 770-6(b) Vertical runs of cable in a shaft to be listed as having fire-resistance characteristics to prevent carrying of fire from floor to floor. (See 770-6(c))

NEC 770-6(b) EX Cables permitted in fireproof shaft having fire stops at each floor or cables enclosed in noncombustible tubing.

NEC 770-6(c) Cables installed in ducts and plenums and other spaces used for environmental air have the same requirements as NEC 300-22 for other wiring methods. (See Exceptions and (FPN))

NEC 770-3 Conductive members transmit light for signaling, communication, and control through an optical fiber.

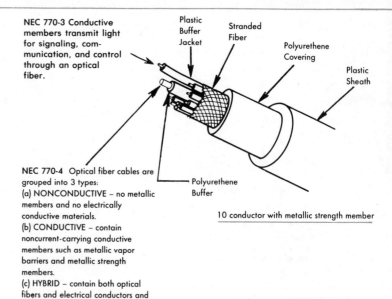

Plastic Buffer Jacket

Stranded Fiber

Polyurethene Covering

Plastic Sheath

Polyurethene Buffer

10 conductor with metallic strength member

NEC 770-4 Optical fiber cables are grouped into 3 types:
(a) NONCONDUCTIVE – no metallic members and no electrically conductive materials.
(b) CONDUCTIVE – contain noncurrent-carrying conductive members such as metallic vapor barriers and metallic strength members.
(c) HYBRID – contain both optical fibers and electrical conductors and are classified as electrical cables per the type of electrical conductors.

OPTICAL FIBER CABLES — ARTICLE 770

NEC 690 New Part H Covers Storage Batteries. 690-71 deals with installation, 690-72 deals with state of charge and 690-73 deals with grounding.

Module

NEC 690-2 A module is the smallest complete, environmentally protected assembly of solar cells, exclusive of tracking, designed to generate DC power under sunlight.

Panel

NEC 690-2 A panel is a collection of modules, secured together, wired, and designed to provide a field-installable unit.

Array

NEC 690-2 An array is an assembly of modules or panels mounted on a support structure.

NEC 690-18 A means is required to disable an array or portions of an array to protect persons against shock while installing, servicing, or replacing of array during daylight hours.

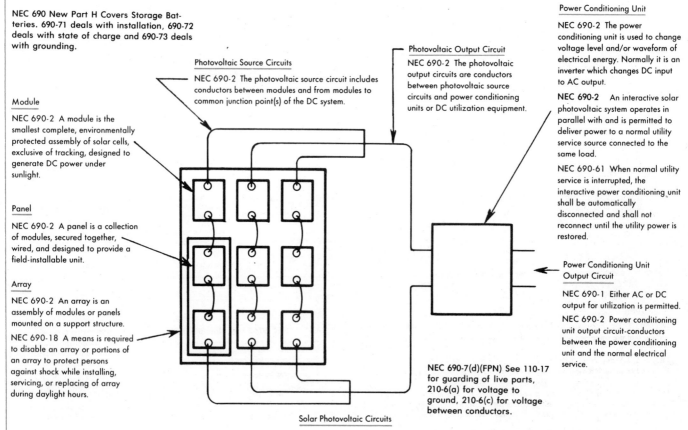

Photovoltaic Source Circuits

NEC 690-2 The photovoltaic source circuit includes conductors between modules and from modules to common junction point(s) of the DC system.

Photovoltaic Output Circuit

NEC 690-2 The photovoltaic output circuits are conductors between photovoltaic source circuits and power conditioning units or DC utilization equipment.

Power Conditioning Unit

NEC 690-2 The power conditioning unit is used to change voltage level and/or waveform of electrical energy. Normally it is an inverter which changes DC input to AC output.

NEC 690-2 An interactive solar photovoltaic system operates in parallel with and is permitted to deliver power to a normal utility service source connected to the same load.

NEC 690-61 When normal utility service is interrupted, the interactive power conditioning unit shall be automatically disconnected and shall not reconnect until the utility power is restored.

Power Conditioning Unit Output Circuit

NEC 690-1 Either AC or DC output for utilization is permitted.

NEC 690-2 Power conditioning unit output circuit-conductors between the power conditioning unit and the normal electrical service.

NEC 690-7(d)(FPN) See 110-17 for guarding of live parts, 210-6(a) for voltage to ground, 210-6(c) for voltage between conductors.

Solar Photovoltaic Circuits

NEC 690-2 The photovoltaic power source is one or more arrays which generate DC power at system voltage and current.

NEC 690-2 The solar photovoltaic system is used to convert solar energy into electrical energy suitable for connection to a utilization load.

NEC 690-7 (d) In one-family and two-family dwellings, live parts in source and output circuits over 150 volts to ground shall not be accessible, while energized, to other than qualified persons.

SOLAR PHOTOVOLTAIC SYSTEMS — ARTICLE 690

TRADE COMPETENCY TEST NO. 4A

Student
Number_____
Instructor's
Name_____

INSTRUCTIONS: The following statements are either True or False. Draw a circle around the **T** if the statement is True, or around the **F** if it is False. The questions are to be answered, as nearly as possible, in accordance with the specific provisions of the NEC.

• True or False •

1. The secondary circuits of 460-volt wound-rotor induction motors do not require overload protection.　　　T　　　F

2. Capacitors used with motors increase motor current.　　　T　　　F

3. A ¾-hp, 115-volt motor on an automatic air-compressor does not need overload protection.　　　T　　　F

4. When the Code speaks of a "transformer," a polyphase bank of single-phase transformers is included in the term.　　　T　　　F

5. Transformers sometimes have overcurrent devices in both primary and secondary circuits.　　　T　　　F

6. The primary windings of potential transformers installed indoors shall normally be protected with fuses.　　　T　　　F

7. The rated-load current of a sealed (hermetic-type) motor-compressor shall be marked on either the equipment or the motor-compressor nameplate.　　　T　　　F

8. An overload in electrical apparatus includes short circuits or ground faults.　　　T　　　F

9. A nonautomatically started, not permanently installed, 1-hp motor that is within sight of the controller location does not normally require running overcurrent protection.　　　T　　　F

10. The overcurrent device for a 30-amp transformer arc welder shall not exceed 30 amps.　　　T　　　F

11. A mogul-base lampholder operating on a 277-volts-to-ground circuit may have a pull-chain fixture switch incorporated in the fixture, provided it is mounted at least 7 ft. from floor.　　　T　　　F

12. The disconnect means for 460-volt capacitors need not open all ungrounded conductors simultaneously.　　　T　　　F

Continued on Next Page

13. The maximum total volt drop for feeders and branch circuits should not exceed 3 percent overall.   T   F

14. An adjustable overload device protecting a 50-degree centigrade motor having a full-load current rating of 20 amps shall be selected to trip at not more than 23 amps.   T   F

15. A 1/40-hp split-phase, automatically-started motor does not require running overcurrent protection.   T   F

16. An inherently short-time intermittent duty motor usually does not require running overload protection.   T   F

17. Motor overload protection shall never be shunted out during the starting period.   T   F

18. Nonflammable fluid-insulated transformers installed indoors and rated at 15,000 volts must be installed in vaults.   T   F

19. Replacement parts may be stored in transformer vaults if at least three feet from operating equipment.   T   F

20. Each transformer shall have on its nameplate the kilowatt rating of the transformer.   T   F

21. A motor branch circuit short-circuit protective device consisting of a nontime-delay fuse shall never have a rating in excess of 400 percent of the full-load motor current.   T   F

22. Mineral-insulated metal-sheathed cable is not approved for continuously moist locations.   T   F

23. Infrared heating units may be supplied from a 50-ampere branch circuit in an industrial occupancy.   T   F

24. In some cases, overload protection for a 50-amp continuous duty motor may be set at 70 amps.   T   F

25. The maximum size of liquid-tight flexible metal conduit is 2½ inches.   T   F

TEAR OUT HERE

TRADE COMPETENCY TEST NO. 4B

Student
Number _____

Instructor's
Name _____

INSTRUCTIONS: In the space provided at the right of each of the following questions, write **a, b, c,** or **d** to indicate which of the alternatives given makes it most nearly correct. The questions are to be answered, as nearly as possible, in accordance with the specific provisions of the NEC.

• Multiple-Choice •

1. Branch-circuit conductors supplying a 50-amp, 3-phase, 230-volt, 40° C induction motor shall be rated at not less than:

   (a) 50 amps, (b) 62.5 amps, (c) 100 amps, (d) 57.5 amps

   1. _____

2. The maximum size of the overload relay used to protect the motor in Question 1 is:

   (a) 50 amps, (b) 70 amps, (c) 100 amps, (d) 57.5 amps

   2. _____

3. The maximum setting of the primary overcurrent device for a 460/230-volt, askarel-insulated, single-phase transformer, without secondary protection, whose primary full-load current is 40 amps shall be:

   (a) 40 amps, (b) 50 amps, (c) 80 amps, (d) 100 amps

   3. _____

4. The rated secondary current of the transformer in Question 3 is:

   (a) 50 amps, (b) 100 amps, (c) 120 amps, (d) 80 amps

   4. _____

5. A 3-hp, 3-phase, 230-volt squirrel-cage motor with no code letter is started at full voltage. The size of a Type TW branch-circuit conductor shall be not less than:

   (a) No. 14, (b) No. 12, (c) No. 10, (d) No. 8

   5. _____

6. High fire point liquid-insulated transformers installed indoors shall be installed in vaults if rated over:

   (a) 5000 volts, (b) 15,000 volts, (c) 25,000 volts, (d) 35,000 volts.

   6. _____

7. The primary current of a transformer arc welder for continuous use is 30 amps. Rated ampacity of supply conductors shall be at least:

   (a) 30 amps, (b) 37.5 amps, (c) 90 amps, (d) 60 amps

   7. _____

Continued on Next Page

TEAR OUT HERE

8. The maximum overcurrent protection of a No. 6 THHN feeder which supplies two welders as in Question 7 shall be:

(a) 67.5 amps, (b) 120 amps, (c) 75 amps, (d) 150 amps

8. _____

9. A 30-hp, 3-phase, 230-volt squirrel-cage motor with Code letter B is started with an autotransformer compensator. The initial rating of a dual element fuse for branch circuit protection shall not exceed:

(a) 100 amps, (b) 125 amps, (c) 150 amps, (d) 175 amps

9. _____

10. To satisfy Code requirements, conductors supplying a capacitor with a nameplate current of 50 amps shall be rated not less than:

(a) 57.5 amps, (b) 62.5 amps, (c) 72.5 amps, (d) 67.5 amps

10. _____

11. A manually-operated resistance welder draws 50 amps according to the nameplate. Circuit conductors are permitted to be chosen on the basis of:

(a) 25 amps, (b) 50 amps, (c) 75 amps, (d) 100 amps

11. _____

12. Overcurrent protection for this welder shall not exceed a maximum value of:

(a) 250 amps, (b) 150 amps, (c) 200 amps, (d) 175 amps

12. _____

13. The minimum size of conductors permitted in a cablebus is:
(a) No. 2, (b) No. 1, (c) No. 1/0, (d) No. 2/0

13. _____

14. The NEC provides that a No. 14 motor-control circuit conductor requires separate overcurrent protection, if the control circuit extends beyond the control equipment enclosure, and the motor branch-circuit protective device exceeds:

(a) 75 amps, (b) 45 amps, (c) 60 amps, (d) 30 amps

14. _____

15. The disconnecting means for a stationary AC motor need not be horsepower rated if the motor is larger than:

(a) 100 hp, (b) 50 hp, (c) 25 hp, (d) 75 hp

15. _____

16. The smallest stationary motor rated 300 volts or less requiring a horsepower rated switch is:

(a) ½ hp, (b) 1 hp, (c) 2 hp, (d) 3 hp

16. _____

National Electrical Code

TRADE COMPETENCY TEST NO. 4C

Student
Number _____

Instructor's
Name _____

INSTRUCTIONS: Complete each of the following statements, by writing the missing word or words in the space provided at the right of each question, to make a true statement. The questions are to be answered, as nearly as possible, in accordance with the specific provisions of the NEC.

1. When the term "in sight" is used in Article 430 it also includes a distance of not more than _____ feet.

    1. _____

2. On an ungrounded circuit _____ thermal overcurrent units are needed to protect a three-phase induction motor from overload.

    2. _____

3. Separate motor-running overcurrent protection shall be based on the full-load current, as listed on the motor _____.

    3. _____

4. A suitable plug and hard-service cord is permitted to supply a fixed or stationary nonmedical X-ray unit, if the branch circuit is rated at no more than _____ amps.

    4. _____

5. Wired office partitions shall not extend from floor to _____.

    5. _____

6. The ampacity of conductors supplying a 20-amp oscillator type induction heating unit shall not be less than _____ amps.

    6. _____

7. A general-use snap switch suitable only for use on AC may be used to disconnect a 230-volt, 2-hp motor whose full-load current rating does not exceed _____ percent of the ampere rating of the switch.

    7. _____

8. Fuses employed as motor overload protective devices must be inserted in each _____ conductor.

    8. _____

9. The minimum length of a pull box for a straight through pull of three No. 4/0 conductors in a 2½-inch conduit is _____ inches.

    9. _____

10. The conductors of mineral-insulated metal-sheathed cable (Type MI) are made of _____ _____.

    10. _____

11. The voltage considered in a solar photovoltaic power source is the rated _____ voltage.

    11. _____

Blueprint Reading

12. Unless otherwise designed and marked, busways shall be supported at least every _____ feet.

12. _____

13. The controller may be an attachment plug and receptacle for portable motors rated at _____ hp or less.

13. _____

14. For stationary motors 2 hp or less, 300 volts or less, a general-use switch may be used as a controller if its rating is at least _____ times the full-load motor current.

14. _____

15. The motor disconnecting means shall disconnect both the motor and the _____ .

15. _____

16. Flexible metal conduit used in a manufactured wiring system shall contain a No. _____ AWG copper grounding conductor.

16. _____

17. Except for polyphase wound-rotor units, AC motors _____ hp and larger must have code letters marked on nameplates.

17. _____

18. Optical fiber cables are used to transmit _____ for control, signaling, and communications through an optical fiber.

18. _____

19. The conductors in a pull box having any dimension over 6 feet shall be _____ in an approved manner.

19. _____

20. The primary overcurrent device, when properly selected, may protect the secondary conductors only if the secondary of the transformer is single-phase and feeds a _____ wire circuit.

20. _____

21. The minimum height of the door sill or curb in an approved transformer vault is _____ inches.

21. _____

22. Sealed refrigeration units have what are known as _____ motors.

22. _____

23. The allowable ampacities of wires used in crane circuits is _____ than for general electrical construction work.

23. _____

24. The three types of optical fiber cables listed in the NEC are the nonconductive, the conductive, and the _____ .

24. _____

25. A busway which is free from contact with combustible material may be reduced in size provided the smaller busway has a current rating of at least ⅓ the rating of the main busway overcurrent device and does not extend more than _____ feet.

25. _____

National Electrical Code

TEAR OUT HERE

Specialized and Hazardous Locations

Classes

This chapter is concerned with locations that, because of their nature, call for higher standards with respect to materials or for methods of installation, than are thought necessary in the general run of electrical wiring. The NEC has grouped dangerous operations under three headings: Class I, Class II and Class III. Class I deals with activities in which fire or explosion may result from gases or vapors. Class II is concerned with the possibility of fire or explosion from dusts. Class III is reserved for combustible fibers.

Each class has two divisions. In the first section, under each category, the particular danger is imminent at any or all times during the normal course of operation. In the second one, the danger is not believed present under normal conditions but is likely to arise from a reasonably foreseeable accidental occurrence. For example, a building which houses a machine for compressing acetylene gas will fall within Class I, Division 1. If high-pressure mains from this machine pass into another building where shut-off valves and pressure-recording instruments are inserted in these lines, this adjacent building comes under the heading Class I, Division 2.

Plants where combustible or electrically-conductive dusts exist are termed Class II, Division 1. Locations in which readily ignitible fibers are merely stored or handled, other than in the actual process of manufacture are assessed as Class III, Division 2. This classification applies mostly to warehouses in which manufactured goods, in bales, are held for shipment to consumers.

Methods for Reducing Hazards

The NEC requirements for proper seals in Class I, Division 1 and 2 locations is of utmost importance. Seals when required are installed within 18 inches of an enclosure for switches, relays, circuit breakers, fuses or other equipment which could produce arcs, sparks or high temperatures. The purpose of these seals is to restrict an explosion to one enclosure. Seals are also required in Class I, Division 1 and 2, areas when a conduit run of 2 inches or larger enters an enclosure or fitting which contains terminals, splices or taps. This seal is provided to limit the total volume that could be exposed to an explosion. Seals are also required in conduits passing from a hazardous to a nonhazardous location. The purpose of this seal is to prevent an explosive mixture from being communicated, through the conduit, from a hazardous to a nonhazardous area.

In Class II areas an alternate method of sealing is provided which permits a horizontal raceway, not less than 10 feet long or a vertical raceway, extending not less than 5 feet downward from a dust-ignition-proof enclosure. All sealing fittings and compound shall be approved for the purpose. At the present time Underwriters Laboratories lists a sealing fitting but does not list a compound except when a compound is used in connection with the explosion-proof seal fittings of a specific manufacturer.

Intrinsically safe equipment and associated wiring which is approved for a specific location is permitted in hazardous areas. This equipment is such that under normal or abnormal conditions it is not capable of releasing sufficient electrical or thermal energy to cause ignition of a specific explosive mixture.

The NEC recommends that wherever possible electrical equipment for hazardous locations be located in less hazardous areas. It also suggests that by adequate, positive-pressure ventilation from a clean source of outside air the hazards may be reduced or hazardous locations limited or eliminated. In many cases the installation of dust-collecting systems can greatly reduce the hazards in a Class II area.

Rigid metal conduit, IMC, or Type MI cable are prescribed as only acceptable wiring methods in majority of hazardous locations. Grounding continuity must be assured by means of bonding and not by double-locknuts or locknut and bushing. Fittings must be explosion-proof, dust-ignition-proof, or dust-tight, depending upon the nature of the particular hazard. Motors within the scope of Class I, Division 1, must be explosion-proof. Under Class I, Division 2, standard motors can be used if they have no sliding contacts. Motors within Classes II and III must be totally enclosed or pipe-ventilated. Flexible cords of the three class areas shall be approved for extra hard usage.

Especially Listed Occupancies

The NEC considers all areas up to a level of 18 inches in a commercial garage as Class I, Division 2, locations. A commercial garage where only exchange of parts and routine maintenance work are performed is considered a nonhazardous location. Aircraft hangars are accorded similar but stricter protection. Space in the immediate vicinity of a gasoline dispensing island is denoted as Class I, Division 1, to a height of 4 feet above grade. Surrounding territory within a radius of 20 feet of the island falls under Class I, Division 2, to a height of 18 inches above grade. Bulk storage plants for gasoline are subject to comparable restrictions.

Finishing processes, where paints, lacquers, or other flammable coatings are ap-

HAZARDOUS LOCATIONS AT A GLANCE

MATERIAL	CLASS I		CLASS II			CLASS III	
	DIVISION 1	DIVISION 2	DIVISION 1	DIVISION 2	MET. DUST	DIVISION 1	DIVISION 2
RACEWAY							
RIGID	✔	✔	✔	✔	✔	✔	✔
IMC	✔	✔	✔	✔	✔	✔	✔
EMT				✔			
OPEN WIRING						Storage Areas Only → ✔	
BOXES—FTGS							
THREADED	✔	✔	✔	✔	✔	✔	✔
EXP PROOF	✔						
D-I PROOF		Where Joints or Terminals → ✔			✔		
DUST TIGHT		Where No Joints or Terminals → ✔					
DUST MIN.				✔		✔	✔
FLEX. CON.							
EXP PROOF	✔						
STANDARD							
EHU CORD	✔	✔	✔	✔	✔	✔	✔
SWITCHES							
EXP PROOF	✔	✔					
G-P HERM		✔					
G-P OIL		✔					
D-I PROOF			✔		✔		
DUST MIN.				✔		✔	✔

D-I PROOF—DUST-IGNITION PROOF
EHU CORD—EXTRA-HARD-USAGE CORD
G-P HERM—GENERAL PURPOSE ENCLOSURE—CONTACTS HERMETICALLY SEALED
G-P OIL—GENERAL PURPOSE ENCLOSURE—CONTACTS IMMERSED IN OIL
DUST MIN. DESIGNED TO MINIMIZE ENTRANCE OF DUST
NOTE—MI CABLE ACCEPTABLE IN ALL CLASSIFICATIONS

National Electrical Code

plied by spraying, dipping, brushing, or similar means, must be carried out under prescribed safeguards for electrical wiring and equipment. Of particular interest here are electrostatic units which incorporate high-voltage transformers.

Anesthetizing locations of hospitals are deemed to be Class I, Division 1, to a height of 5 feet above floor level. Special rules apply to lighting and surgical equipment within the operating room, isolating transformers being generally required. Gas storage rooms are designated as Class I, Division 1, throughout.

Theatres and motion picture studios are given adequate consideration in the NEC.

Unless theatres are provided with sufficient emergency lighting, there is an ever-present likelihood of panic in case of fire. And fires may occur suddenly from temporary wiring used in connection with stage scenery or from projection room equipment. Motion picture studios also suffer threat of fire arising from temporary wiring, arc lamps, and portable devices.

Radio or television studios and receiving stations must be guarded from danger offered by high-voltage electronic wiring systems or equipment. Other subjects treated in this section are allied in some degree to the matters outlined here.

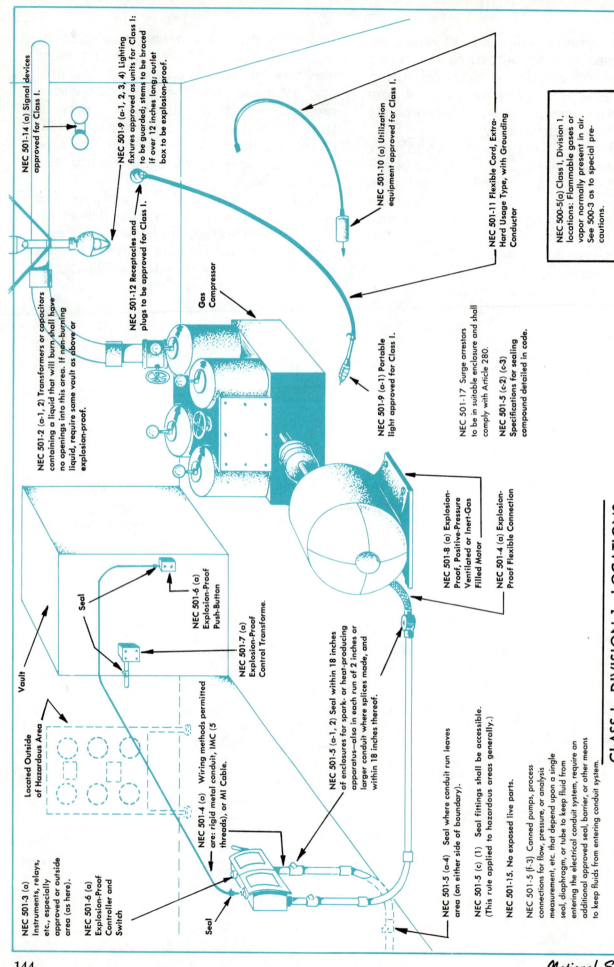

NEC 501-14 (a) Signal devices approved for Class I.

NEC 501-9 (a-1, 2, 3, 4) Lighting fixtures approved as units for Class I: to be guarded; stems to be braced if over 12 inches long; outlet box to be explosion-proof.

NEC 501-10 (a) Utilization equipment approved for Class I.

NEC 501-11 Flexible Cord, Extra-Hard Usage Type, with Grounding Conductor

NEC 500-5(a) Class I, Division 1, locations: Flammable gases or vapor normally present in air. See 500-3 as to special precautions.

NEC 501-2 (a-1, 2) Transformers or capacitors containing a liquid that will burn shall have no openings into this area. If non-burning liquid, require same vault as above or explosion-proof.

NEC 501-12 Receptacles and plugs to be approved for Class I.

Gas Compressor

NEC 501-9 (a-1) Portable light approved for Class I.

NEC 501-17 Surge arrestors to be in suitable enclosure and shall comply with Article 280.

NEC 501-5 (c-2) (c-3) Specifications for sealing compound detailed in code.

NEC 501-8 (a) Explosion-Proof, Positive-Pressure Ventilated or Inert-Gas Filled Motor

NEC 501-4 (a) Explosion-Proof Flexible Connection

NEC 501-3 (a) Instruments, relays, etc., especially approved or outside area (as here).

NEC 501-6 (a) Explosion-Proof Controller and Switch

Vault

Located Outside of Hazardous Area

Seal

NEC 501-6 (a) Explosion-Proof Push-Button

NEC 501-7 (a) Explosion-Proof Control Transforme.

NEC 501-4 (a) Wiring methods permitted are: rigid metal conduit, IMC (5 threads), or MI Cable.

NEC 501-5 (a-1, 2) Seal within 18 inches of enclosures for spark- or heat-producing apparatus—also in each run of 2 inches or larger conduit where splices made, and within 18 inches thereof.

NEC 501-5 (a-4) Seal where conduit run leaves area (on either side of boundary).

NEC 501-5 (c) (1) Seal fittings shall be accessible. (This rule applied to hazardous areas generally.)

NEC 501-15. No exposed live parts.

NEC 501-5 (f-3) Canned pumps, process connections for flow, pressure, or analysis measurement, etc. that depend upon a single seal, diaphragm, or tube to keep fluid from entering the electrical conduit system, require an additional approved seal, barrier, or other means to keep fluids from entering conduit system.

Seal

CLASS I, DIVISION 1, LOCATIONS

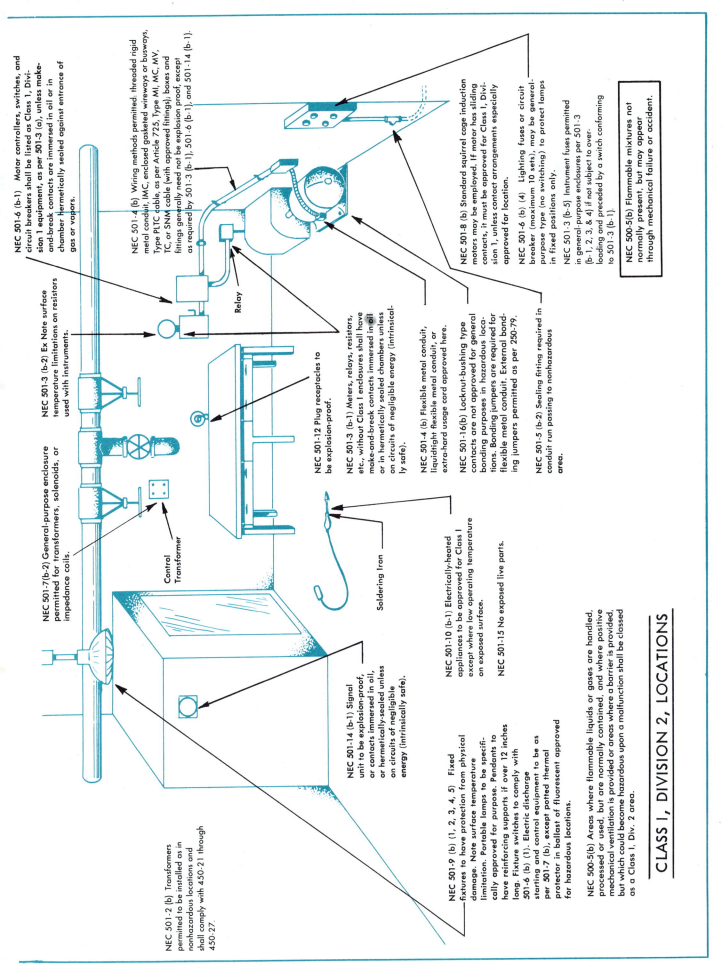

NEC 501-6 (b-1) Motor controllers, switches, and circuit breakers shall be listed as Class 1, Division 1 equipment, as per 501-3 (a), unless make-and-break contacts are immersed in oil or in chamber hermetically sealed against entrance of gas or vapors.

NEC 501-4 (b) Wiring methods permitted: threaded rigid metal conduit, IMC, enclosed gasketed wireways or busways, Type PLTC cable, as per Article 725, Type MI, MC, MV, TC, or SNM cable (with approved fittings); boxes and fittings generally need not be explosion proof, except as required by 501-3 (b-1), 501-6 (b-1), and 501-14 (b-1).

NEC 501-8 (b) Standard squirrel cage induction motors may be employed. If motor has sliding contacts, it must be approved for Class 1, Division 1, unless contact arrangements especially approved for location.

NEC 501-6 (b) (4) Lighting fuses or circuit breaker (maximum 10 sets), may be general-purpose type (no switching) to protect lamps in fixed positions only.

NEC 501-3 (b-5) Instrument fuses permitted in general-purpose enclosures per 501-3 (b-1, 2, 3, & 4) if not subject to over-loading and preceded by a switch conforming to 501-3 (b-1).

NEC 500-5(b) Flammable mixtures not normally present, but may appear through mechanical failure or accident.

NEC 501-3 (b-2) Ex Note surface temperature limitations on resistors used with instruments.

Relay

NEC 501-7(b-2) General-purpose enclosure permitted for transformers, solenoids, or impedance coils.

Control Transformer

NEC 501-12 Plug receptacles to be explosion-proof.

NEC 501-3 (b-1) Meters, relays, resistors, etc., without Class 1 enclosures shall have make-and-break contacts immersed in oil or in hermetically sealed chambers unless on circuits of negligible energy (intrinsically safe).

NEC 501-4 (b) Flexible metal conduit, liquidtight flexible metal conduit, or extra-hard usage cord approved here.

NEC 501-16(b) Locknut-bushing type contacts are not approved for general bonding purposes in hazardous locations. Bonding jumpers are required for flexible metal conduit. External bonding jumpers permitted as per 250-79.

NEC 501-5 (b-2) Sealing fitting required in conduit run passing to nonhazardous area.

NEC 501-10 (b-1) Electrically-heated appliances to be approved for Class 1 except where low operating temperature on exposed surface.

NEC 501-15 No exposed live parts.

Soldering Iron

NEC 501-14 (b-1) Signal unit to be explosion-proof, or contacts immersed in oil, or hermetically-sealed unless on circuits of negligible energy (intrinsically safe).

NEC 501-9 (b) (1, 2, 3, 4, 5) Fixed fixtures to have protection from physical damage. Note surface temperature limitation. Portable lamps to be specifically approved for purpose. Pendants to have reinforcing supports if over 12 inches long. Fixture switches to comply with 501-6 (b) (1). Electric discharge starting and control equipment to be as per 501-7 (b), except potted thermal protector in ballast of fluorescent approved for hazardous locations.

NEC 500-5(b) Areas where flammable liquids or gases are handled, processed or used, but are normally contained, and where positive mechanical ventilation is provided or areas where a barrier is provided, but which could become hazardous upon a malfunction shall be classed as a Class I, Div. 2 area.

NEC 501-2 (b) Transformers permitted to be installed as in nonhazardous locations and shall comply with 450-21 through 450-27.

CLASS I, DIVISION 2, LOCATIONS

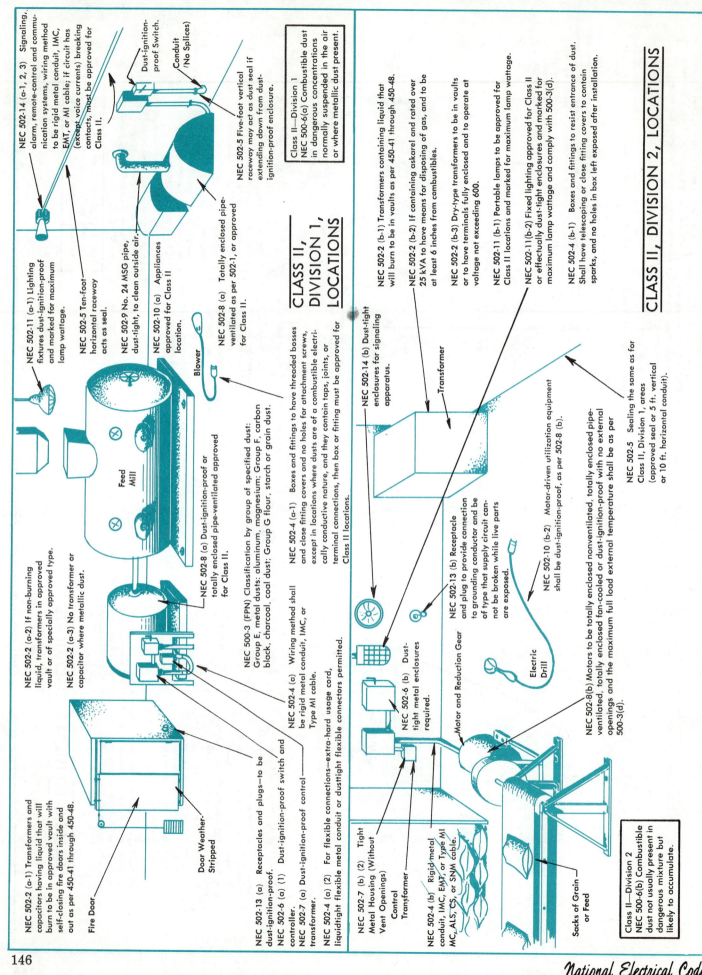

NEC 502-14 (a-1, 2, 3) Signaling, alarm, remote-control and communication systems; wiring method to be rigid metal conduit, IMC, EMT, or MI cable; if circuit has (except voice currents) breaking contacts, must be approved for Class II.

Dust-ignition-proof Switch.

Conduit (No Splices)

NEC 502-5 Five-foot vertical raceway may act as dust seal if extending down from dust-ignition-proof enclosure.

Class II—Division 1
NEC 500-6(a) Combustible dust in dangerous concentrations normally suspended in the air or where metallic dust present.

NEC 502-11 (a-1) Lighting fixtures dust-ignition-proof and marked for maximum lamp wattage.

NEC 502-5 Ten-foot horizontal raceway acts as seal.

NEC 502-9 No. 24 MSG pipe, dust-tight, to clean outside air.

NEC 502-10 (a) Appliances approved for Class II location.

NEC 502-8 (a) Totally enclosed pipe-ventilated as per 502-1, or approved for Class II.

Blower

Feed Mill

NEC 502-8 (a) Dust-ignition-proof or totally enclosed pipe-ventilated approved for Class II.

NEC 500-3 (FPN) Classification by group of specified dust: Group E, metal dusts: aluminum, magnesium; Group F, carbon black, charcoal, coal dust; Group G flour, starch or grain dust.

NEC 502-4 (a) Wiring method shall be rigid metal conduit, IMC, or Type MI cable.

NEC 502-4 (a-1) Boxes and fittings to have threaded bosses and close fitting covers and no holes for attachment screws, except in locations where dusts are of a combustible electrically conductive nature, and they contain taps, joints, or terminal connections, then box or fitting must be approved for Class II locations.

NEC 502-2 (a-1) Transformers and capacitors having liquid that will burn to be in approved vault with self-closing fire doors inside and out as per 450-41 through 450-48.

Fire Door

NEC 502-2 (a-2) If non-burning liquid, transformers in approved vault or of specially approved type.

NEC 502-2 (a-3) No transformer or capacitor where metallic dust.

NEC 502-13 (a) Receptacles and plugs—to be dust-ignition-proof.

NEC 502-7 (a) Dust-ignition-proof control transformer.

NEC 502-6 (a) (1) Dust-ignition-proof switch and controller.

NEC 502-4 (a) (2) For flexible connections—extra-hard usage cord, liquidtight flexible metal conduit or dusttight flexible connectors permitted.

Door Weather-Stripped

NEC 502-7 (b) (2) Tight Metal Housing (Without Vent Openings) Control Transformer

NEC 502-4 (b) Rigid metal conduit, IMC, EMT, or Type MI MC, ALS, CS, or SNM cable.

NEC 502-6 (b) Dust-tight metal enclosures required.

Motor and Reduction Gear

Sacks of Grain or Feed

Class II—Division 2
NEC 500-6(b) Combustible dust not usually present in dangerous mixture but likely to accumulate.

CLASS II, DIVISION 1, LOCATIONS

NEC 502-2 (b-1) Transformers containing liquid that will burn to be in vaults as per 450-41 through 450-48.

NEC 502-2 (b-2) If containing askarel and rated over 25 kVA to have means for disposing of gas, and to be at least 6 inches from combustibles.

NEC 502-2 (b-3) Dry-type transformers to be in vaults or to have terminals fully enclosed and to operate at voltage not exceeding 600.

NEC 502-11 (b-1) Portable lamps to be approved for Class II locations and marked for maximum lamp wattage.

NEC 502-11(b-2) Fixed lighting approved for Class II or effectually dust-tight enclosures and marked for maximum lamp wattage and comply with 500-3(d).

NEC 502-4 (b-1) Boxes and fittings to resist entrance of dust. Shall have telescoping or close fitting covers to contain sparks, and no holes in box left exposed after installation.

NEC 502-14 (b) Dust-tight enclosures for signaling apparatus.

Transformer

NEC 502-13 (b) Receptacle and plug to provide connection to grounding conductor and be of type that supply circuit cannot be broken while live parts are exposed.

Electric Drill

NEC 502-10 (b-2) Motor-driven utilization equipment shall be dust-ignition-proof, as per 502-8 (b).

NEC 502-8(b) Motors to be totally enclosed nonventilated, totally enclosed pipe-ventilated, totally enclosed fan-cooled or dust-ignition-proof with no external openings and the maximum full load external temperature shall be as per 500-3(d).

NEC 502-5 Sealing the same as for Class II, Division 1, areas (approved seal or 5 ft. vertical or 10 ft. horizontal conduit).

CLASS II, DIVISION 2, LOCATIONS

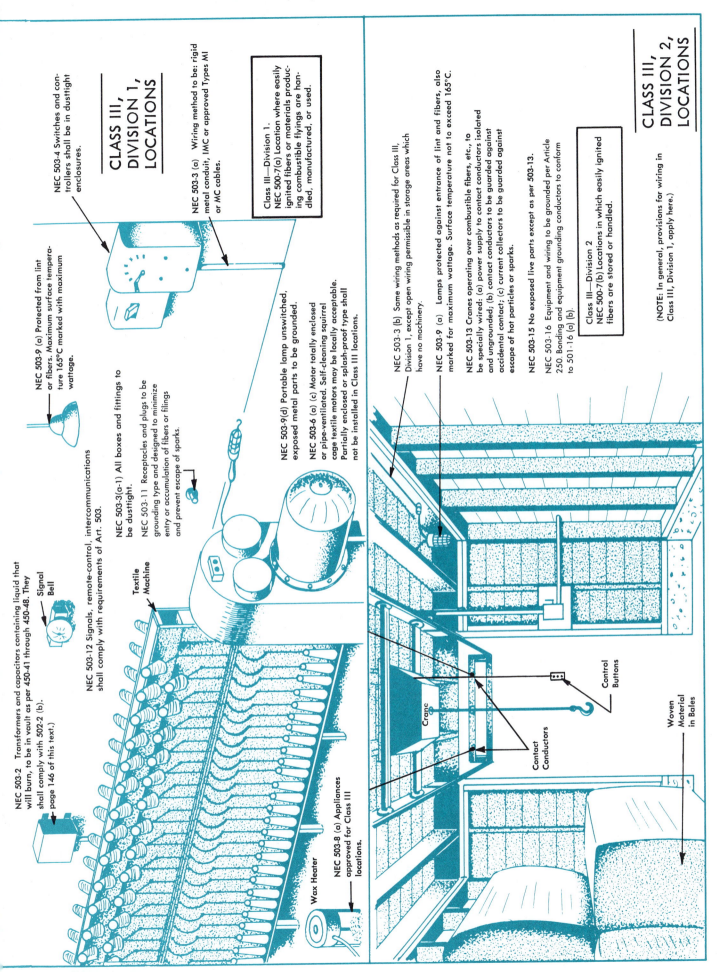

CLASS III, DIVISION 1, LOCATIONS

NEC 503-4 Switches and controllers shall be in dusttight enclosures.

NEC 503-3 (a) Wiring method to be: rigid metal conduit, IMC or approved Types MI or MC cables.

Class III—Division 1.
NEC 500-7(a) Location where easily ignited fibers or materials producing combustible flyings are handled, manufactured, or used.

NEC 503-9 (a) Protected from lint or fibers. Maximum surface temperature 165°C marked with maximum wattage.

Signal Bell

NEC 503-3(a-1) All boxes and fittings to be dusttight.

NEC 503-11 Receptacles and plugs to be grounding type and designed to minimize entry or accumulation of fibers or filings and prevent escape of sparks.

NEC 503-12 Signals, remote-control, intercommunications shall comply with requirements of Art. 503.

Textile Machine

NEC 503-9(d) Portable lamp unswitched, exposed metal parts to be grounded.

NEC 503-6 (a) (c) Motor totally enclosed or pipe-ventilated. Self-cleaning squirrel cage textile motors may be locally acceptable. Partially enclosed or splash-proof type shall not be installed in Class III locations.

NEC 503-2 Transformers and capacitors containing liquid that will burn, to be in vault as per 450-41 through 450-48. They shall comply with 502-2 (b). page 146 of this text.)

Wax Heater

NEC 503-8 (a) Appliances approved for Class III locations.

CLASS III, DIVISION 2, LOCATIONS

NEC 503-3 (b) Same wiring methods as required for Class III, Division 1, except open wiring permissible in storage areas which have no machinery.

NEC 503-9 (a) Lamps protected against entrance of lint and fibers, also marked for maximum wattage. Surface temperature not to exceed 165°C.

NEC 503-13 Cranes operating over combustible fibers, etc., to be specially wired: (a) power supply to contact conductors isolated and ungrounded; (b) contact conductors to be guarded against accidental contact; (c) current collectors to be guarded against escape of hot particles or sparks.

NEC 503-15 No exposed live parts except as per 503-13.

NEC 503-16 Equipment and wiring to be grounded per Article 250. Bonding and equipment grounding conductors to conform to 501-16 (a) (b).

Class III—Division 2
NEC 500-7(b) Locations in which easily ignited fibers are stored or handled.

(NOTE: In general, provisions for wiring in Class III, Division 1, apply here.)

Control Buttons

Crane

Contact Conductors

Woven Material in Bales

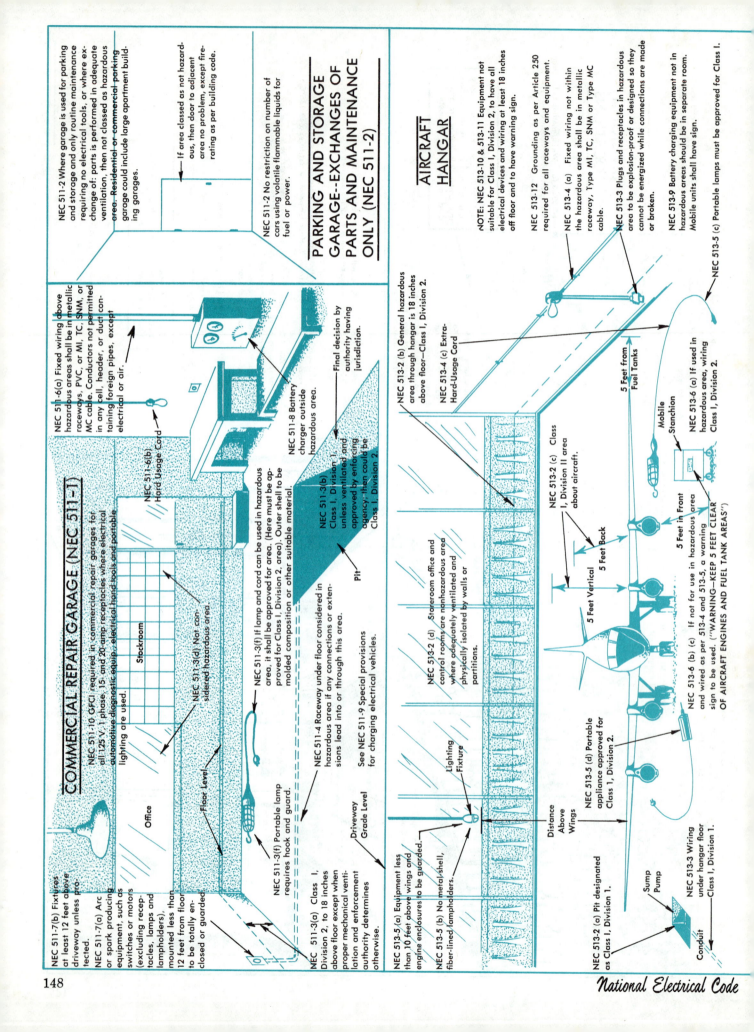

COMMERCIAL REPAIR GARAGE (NEC 511-1)

NEC 511-7(b) Fixtures at least 12 feet above driveway unless protected.

NEC 511-7(a) Arc or spark producing equipment, such as switches or motors (excluding receptacles, lamps and lampholders), mounted less than 12 feet from floor to be totally enclosed or guarded.

NEC 511-3(a) Class I, Division 2, to 18 inches above floor except when proper mechanical ventilation and enforcement authority determines otherwise.

NEC 511-6(a) Fixed wiring above hazardous areas shall be in metallic raceways, PVC, or MI, TC, SNM, or MC cable. Conductors not permitted in any cell, header, or duct containing foreign pipes, except electrical or air.

NEC 511-6(b) Hard Usage Cord

NEC 511-8 Battery charger outside hazardous area.

Final decision by authority having jurisdiction.

NEC 511-3(b) Class I, Division I, unless ventilated and approved by enforcing agency, then could be Class I, Division 2.

Pit

NEC 511-3(f) If lamp and cord can be used in hazardous area, it shall be approved for area. (Here must be approved for Class I, Division 2, area). Outer shell to be molded composition or other suitable material.

NEC 511-3(d) Not considered hazardous area.

Stockroom

Office

Floor Level

Driveway Grade Level

NEC 511-4 Raceway under floor considered in hazardous area if any connections or extensions lead into or through this area.

See NEC 511-9 Special provisions for charging electrical vehicles.

NEC 511-3(f) Portable lamp requires hook and guard.

NEC 511-10 GFCI required in commercial repair garages for all 125 V, 1 phase, 15- and 20-amp receptacles where electrical automotive diagnostic equip., electrical hand tools and portable lighting are used.

PARKING AND STORAGE GARAGE--EXCHANGES OF PARTS AND MAINTENANCE ONLY (NEC 511-2)

NEC 511-2 Where garage is used for parking and storage and only routine maintenance requiring no electrical tools, or where exchange of: parts is performed in adequate ventilation, then not classed as hazardous area. Residential or commercial parking garage could include large apartment building garages.

NEC 511-2 No restriction on number of cars using volatile flammable liquids for fuel or power.

If area classed as not hazardous, then door to adjacent area no problem, except fire-rating as per building code.

AIRCRAFT HANGAR

NEC 513-2 (b) General hazardous area through hangar is 18 inches above floor—Class I, Division 2.

NEC 513-4 (c) Extra-Hard-Usage Cord

NOTE: NEC 513-10 & 513-11 Equipment not suitable for Class I, Division 2, to have all electrical devices and wiring at least 18 inches off floor and to have warning sign.

NEC 513-12 Grounding as per Article 250 required for all raceways and equipment.

NEC 513-4 (a) Fixed wiring not within the hazardous area shall be in metallic raceway, Type MI, TC, SNM or Type MC cable.

NEC 513-3 Plugs and receptacles in hazardous area to be explosion-proof or designed so they cannot be energized while connections are made or broken.

NEC 513-9 Battery charging equipment not in hazardous areas should be in separate room. Mobile units shall have sign.

NEC 513-5 (c) Portable lamps must be approved for Class I.

NEC 513-6 (a)

5 Feet from Fuel Tanks

Mobile Stanchion

NEC 513-6 (a) If used in hazardous area, wiring Class I, Division 2.

NEC 513-2 (c) Class I, Division II area about aircraft.

5 Feet Back

5 Feet in Front

5 Feet Vertical

NEC 513-6 (b) (c) If not for use in hazardous area and wired as per 513-4 and 513-5, a warning sign to be used. ("WARNING—KEEP 5 FEET CLEAR OF AIRCRAFT ENGINES AND FUEL TANK AREAS")

NEC 513-2 (d) Storeroom office and control rooms are nonhazardous area where adequately ventilated and physically isolated by walls or partitions.

Lighting Fixture

NEC 513-5 (a) Equipment less than 10 feet above wings and engine enclosures to be guarded.

NEC 513-5 (b) No metal-shell, fiber-lined lampholders.

NEC 513-5 (d) Portable appliance approved for Class I, Division 2.

Distance Above Wings

NEC 513-2 (a) Pit designated as Class I, Division 1.

Sump Pump

NEC 513-3 Wiring under hangar floor—Class I, Division 1.

Conduit

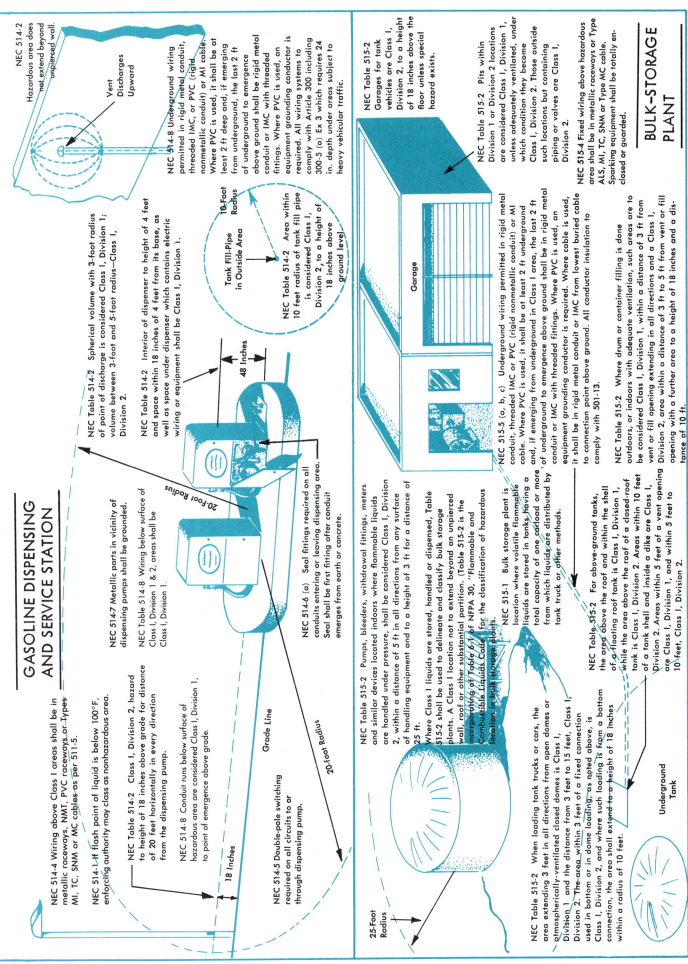

GASOLINE DISPENSING AND SERVICE STATION

NEC 514-2 Hazardous area does not extend beyond unpierced wall.

Vent Discharges Upward

NEC 514-8 Underground wiring permitted in rigid metal conduit, threaded IMC, or PVC (rigid nonmetallic conduit) or MI cable. Where PVC is used, it shall be at least 2 ft deep and, if emerging from underground, the last 2 ft of underground to emergence above ground shall be rigid metal conduit or IMC with threaded fittings. Where PVC is used, an equipment grounding conductor is required. All wiring systems to comply with Article 300 including 300-5 (a) Ex 3 which requires 24 in. depth under areas subject to heavy vehicular traffic.

NEC Table 514-2   Spherical volume with 3-foot radius of point of discharge is considered Class I, Division 1; volume between 3-foot and 5-foot radius—Class I, Division 2.

NEC Table 514-2   Interior of dispenser to height of 4 feet and space within 18 inches of 4 feet from its base, as well as space under dispenser which contains electric wiring or equipment shall be Class I, Division 1.

10-Foot Radius

Tank Fill-Pipe in Outside Area

NEC Table 514-2   Area within 10 feet radius of tank fill pipe is considered Class I, Division 2, to a height of 18 inches above ground level

48 Inches

20-Foot Radius

NEC 514-4 Wiring above Class I areas shall be in metallic raceways, NMT, PVC raceways or Types MI, TC, SNM or MC cables as per 511-5.

NEC 514-1 If flash point of liquid is below 100°F, enforcing authority may class as nonhazardous area.

NEC Table 514-2   Class I, Division 2, hazard to height of 18 inches above grade for distance of 20 feet horizontally in every direction from the dispensing pump.

NEC 514-8 Conduit runs below surface of hazardous area are considered Class I, Division 1, to point of emergence above grade.

18 Inches

Grade Line

20-Foot Radius

NEC 514-7 Metallic parts in vicinity of dispensing pumps shall be grounded.

NEC Table 514-8 Wiring below surface of Class I, Division 1 & 2, areas shall be Class I, Division 1.

NEC 514-6 (a)   Seal fittings required on all conduits entering or leaving dispensing area. Seal shall be first fitting after conduit emerges from earth or concrete.

NEC 514-5 Double-pole switching required on all circuits to or through dispensing pump.

NEC Table 515-2   Pumps, bleeders, withdrawal fittings, meters and similar devices located indoors where flammable liquids are handled under pressure, shall be considered Class I, Division 2, within a distance of 5 ft in all directions from any surface of handling equipment and to a height of 3 ft for a distance of 25 ft.

Where Class I liquids are stored, handled or dispensed, Table 515-2 shall be used to delineate and classify bulk storage plants. A Class I location not to extend beyond an unpierced wall, roof or other substantial partition. (Table 515-2 is the incorporation of Table 6-1 of NFPA 30, "Flammable and Combustible Liquids Code" for the classification of hazardous location in bulk storage plants.

NEC 515-1   Bulk storage plant is location where volatile flammable liquids are stored in tanks having a total capacity of one carload or more from which liquids are distributed by tank truck or other methods.

NEC Table 515-2   For above-ground tanks, the area above the roof and within the shell of a floating roof tank is Class I, Division 1, while the area above the roof of a closed-roof tank is Class I, Division 2. Areas within 10 feet of a tank shell and inside a dike are Class I, Division 2. Areas within 5 feet of a vent opening are Class I, Division 1, and within 5 feet to 10 feet, Class I, Division 2.

NEC Table 515-2   When loading tank trucks or cars, the area extending 3 feet in all directions from open domes or atmospherically-ventilated closed domes is Class I, Division 1 and the distance from 3 feet to 15 feet, Class I, Division 2. The area within 3 feet of a fixed connection used in bottom or in dome loading, as noted above, is Class I, Division 2, and where such loading is from a bottom connection, the area shall extend to a height of 18 inches within a radius of 10 feet.

25-Foot Radius

Underground Tank

Garage

NEC Table 515-2   Garages for tank vehicles are Class I, Division 2, to a height of 18 inches above the floor unless special hazard exists.

NEC Table 515-2   Pits within Division 1 or Division 2 locations are considered Class I, Division 1, unless adequately ventilated, under which condition they become Class I, Division 2. Those outside such locations but containing piping or valves are Class I, Division 2.

NEC 515-4 Fixed wiring above hazardous area shall be in metallic raceways or Type ALS, MI, TC, SNM or Type MC cable. Sparking equipment shall be totally enclosed or guarded.

NEC 515-5 (a, b, c)   Underground wiring permitted in rigid metal conduit, threaded IMC or PVC (rigid nonmetallic conduit) or MI cable. Where PVC is used, it shall be at least 2 ft underground and, if emerging from underground in Class I area, the last 2 ft of underground to emergence above ground shall be in rigid metal conduit or IMC with threaded fittings. Where PVC is used, an equipment grounding conductor is required. Where cable is used, it shall be in rigid metal conduit or IMC from lowest buried cable to connection point above ground. All conductor insulation to comply with 501-13.

NEC Table 515-2   Where drum or container filling is done outdoors, or indoors with adequate ventilation, such areas are to be considered Class I, Division 1, within a distance of 3 ft from vent or fill opening extending in all directions and a Class I, Division 2, area within a distance of 3 ft to 5 ft from vent or fill opening with a further area to a height of 18 inches and a distance of 10 ft.

BULK-STORAGE PLANT

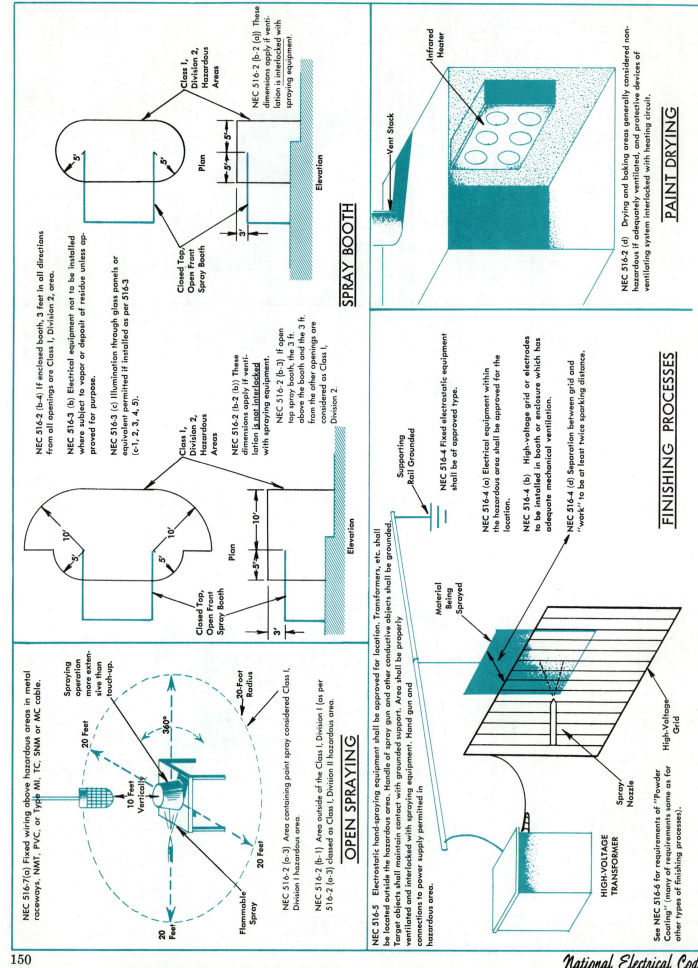

Class I, Division 2, Hazardous Areas

NEC 516-2 (b-2 (a)) These dimensions apply if ventilation is interlocked with spraying equipment.

5'

5'

5'

5'

3'

Plan

Elevation

Closed Top, Open Front Spray Booth

SPRAY BOOTH

Infrared Heater

Vent Stack

NEC 516-2 (d) Drying and baking areas generally considered non-hazardous if adequately ventilated, and protective devices of ventilating system interlocked with heating circuit.

PAINT DRYING

NEC 516-2 (b-4) If enclosed booth, 3 feet in all directions from all openings are Class I, Division 2, area.

NEC 516-3 (b) Electrical equipment not to be installed where subject to vapor or deposit of residue unless approved for purpose.

NEC 516-3 (c) Illumination through glass panels or equivalent permitted if installed as per 516-3 (c-1, 2, 3, 4, 5).

Class I, Division 2, Hazardous Areas

NEC 516-2 (b-2 (b)) These dimensions apply if ventilation is not interlocked with spraying equipment.

NEC 516-2 (b-3) If open top spray booth, the 3 ft. above the booth and the 3 ft. from the other openings are considered as Class I, Division 2.

10'

10'

5'

5'

5'

10'

3'

Plan

Elevation

Closed Top, Open Front Spray Booth

NEC 516-7(a) Fixed wiring above hazardous areas in metal raceways, NMT, PVC, or Type MI, TC, SNM or MC cable.

Spraying operation more extensive than touch-up.

360°

20 Feet

20-Foot Radius

10 Feet Vertically

Flammable Spray

20 Feet

20 Feet

NEC 516-2 (a-3) Area containing paint spray considered Class I, Division I hazardous area.

NEC 516-2 (b-1) Area outside of the Class I, Division I (as per 516-2 (a-3)) classed as Class I, Division II hazardous area.

OPEN SPRAYING

NEC 516-5 Electrostatic hand-spraying equipment shall be approved for location. Transformers, etc., shall be located outside the hazardous area. Handle of spray gun and other conductive objects shall be grounded. Target objects shall maintain contact with grounded support. Area shall be properly ventilated and interlocked with spraying equipment. Hand gun and connections to power supply permitted in hazardous area.

Supporting Rail Grounded

NEC 516-4 Fixed electrostatic equipment shall be of approved type.

NEC 516-4 (a) Electrical equipment within the hazardous area shall be approved for the location.

NEC 516-4 (b) High-voltage grid or electrodes to be installed in booth or enclosure which has adequate mechanical ventilation.

NEC 516-4 (d) Separation between grid and "work" to be at least twice sparking distance.

Material Being Sprayed

Spray Nozzle

High-Voltage Grid

HIGH-VOLTAGE TRANSFORMER

See NEC 516-6 for requirements of "Powder Coating" (many of requirements same as for other types of finishing processes).

FINISHING PROCESSES

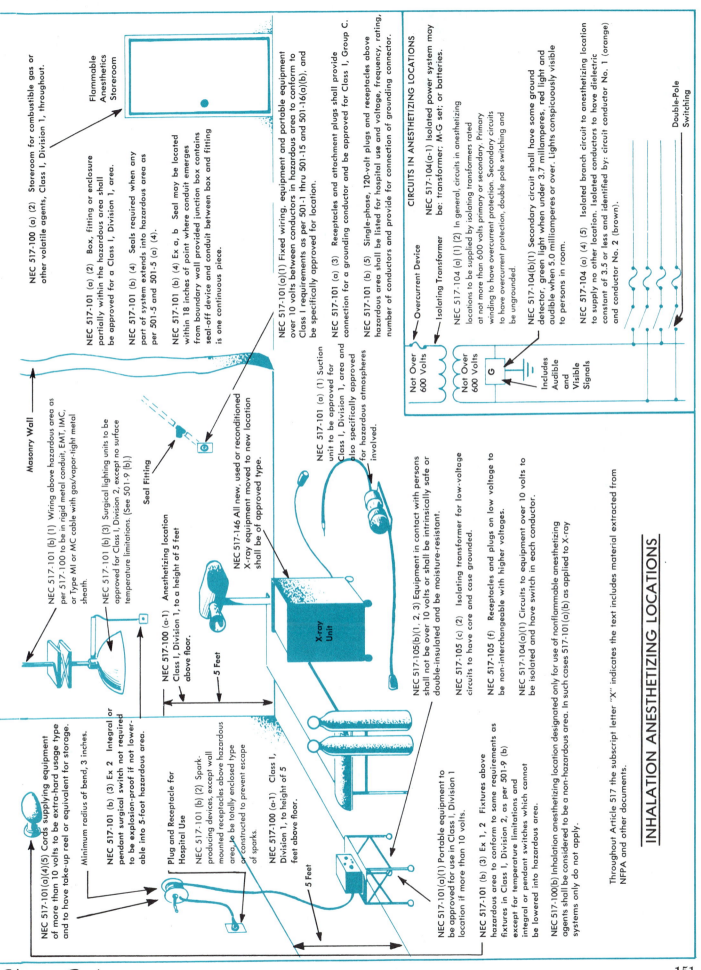

NEC 517-100 (a) (2) Storeroom for combustible gas or other volatile agents, Class I, Division 1, throughout.

Flammable Anesthetics Storeroom

NEC 517-101 (a) (2) Box, fitting or enclosure partially within the hazardous area shall be approved for a Class I, Division 1, area.

NEC 517-101 (b) (4) Seals required when any part of system extends into hazardous area as per 501-5 and 501-5 (a) (4).

NEC 517-101 (b) (4) Ex a, b Seal may be located within 18 inches of point where conduit emerges from boundary wall provided junction box contains seal-off device and conduit between box and fitting is one continuous piece.

NEC 517-101(a)(1) Fixed wiring, equipment and portable equipment over 10 volts between conductors in hazardous area to conform to Class I requirements as per 501-1 thru 501-15 and 501-16(a)(b), and be specifically approved for location.

NEC 517-101 (a) (3) Receptacles and attachment plugs shall provide connection for a grounding conductor and be approved for Class I, Group C.

NEC 517-101 (b) (5) Single-phase, 120-volt plugs and receptacles above hazardous area shall be listed for hospital use and voltage, frequency, rating, number of conductors and provide for connection of grounding connector.

CIRCUITS IN ANESTHETIZING LOCATIONS

NEC 517-104(a-1) Isolated power system may be: transformer; M-G set; or batteries.

Overcurrent Device

Isolating Transformer

NEC 517-104 (a) (1) (2) In general, circuits in anesthetizing locations to be supplied by isolating transformers rated at not more than 600 volts primary or secondary. Primary winding to have overcurrent protection. Secondary circuits to have overcurrent protection, double pole switching and be ungrounded.

NEC 517-104(b)(1) Secondary circuit shall have some ground detector, green light when under 3.7 millamperes, red light and audible when 5.0 millamperes or over. Lights conspicuously visible to persons in room.

NEC 517-104 (a) (4) (5) Isolated branch circuit to anesthetizing location to supply no other location. Isolated conductors to have dielectric constant of 3.5 or less and identified by: circuit conductor No. 1 (orange) and conductor No. 2 (brown).

Not Over 600 Volts

Not Over 600 Volts

G

Includes Audible and Visible Signals

Double-Pole Switching

Masonry Wall

NEC 517-101 (b) (1) Wiring above hazardous area as per 517-100 to be in rigid metal conduit, EMT, IMC, or Type MI or MC cable with gas/vapor-tight metal sheath.

NEC 517-101 (b) (3) Surgical lighting units to be approved for Class I, Division 2, except no surface temperature limitations. (See 501-9 (b).)

Seal Fitting

NEC 517-101 (a) (1) Suction unit to be approved for Class I, Division 1, area and also specifically approved for hazardous atmospheres involved.

NEC 517-100 (a-1) Anesthetizing location Class I, Division 1, to a height of 5 feet above floor.

5 Feet

NEC 517-146 All new, used or reconditioned X-ray equipment moved to new location shall be of approved type.

X-ray Unit

NEC 517-105(b)(1, 2, 3) Equipment in contact with persons shall not be over 10 volts or shall be intrinsically safe or double-insulated and be moisture-resistant.

NEC 517-105 (c) (2) Isolating transformer for low-voltage circuits to have core and case grounded.

NEC 517-105 (f) Receptacles and plugs on low voltage to be non-interchangeable with higher voltages.

NEC 517-104(a)(1) Circuits to equipment over 10 volts to be isolated and have switch in each conductor.

NEC 517-101(a)(4)(5) Cords supplying equipment of more than 10 volts to be extra-hard usage type and to have take-up reel or equivalent for storage.

Minimum radius of bend, 3 inches.

NEC 517-101 (b) (3) Ex 2 Integral or pendant surgical switch not required to be explosion-proof if not lowerable into 5-foot hazardous area.

Plug and Receptacle for Hospital Use

NEC 517-101 (b) (2) Spark-producing devices, except wall mounted receptacles above hazardous area, to be totally enclosed type or constructed to prevent escape of sparks.

NEC 517-100 (a-1) Class I, Division 1, to height of 5 feet above floor.

5 Feet

NEC 517-101(a)(1) Portable equipment to be approved for use in Class I, Division 1 location if more than 10 volts.

NEC 517-101 (b) (3) Ex 1, 2 Fixtures above hazardous area to conform to same requirements as fixtures in Class I, Division 2, as per 501-9 (b) except for temperature limitations and integral or pendant switches which cannot be lowered into hazardous area.

NEC 517-100(b) Inhalation anesthetizing location designated only for use of nonflammable anesthetizing agents shall be considered to be a non-hazardous area. In such cases 517-101(a)(b) as applied to X-ray systems only do not apply.

Throughout Article 517 the subscript letter "X" indicates the text includes material extracted from NFPA and other documents.

INHALATION ANESTHETIZING LOCATIONS

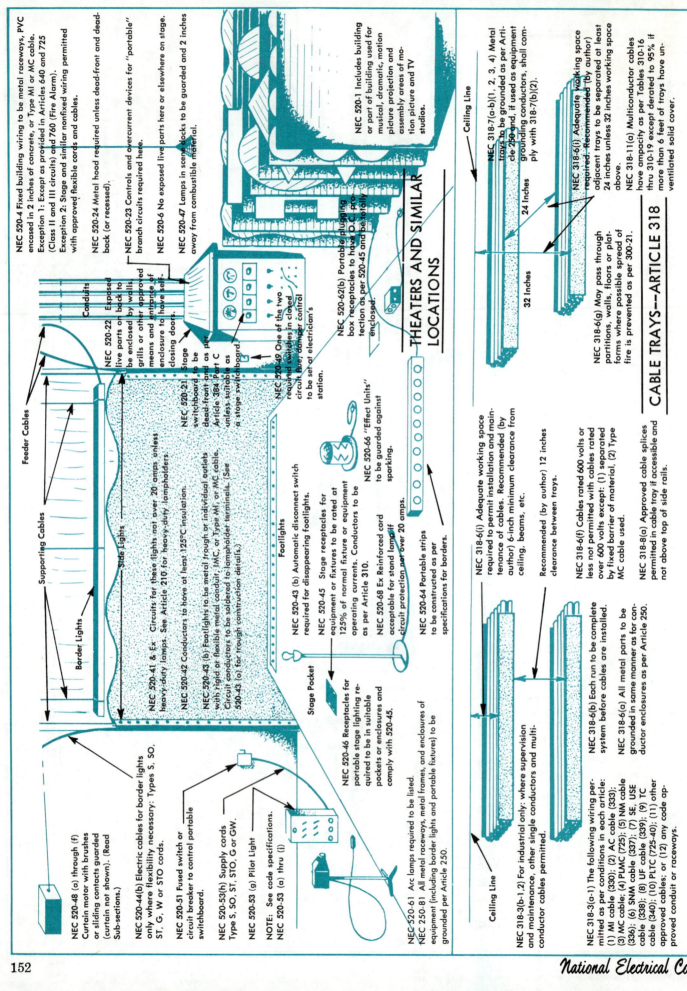

NEC 520.4 Fixed building wiring to be metal raceways, PVC encased in 2 inches of concrete, or Type MI or MC cable. Exception 1: Except as provided in Articles 640 and 725 (Class II and III circuits) and 760 (Fire Alarm). Exception 2: Stage and similar nonfixed wiring permitted with approved flexible cords and cables.

NEC 520.24 Metal hood required unless dead-front and dead-back (or recessed).

NEC 520.23 Controls and overcurrent devices for "portable" branch circuits required here.

NEC 520.6 No exposed live parts here or elsewhere on stage.

NEC 520.47 Lamps in scene docks to be guarded and 2 inches away from combustible material.

NEC 520.1 Includes building or part of building used for musical, dramatic, motion picture projection and assembly areas of motion picture and TV studios.

THEATERS AND SIMILAR LOCATIONS

Conduits

Exposed live parts or back to be enclosed by walls, grills or other approved means and entrance to have self-closing doors.

NEC 520.22 Stage switchboard to be dead-front and as per Article 384 Part C unless suitable as a stage switchboard.

NEC 520.21 switchboard

NEC 520.62(b) Portable plugging box receptacles to have O.C. Protection as per 520.45 and be totally enclosed.

NEC 520.49 One of the two required switches in closed circuit flue; damper control to be set at electrician's station.

Feeder Cables

Supporting Cables

Side Lights

Border Lights

NEC 520.66 "Effect Units" to be guarded against sparking.

NEC 520.43 (b) Automatic disconnect switch required for disappearing footlights.

NEC 520.45 Stage receptacles for equipment or fixtures to be rated at 125% of normal fixture or equipment operating currents. Conductors to be as per Article 310.

NEC 520.68 Ex Reinforced cord acceptable for stand lamp if circuit protection not over 20 amps.

NEC 520.64 Portable strips to be constructed as per specifications for borders.

NEC 520.41 & Ex Circuits for these lights not over 20 amps unless heavy-duty lamps. See Article 210 for heavy-duty lampholders.

NEC 520.42 Conductors to have at least 125°C insulation.

NEC 520.43 (b) Footlights to be metal trough or individual outlets with rigid or flexible metal conduit, IMC, or Type MI, or MC cable. Circuit conductors to be soldered to lampholder terminals. (See 520.43 (a) for trough construction details.)

Footlights

Stage Pocket

NEC 520.46 Receptacles for portable stage lighting required to be in suitable pockets or enclosures and comply with 520.45.

NEC 520.48 (a) through (f) Curtain motor with brushes or sliding contacts guarded (curtain not shown). (Read Sub-sections.)

NEC 520.44(b) Electric cables for border lights only where flexibility necessary: Types S, SO, ST, G, W or STO cords.

NEC 520.51 Fused switch or circuit breaker to control portable switchboard.

NEC 520.53(h) Supply cords Type S, SO, ST, STO, G or GW.

NEC 520.53 (g) Pilot Light

NOTE: See code specifications. NEC 520.53 (a) thru (i)

NEC 520.61 Arc lamps required to be listed. NEC 250.81 All metal raceways, metal frames, and enclosures of equipment (including border lights and portable fixtures) to be grounded per Article 250.

CABLE TRAYS--ARTICLE 318

Ceiling Line

NEC 318-7(a-b)(1, 2, 3, 4) Metal trays to be grounded as per Article 250 and, if used as equipment grounding conductors, shall comply with 318-7(b)(2).

NEC 318-6(i) Adequate working space required. Recommended (by author) adjacent trays to be separated at least 24 inches unless 32 inches working space above.

NEC 318-11(a) Multiconductor cables have ampacity as per Tables 310-16 thru 310-19 except derated to 95% if more than 6 feet of trays have un-ventilated solid cover.

24 Inches

32 Inches

NEC 318-6(g) May pass through partitions, walls, floors or platforms where possible spread of fire is prevented as per 300-21.

NEC 318-6(i) Adequate working space required to permit installation and maintenance of cables. Recommended (by author) 6-inch minimum clearance from ceiling, beams, etc.

Recommended (by author) 12 inches clearance between trays.

NEC 318-6(f) Cables rated 600 volts or less not permitted with cables rated over 600 volts except: (1) separated by fixed barrier of material, (2) Type MC cable used.

NEC 318-8(a) Approved cable splices permitted in cable tray if accessible and not above top of side rails.

Ceiling Line

NEC 318-6(b) Each run to be complete system before cables are installed.

NEC 318-6(a) All metal parts to be grounded in same manner as for conductor enclosures as per Article 250.

NEC 318-3(b-1,2) For industrial only: where supervision and maintenance, other single conductors and multi-conductor cables permitted.

NEC 318-3(a-1) The following wiring permitted as per conditions in each article: (1) MI cable (330); (2) AC cable (333); (3) MC cable; (4) PLMC (725); (5) NM cable (336); (6) SNM cable (337); (7) SE, USE cable (338); (8) UF cable (339); (9) TC cable (340); (10) PLTC (725-40); (11) other approved cables; or (12) any code approved conduit or raceways.

CABLE CONNECTORS -- STAGE WIRING

Line

Load

NEC 520-67 The female half of the connector shall be attached to the line side of the cord or cable. Cable connectors constructed so that tension on cord or cable will not be transmitted to connections. Connectors to be rated in Amps and designed so that different rated males and females cannot be properly connected together and be polarized.

Flue-damper control device designed for full circuit voltage.

Control circuit normally closed.

Flue damper device to be located in loft above scenery and enclosed in suitable metal box with tight, self-closing door.

At least two switches in series, one at electrician's station.

NEC 520-53 (e) Dimmer terminals to have enclosures and faceplate and be so arranged that accidental contact between live parts and faceplate is improbable.

NEC 520-49 Requirements for flue damper control device and its circuit.

FLUE DAMPER CONTROL -- STAGE

NEC 530-13 Ex A single switch may be used to disconnect all contactors on a location board which is not more than 6 feet away.

NEC 530-19 See Table 530-19 (a) for feeder demand factors for stage set lighting.

NEC 530-18 (b) Overcurrent devices for feeders may be set at not over 400% of conductor rating.

NEC 530-20 Pendant and portable lamps and special portable equipment need not be grounded if operating supply voltage not over 150 volts to ground.

NEC 530-18 (d) Where plugging box does not have overcurrent devices, every cord or cable smaller than No. 8 supplied through it shall have its own overcurrent device.

MOTION PICTURE AND TV STUDIOS

Disconnect Switch

Feeder from Substation

Feeder to Location Board

NEC 530-18 (e) 20-Amp Circuits

Location Board

Plugging Box

Cords or Cables

NEC 530-11, 530-12, 530-13 Studio wiring requirements, generally, follow those for theater stage.

FESTOONS -- STAGE WIRING

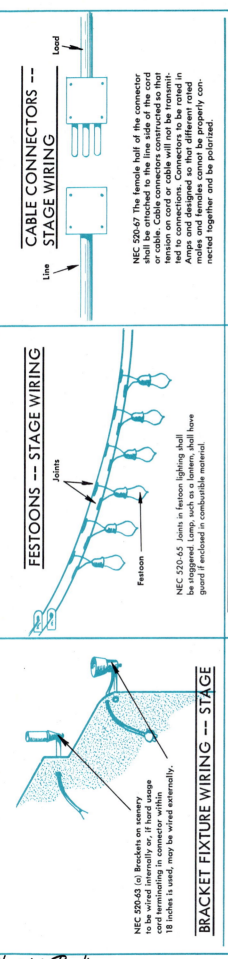

Joints

Festoon

NEC 520-65 Joints in festoon lighting shall be staggered. Lamp, such as a lantern, shall have guard if enclosed in combustible material.

AUTOTRANSFORMER-TYPE DIMMERS

Autotransformer Dimmer

Line

Load

NEC 520-25 (c) Voltage between conductors not to exceed 150 volts. Grounded conductor to be common to both load and supply.

STAGE DETAILS

BRACKET FIXTURE WIRING -- STAGE

NEC 520-63 (a) Brackets on scenery to be wired internally or, if hard usage cord terminating in connector within 18 inches is used, may be wired externally.

RESISTANCE- OR REACTOR-TYPE DIMMERS -- STAGE WIRING

Dimmer

Overcurrent Device

Switch

NEC 520-25 (a) Where installed in ungrounded conductors, each dimmer shall be protected by an overcurrent device set at a value not exceeding 125% of its rating.

NEC 520-25 (b) Resistance or reactor (series) dimmers may be placed in either line wire and shall comply with 380-1. Dimmers in grounded neutral shall not open the circuit.

NEC 520-73 Lamps and receptacles to be controlled by wall switches with pilot light.

NEC 110-2 Equipment (including dressing room lighting) shall be approved by authority enforcing code. (Also see 90-6 and Article 100—Definitions, approved for the purpose.)

NEC 520-71 Pendant lampholders not permitted in dressing rooms.

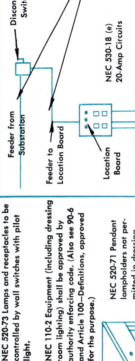

Dressing Tables

THEATER DRESSING ROOMS

NEC 520-72 Lamps within 8 feet of floor to have riveted, open-end guards or otherwise locked in place.

Wall Switch with Pilot Light

NEC 520-73 Switch controlling receptacles to have pilot light.

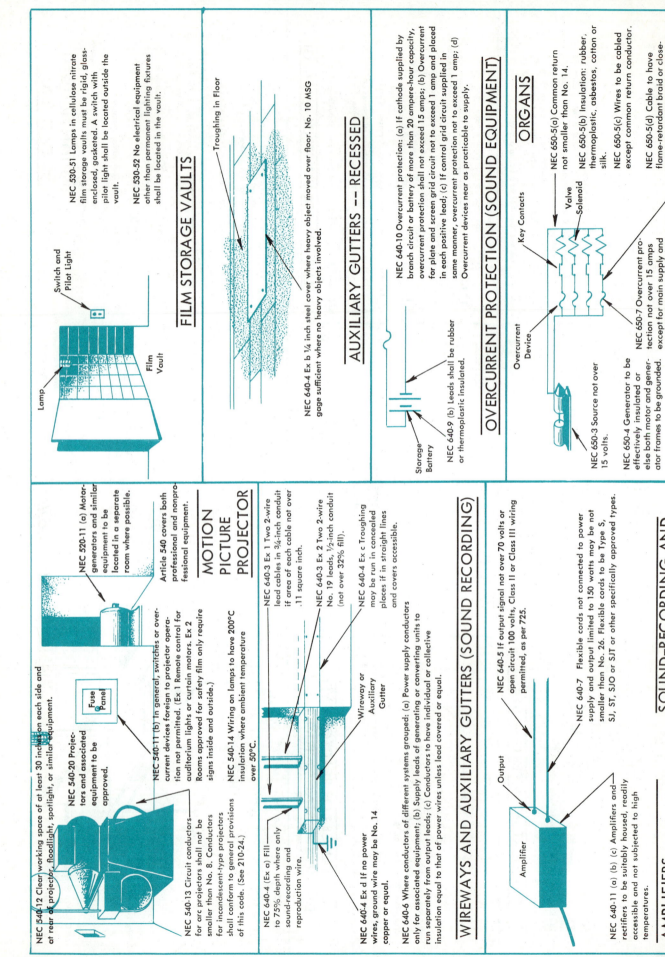

NEC 530-51 Lamps in cellulose nitrate film storage vaults must be rigid, glass-enclosed, gasketed. A switch with pilot light shall be located outside the vault.

NEC 530-52 No electrical equipment other than permanent lighting fixtures shall be located in the vault.

Switch and Pilot Light

Troughing in Floor

FILM STORAGE VAULTS

Lamp

Film Vault

NEC 640-4 Ex b ¼ inch steel cover where heavy object moved over floor. No. 10 MSG gage sufficient where no heavy objects involved.

AUXILIARY GUTTERS -- RECESSED

NEC 640-10 Overcurrent protection: (a) If cathode supplied by branch circuit or battery of more than 20 ampere-hour capacity, overcurrent protection shall not exceed 15 amps; (b) Overcurrent for plate and screen grid circuit not to exceed 1 amp and placed in each positive lead; (c) If control grid circuit supplied in same manner, overcurrent protection not to exceed 1 amp; (d) Overcurrent devices near as practicable to supply.

NEC 640-9 (b) Leads shall be rubber or thermoplastic insulated.

Storage Battery

OVERCURRENT PROTECTION (SOUND EQUIPMENT)

ORGANS

NEC 650-5(a) Common return not smaller than No. 14.

NEC 650-5(b) Insulation: rubber, thermoplastic, asbestos, cotton or silk.

NEC 650-5(c) Wires to be cabled except common return conductor.

NEC 650-5(d) Cable to have flame-retardant braid or close-wound fireproof tape unless in metal raceways.

Key Contacts

Valve Solenoid

Overcurrent Device

NEC 650-7 Overcurrent protection not over 15 amps except for main supply and common return conductors.

NEC 650-5(a) Conductors not less than No. 26.

NEC 650-3 Source not over 15 volts.

NEC 650-4 Generator to be effectively insulated or else both motor and generator frames to be grounded.

NEC 540-12 Clear working space of at least 30 inches on each side and at rear of projector, floodlight, spotlight, or similar equipment.

NEC 520-11 (a) Motor-generators and similar equipment to be located in a separate room where possible.

Article 540 covers both professional and nonprofessional equipment.

MOTION PICTURE PROJECTOR

NEC 540-11 (b) In general, switches or over-current devices foreign to projector operation not permitted. (Ex 1 Remote control for auditorium lights or curtain motors. Ex 2 Rooms approved for safety film only require signs inside and outside.)

NEC 540-14 Wiring on lamps to have 200°C insulation where ambient temperature over 50°C.

NEC 540-20 Projectors and associated equipment to be approved.

Fuse Panel

NEC 540-13 Circuit conductors for arc projectors shall not be smaller than No. 8. Conductors for incandescent-type projectors shall conform to general provisions of this code. (See 210-24.)

NEC 640-4 (Ex a) Fill to 75% depth where only sound-recording and reproduction wire.

NEC 640-4 Ex d If no power wires, ground wire may be No. 14 copper or equal.

NEC 640-3 Ex 1 Two 2-wire lead cables in ¾-inch conduit if area of each cable not over .11 square inch.

NEC 640-3 Ex 2 Two 2-wire No. 19 leads, ½-inch conduit (not over 32% fill).

NEC 640-4 Ex c Troughing may be run in concealed places if in straight lines and covers accessible.

Wireway or Auxiliary Gutter

NEC 640-6 Where conductors of different systems grouped: (a) Power supply conductors only for associated equipment; (b) Supply leads of generating or converting units to run separately from output leads; (c) Conductors to have individual or collective insulation equal to that of power wires unless lead covered or equal.

WIREWAYS AND AUXILIARY GUTTERS (SOUND RECORDING)

NEC 640-5 If output signal not over 70 volts or open circuit 100 volts, Class II or Class III wiring permitted, as per 725.

Output

NEC 640-7 Flexible cords not connected to power supply and output limited to 150 watts may be not smaller than No. 26. Flexible cords to be Type S, SJ, ST, SJO or other specifically approved types.

SOUND-RECORDING AND SIMILAR EQUIPMENT

Amplifier

NEC 640-11 (a) (b) (c) Amplifiers and rectifiers to be suitably housed, readily accessible and not subjected to high temperatures.

AMPLIFIERS (SOUND EQUIPMENT)

NEC 700-1 Emergency systems are normally installed in building where large numbers of persons assemble or occupy such buildings, such as theaters, assembly halls, hotels, sports arenas, hospitals, and similar institutions. Emergency systems also provide power for hospital equipment that is essential to maintain life and for illumination and power for operating rooms, fire alarm systems, fire pumps, critical public address systems, industrial processes, where interruption would cause hazards, and similar functions (See Life Safety Code NFPA 101 1981 (ANSI) for critical emergency illumination and exit lights and lighting.)

Emergency Circuit

NEC 700-20 Only authorized persons have access to emergency switch. Series, 3- or 4-way switches not permitted. (See Ex 1, 2.)

NEC 700-22 Exterior emergency light may be switched by light-actuated devices.

NEC 700-25 Branch-circuit overcurrent devices accessible to authorized persons.

NEC 700-16 Emergency lighting includes exit lights and illuminated exit signs.

NEC 700-1 Article 700 to apply when emergency systems or circuits are legally required by Municipal, State, Federal, or other codes or by any governmental agency having jurisdiction.

NEC 700-7 Audible and visual derangement signals recommended to insure constant readiness of emergency system. (See 700-7(a, b, c, d, e))

NEC 700-9 Wiring to be entirely separate from all other wiring and boxes and raceways to be identified. (See Ex. 1, 2, 3, 4, 5.)

NEC 700-15 Only emergency light outlet may be served by emergency lighting system.

EMERGENCY SYSTEMS

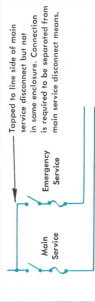

Supply

NEC 720-1 Wiring under 50 volts AC or DC except as per 650, 725 and 760. Battery circuits as per Article 480.

NEC 720-5 Lampholders rated at least 660 watts.

NEC 720-6 Rating of receptacles not less than 15 amps.

NEC 720-7 Twenty-amp receptacles for kitchen and laundry and other locations where portable appliances could be used.

NEC 720-4 Conductors not smaller than No. 12 AWG copper; not smaller than No. 10 AWG copper if supplying more than one appliance receptacle.

NEC 720-10 Grounding as per 250-5 (a) and 250.45.

CIRCUITS AND EQUIPMENT LESS THAN 50 VOLTS

Normal Service Supply

Emergency Source of Power

Automatic Transfer Equipment

NEC 230-83 All ungrounded conductors of one source to be disconnected before conductors of second source is connected.

Emergency System

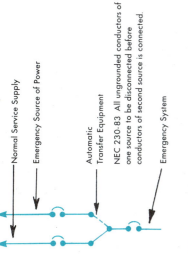

TRANSFER SWITCHES

NEC 700-6 Transfer equipment to be automatic and identified for use or approved by authority having jurisdiction. To be designed and installed so that connection of both normal and emergency power is not possible.

See 250-5 (d) for grounding of separately derived systems.

NEC 700-18 Branch circuits for emergency use to be automatically transferred upon failure of normal power supply.

UNIT EQUIPMENT

Junction Box for Conduit Connection

State of Charge Indicating Discs

NEC 700-12 (f) Unit equipment to be listed for emergency use.

NEC 700-12 (f) Unit equipment for emergency lighting shall consist of: (1) rechargeable battery; (2) means to charge battery; (3) one or more lamps on unit (terminals are permitted for remote lamp location); (4) a relay automatically energized by failure of general lighting system. Battery shall be able to maintain at least 87½% of nominal voltage at full lamp load for at least 1½ hours or maintain at least 60% of rated illumination for at least 1½ hours. Unit must be approved for emergency service.

Unit not to be portable and to be wired as per Chapter 3. Flexible cord and plug permitted if cord not over 3 feet long. Unit wired to same branch circuit as other normal lighting in area and connected ahead of any local switches. Remote location unit equipment lamps shall be wired as per Chapter 3.

EMERGENCY SYSTEMS

Storage Battery

NEC 700-12 (a) Storage batteries permitted for emergency service if capacity is adequate to supply and maintain 87½% of system voltage when supplying full emergency load for 1½ hours. (See Article 480 for installation of storage batteries.)

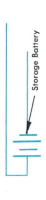

STORAGE BATTERY

Prime Mover

G — Generator Set
E

NEC 700-12 (b) (1) A generator with automatic transfer and automatically-started prime mover with 15-minute time delay before restoring normal power is permitted for emergency power.

NEC 700-12 (b) (2) Where internal combustion engine is prime mover, a 2-hour fuel supply is required.

NEC 700-12 (b) (3) Where prime mover fuel is natural gas from utility, a second fuel supply is required with automatic transfer to second fuel supply. (See 700-12 (b) (3) Ex.)

NEC 700-12 (b) (4) Where battery is used to start prime mover, it shall be suitable for the purpose and have automatic charging means.

GENERATOR SET

SEPARATE SERVICE

Service No. 1   Service No. 2

NEC 700-12 (d) Two distinct services, with separate service drops or laterals, widely separated physically and electrically, is permitted. Each service to be treated as a separate service and installed as per Article 230.

Tapped to line side of main service disconnect but not in same enclosure. Connection is required to be separated from main service disconnect means.

Emergency Service

Main Service

NEC 700-12 (e) Connection ahead of service disconnect (but not within) is permitted if sufficiently separated from service disconnect means.

CONNECTION AHEAD OF DISCONNECTING MEANS

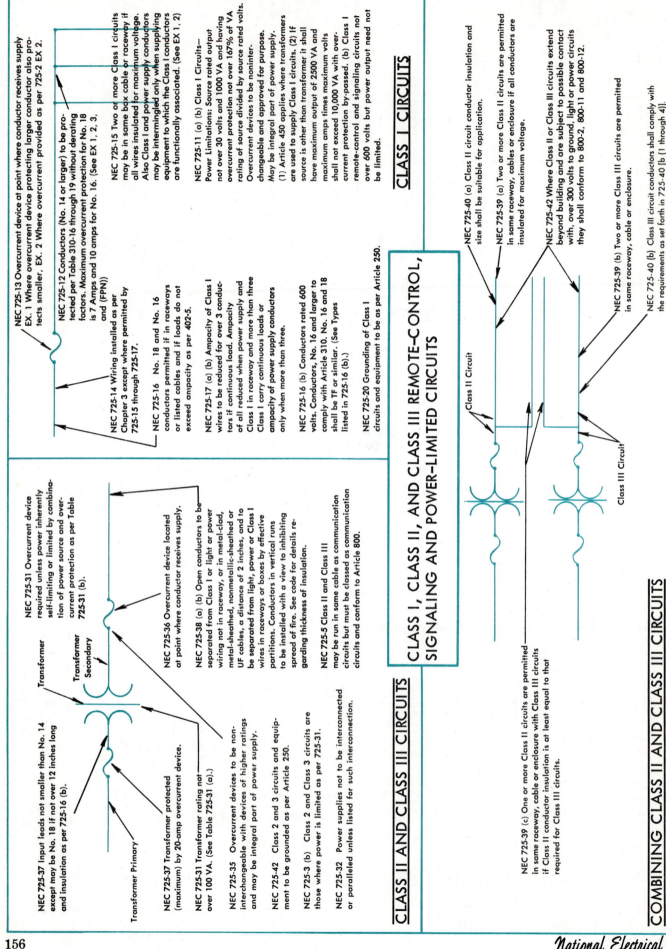

NEC 725-13 Overcurrent device at point where conductor receives supply EX. 1 Where overcurrent device protecting larger conductor also protects smaller, EX. 2 Where overcurrent provided as per 725-2 EX. 2.

NEC 725-12 Conductors (No. 14 or larger) to be protected per Table 310-16 through 19 without derating factors. Maximum overcurrent protection for No. 18 is 7 Amps and 10 amps for No. 16. (See EX 1, 2, 3, and (FPN))

NEC 725-14 Wiring installed as per Chapter 3 except where permitted by 725-15 through 725-17.

NEC 725-16 No. 18 and No. 16 conductors permitted if in raceways or listed cables and if loads do not exceed ampacity as per 402-5.

NEC 725-17 (a) (b) Ampacity of Class I wires to be reduced for over 3 conductors if continuous load. Ampacity of all reduced when power supply and Class I in raceway and more than three Class I carry continuous loads or ampacity of power supply conductors only when more than three.

NEC 725-16 (b) Conductors rated 600 volts. Conductors, No. 16 and larger to comply with Article 310. No. 16 and 18 shall be TF or similar. (See Types listed in 725-16 (b).)

NEC 725-20 Grounding of Class I circuits and equipment to be as per Article 250.

NEC 725-15 Two or more Class I circuits may be in same box cable or raceway if all wires insulated for maximum voltage. Also Class I and power supply conductors may be intermingled only when supplying equipment to which the Class I conductors are functionally associated. (See EX 1, 2)

NEC 725-11 (a) (b) Class I Circuits— Power Limitations: Source rated output not over 30 volts and 1000 VA and having overcurrent protection not over 167% of VA rating of source divided by source rated volts. Overcurrent devices to be noninterchangeable and approved for purpose. May be integral part of power supply. (1) Article 450 applies where transformers are used to supply Class I circuits. (2) If source is other than transformer it shall have maximum output of 2500 VA and maximum amps times maximum volts shall not exceed 10,000 VA with overcurrent protection by-passed. (b) Class I remote-control and signaling circuits not over 600 volts but power output need not be limited.

CLASS I CIRCUITS

NEC 725-40 (a) Class II circuit conductor insulation and size shall be suitable for application.

NEC 725-39 (a) Two or more Class II circuits are permitted in same raceway, cables or enclosure if all conductors are insulated for maximum voltage.

NEC 725-42 Where Class II or Class III circuits extend beyond building and are subject to possible contact with, over 300 volts to ground, light or power circuits they shall conform to 800-2, 800-11 and 800-12.

NEC 725-39 (b) Two or more Class III circuits are permitted in same raceway, cable or enclosure.

NEC 725-40 (b) Class III circuit conductors shall comply with the requirements as set forth in 725-40 [b (1 through 4)].

Class II Circuit

Class III Circuit

CLASS I, CLASS II, AND CLASS III REMOTE-CONTROL, SIGNALING AND POWER-LIMITED CIRCUITS

NEC 725-37 Overcurrent device required unless power inherently self-limiting or limited by combination of power source and overcurrent protection as per Table 725-31 (b).

NEC 725-37 Input leads not smaller than No. 14 except may be No. 18 if not over 12 inches long and insulation as per 725-16 (b).

Transformer Primary

Transformer

Transformer Secondary

NEC 725-36 Overcurrent device located at point where conductor receives supply.

NEC 725-38 (a) (b) Open conductors to be separated from Class I or light or power wiring not in raceway, or in metal-clad, metal-nonmetallic-sheathed or UF cables, a distance of 2 inches, and to be separated from light, power or Class I wires in raceways or boxes by effective partitions. Conductors in vertical runs to be installed with a view to inhibiting spread of fire. See code for details regarding thickness of insulation.

NEC 725-5 Class II and Class III may be run in same cable as communication circuits but must be classed as communication circuits and conform to Article 800.

NEC 725-37 Transformer protected (maximum) by 20-amp overcurrent device.

NEC 725-31 Transformer rating not over 100 VA. (See Table 725-31 (a).)

NEC 725-35 Overcurrent devices to be noninterchangeable with devices of higher ratings and may be integral part of power supply.

NEC 725-42 Class 2 and 3 circuits and equipment to be grounded as per Article 250.

NEC 725-3 (b) Class 2 and Class 3 circuits are those where power is limited as per 725-31.

NEC 725-32 Power supplies not to be interconnected or paralleled unless listed for such interconnection.

CLASS II AND CLASS III CIRCUITS

NEC 725-39 (c) One or more Class II circuits are permitted in same raceway, cable or enclosure with Class III circuits if Class II conductor insulation is at least equal to that required for Class III circuits.

COMBINING CLASS II AND CLASS III CIRCUITS

SIGNS -- GENERAL RULES -- 600 VOLTS OR LESS

NEC 600-4 All signs to be listed and installed as per listing unless otherwise permitted.

NEC 600-8 See specified details for sign enclosures (metal galvanized or equal).

NEC 600-8 (b) Flashers in separate box or compartment.

NEC 600-21 (b) Conductors not smaller than No. 14. Ex 1 if listed in Table 402-3, No. 18 permitted for: (a) portable signs, (b) short leads to lampholders or ballasts, (c) permanent leads not over 8 ft long to discharge lampholders or ballasts if in channels, (d) multiple conductor incandescent signs of not over 250 watts. Ex 2 Not smaller than No. 20 permitted as short, permanent leads to synchronous motors.

NEC 600-5 Grounding as per general rules except: (1) Isolated and accessible only to authorized persons; (2) Isolated outline lighting may use No. 14 for bonding if protected from physical damage and grounded as per Article 250.

NEC 600-21 (a) Standard wiring methods except NM and NMC not permitted.

NEC 600-2 Disconnect required.

NEC 600-2 (a) Disconnect in sight of sign except, if controller is used, disconnect shall be in sight of controller and disconnect to be lockable.

NEC 600-6 (a) The load on a sign branch circuit shall not exceed 20 amps, when lamps, ballasts, and transformers, or combinations thereof. When it supplies exclusively electric discharge lighting transformers, the load may not exceed 30 amps.

NEC 600-22 Lampholders to be of unswitched type.

NEC 600-21 (e) Conductors to be connected to lampholders by solder or wire connectors and treated to prevent corrosion except pin-type lampholders used with stranded wire.

NEC 600-2 (b) Devices such as switches or flashers to have twice ampere rating of transformer, or be approved for purpose, except AC general-use snap switches permitted on AC inductive loads not exceeding switch rating.

SIGNS--OVER 600 VOLTS

NEC 600-31 (a) Conductors over 600 volts shall be installed in flexible metal conduit, liquidtight flexible metal conduit, EMT, rigid nonmetallic conduit, rigid or intermediate metal conduit, or as concealed conductors on insulators, or as Type MC cable.

NEC 600-31 (c) No sharp bends.

Raceway

NEC 600-31 (b) Conductors shall not be smaller than No. 14. Exception permits No. 18 or larger for (a) permanent leads not over 8 ft long to discharge lampholders or ballasts if in channels; (b) permanent leads not over 8 ft long from line ends of tubing to transformer secondary windings connected within transformer enclosure for show window displays or portable signs.

NEC 600-31 (e) Where conductor covered with lead or metallic raceway, insulation shall extend beyond raceway for distances specified in paragraphs (1), (2), and (3) of this section.

NEC 600-31 (e-4) Metal raceway having a single conductor from a transformer secondary shall not exceed 20 feet in length.

NEC 600-32 (a) (b) Transformer voltage (open circuit) not to exceed 15,000 (plus 1,000 for test); end-grounded transformer 7,500 (plus 500 for test); open core and coil 5,000 volts (plus 500 for test).

NEC 600-32 (b) Maximum rating 4,500 VA and 30 milliamps for outline lighting transformer. See Exception.

NEC 600-32 (d) Secondaries not to be connected in parallel, and in series (Ex 1) to form midpoint-grounded system.

NEC 600-32 (e) Transformers to be accessible.

ELECTRIC-DISCHARGE TUBING

Tubing Normally Formed in Letters or Numbers

Electric-Discharge Tubing Open High-Voltage Conductors Mounted on Approved Insulators

Glass Support

Where over 7,500 volts, require ¼-inch space from any surface.

NEC 600-33 (a, b, c) Electric-discharge tubing shall be of such design and length not to cause overvolt on transformer; adequate supports to be of noncombustible, nonabsorbent material, and tubing not to be exposed to physical damage or be in contact with flammable material.

TERMINALS FOR NEON TUBING

NEC 600-33 (c) Glass tubing to be free from contact with combustible materials. If voltage in excess of 7,500 volts, tubing to be maintained ¼ inch from nearest surface.

NEC 600-34 (a, b) Terminals to be isolated; if enclosed, 1½-inch separation from grounded metal.

NEC 600-34 (d) Where voltage over 7,500, bushings required unless electrode receptacles provided. Electrode terminal assemblies require support within 6 inches of electrode terminals.

NEC 600-34 (c) Electrode receptacles noncombustible, nonabsorbent, insulating material.

NEC 600-34 (f) Bushing seals may be used.

NOTE: See NEC 600-8 Specification for Enclosures.

1½"

No. 14

Not Over 6 Inches to Support

Seal

OUTSIDE SIGN OR OUTLINE LIGHTING

NEC 600-22 Where lampholders are used they shall not be of the miniature type and shall be constructed and installed to prevent turning. Bodies to be constructed of suitable insulating material and screw-shell connected to grounded circuit conductor.

NEC 600-23 Conductors within sign and outline lighting troughs to be mechanically secure.

NEC 600-8 Conductors in sign and outline troughs to be enclosed in metal or other noncombustible enclosure.

CONCEALED CONDUCTORS ON INSULATIONS -- INDOORS (OVER 600 VOLTS)

Secondary Conductors

Insulators

Channel Lined with Noncombustible Material

Primary Conductors

NEC 600-31 (d) Concealed conductors installed in channels indoors to have 1-inch clearances for voltages of 10,000 or less, and 1½-inch clearance for voltages over 10,000. Conductors not permitted outside enclosure. Insulator constructed with noncombustible, nonabsorbent material (normally made of glass).

ELECTRIC SIGNS & OUTLINE LIGHTING

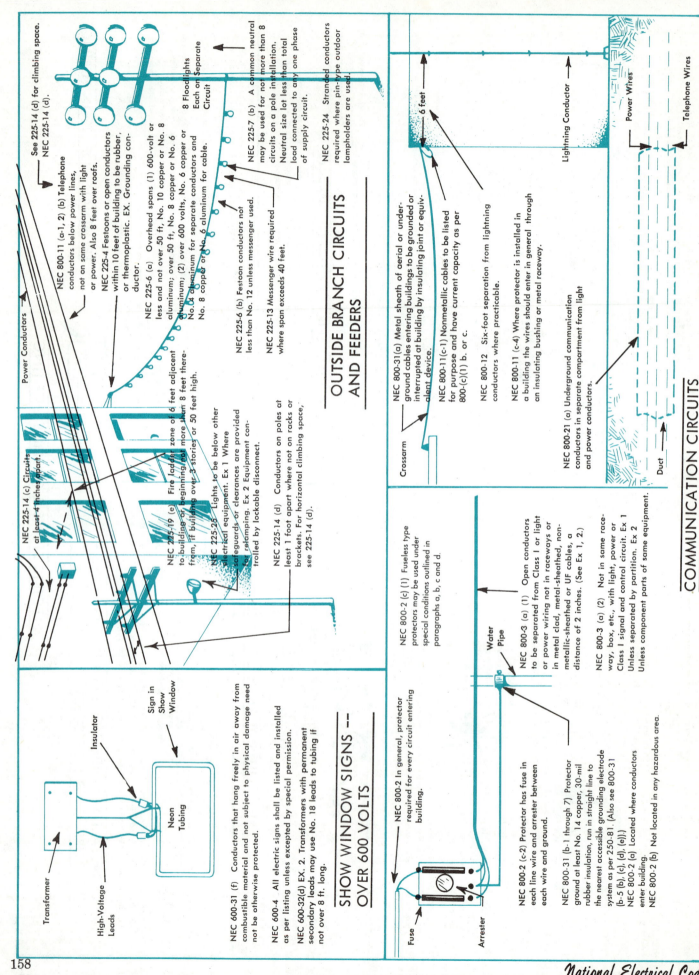

See 225-14 (d) for climbing space.
NEC 225-14 (d).

Power Conductors

NEC 800-11 (a-1, 2) (b) Telephone conductors below power lines, not on same crossarm with light or power. Also 8 feet over roofs.

NEC 225-4 Festoons or open conductors within 10 feet of building to be rubber, or thermoplastic. EX. Grounding conductor.

NEC 225-6 (a) Overhead spans (1) 600-volt or less and not over 50 ft, No. 10 copper or No. 8 aluminum; over 50 ft, No. 8 copper or No. 6 aluminum; (2) over 600 volts, No. 6 copper or No.4 aluminum for separate conductors and No. 8 copper or No. 6 aluminum for cable.

NEC 225-6 (b) Festoon conductors not less than No. 12 unless messenger used.

NEC 225-13 Messenger wire required where span exceeds 40 feet.

8 Floodlights Each on Separate Circuit

NEC 225-7 (b) A common neutral may be used for not more than 8 circuits on a pole installation. Neutral size lot less than total load connected to any one phase of supply circuit.

NEC 225-24 Stranded conductors required where pin-type outdoor lampholders are used.

NEC 225-14 (c) Circuits at least 4 inches apart.

NEC 225-19 (e) Fire ladder zone of 6 feet adjacent to building or beginning not more than 8 feet therefrom, if building over 3 stories or 50 feet high.

NEC 225-25 Lights to be below other electrical equipment. Ex 1 Where safeguards or clearances are provided for relamping. Ex 2 Equipment controlled by lockable disconnect.

NEC 225-14 (d) Conductors on poles at least 1 foot apart where not on racks or brackets. For horizontal climbing space, see 225-14 (d).

OUTSIDE BRANCH CIRCUITS AND FEEDERS

Crossarm

NEC 800-31(a) Metal sheath of aerial or underground cables entering buildings to be grounded or interrupted at building by insulating joint or equivalent device.

NEC 800-11(c-1) Nonmetallic cables to be listed for purpose and have current capacity as per 800-(c)(1) b. or c.

NEC 800-12 Six-foot separation from lightning conductors where practicable.

NEC 800-11 (c-4) Where protector is installed in a building the wires should enter in general through an insulating bushing or metal raceway.

NEC 800-21 (a) Underground communication conductors in separate compartment from light and power conductors.

6 feet

Lightning Conductor

Power Wires

Telephone Wires

Duct

COMMUNICATION CIRCUITS

Transformer

Insulator

High-Voltage Leads

Sign in Show Window

Neon Tubing

NEC 600-31 (f) Conductors that hang freely in air away from combustible material and not subject to physical damage need not be otherwise protected.

NEC 600-4 All electric signs shall be listed and installed as per listing unless excepted by special permission.

NEC 600-32(d) EX. 2. Transformers with permanent secondary leads may use No. 18 leads to tubing if not over 8 ft. long.

SHOW WINDOW SIGNS -- OVER 600 VOLTS

Water Pipe

Fuse

Arrester

NEC 800-2 (c) (1) Fuseless type protectors may be used under special conditions outlined in paragraphs a, b, c and d.

NEC 800-3 (a) (1) Open conductors to be separated from Class 1 or light or power wiring not in raceways or in metal clad, metal-sheathed, nonmetallic-sheathed or UF cables, a distance of 2 inches. (See Ex 1, 2.)

NEC 800-3 (a) (2) Not in same raceway, box, etc., with light, power or Class 1 signal and control circuit. Ex 1 Unless separated by partition. Ex 2 Unless component parts of same equipment.

NEC 800-2 In general, protector required for every circuit entering building.

NEC 800-2 (c-2) Protector has fuse in each line wire and arrester between each wire and ground.

NEC 800-31 (b-1 through 7) Protector ground at least No. 14 copper, 30-mil rubber insulation, run in straight line to the nearest accessible grounding electrode system as per 250-81. (Also see 800-31 (b-5 (b), (c), (d), (e)].)
NEC 800-2 (a) Located where conductors enter building.
NEC 800-2 (b) Not located in any hazardous area.

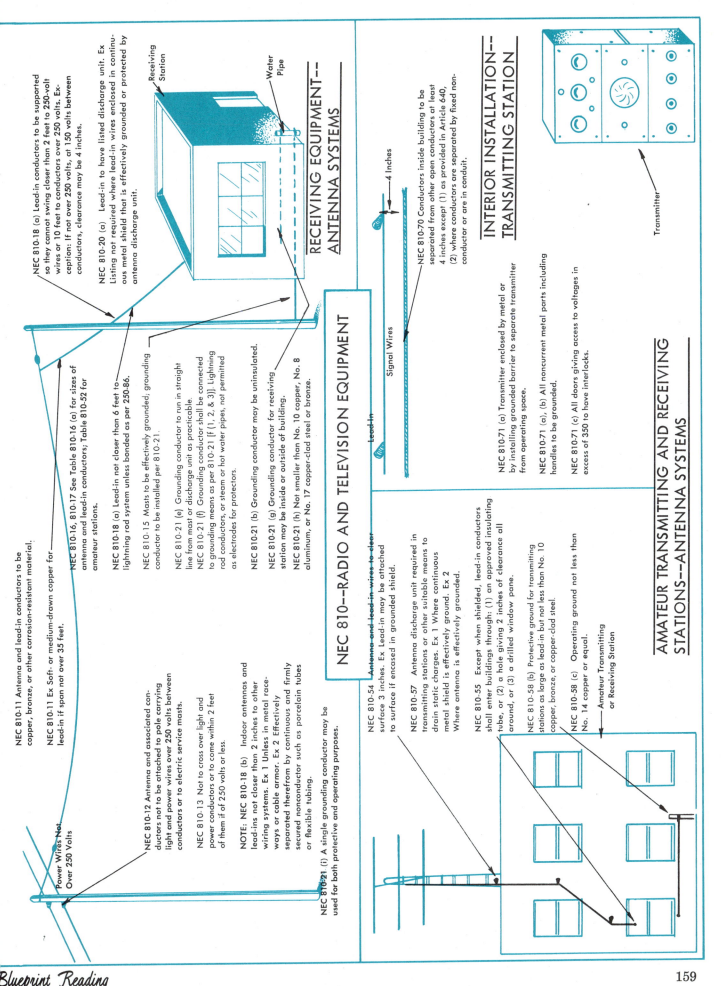

NEC 810-18 (a) Lead-in conductors to be supported so they cannot swing closer than 2 feet to 250-volt wires or 10 feet to conductors over 250 volts. Exception: If not over 250 volts, at 150 volts between conductors, clearance may be 4 inches.

NEC 810-20 (a) Lead-in to have listed discharge unit. Ex Listing not required where lead-in wires enclosed in continuous metal shield that is effectively grounded or protected by antenna discharge unit.

Receiving Station

Water Pipe

RECEIVING EQUIPMENT-- ANTENNA SYSTEMS

NEC 810-11 Antenna and lead-in conductors to be copper, bronze, or other corrosion-resistant material.

NEC 810-11 Ex Soft- or medium-drawn copper for lead-in if span not over 35 feet.

NEC 810-16, 810-17 See Table 810-16 (a) for sizes of antenna and lead-in conductors; Table 810-52 for amateur stations.

NEC 810-18 (a) Lead-in not closer than 6 feet to lightning rod system unless bonded as per 250-86.

NEC 810-15 Masts to be effectively grounded; grounding conductor to be installed per 810-21.

NEC 810-21 (e) Grounding conductor to run in straight line from mast or discharge unit as practicable.

NEC 810-21 (f) Grounding conductor shall be connected to grounding means as per 810-21 [f (1, 2, & 3)]. Lightning rod conductors, or steam or hot water pipes, not permitted as electrodes for protectors.

NEC 810-21 (b) Grounding conductor may be uninsulated.

NEC 810-21 (g) Grounding conductor for receiving station may be inside or outside of building.

NEC 810-21 (h) Not smaller than No. 10 copper, No. 8 aluminum, or No. 17 copper-clad steel or bronze.

NEC 810-12 Antenna and associated conductors not to be attached to pole carrying light and power wires over 250 volts between conductors or to electric service masts.

NEC 810-13 Not to cross over light and power conductors or to come within 2 feet of them if of 250 volts or less.

NOTE: NEC 810-18 (b) Indoor antennas and lead-ins not closer than 2 inches to other wiring systems. Ex 1 Unless in metal raceways or cable armor. Ex 2 Effectively separated therefrom by continuous and firmly secured nonconductor such as porcelain tubes or flexible tubing.

NEC 810-21 (i) A single grounding conductor may be used for both protective and operating purposes.

Power Wires Not Over 250 Volts

NEC 810--RADIO AND TELEVISION EQUIPMENT

Lead-In

Signal Wires

4 Inches

NEC 810-70 Conductors inside building to be separated from other open conductors at least 4 inches except (1) as provided in Article 640, (2) where conductors are separated by fixed nonconductor or are in conduit.

INTERIOR INSTALLATION-- TRANSMITTING STATION

Transmitter

NEC 810-71 (a) Transmitter enclosed by metal or by installing grounded barrier to separate transmitter from operating space.

NEC 810-71 (a), (b) All noncurrent metal parts including handles to be grounded.

NEC 810-71 (c) All doors giving access to voltages in excess of 350 to have interlocks.

AMATEUR TRANSMITTING AND RECEIVING STATIONS--ANTENNA SYSTEMS

NEC 810-54 Antenna and lead-in wires to clear surface 3 inches. Ex Lead-in may be attached to surface if encased in grounded shield.

NEC 810-57 Antenna discharge unit required in transmitting stations or other suitable means to drain static charges. Ex 1 Where continuous metal shield is effectively ground. Ex 2 Where antenna is effectively grounded.

NEC 810-55 Except when shielded, lead-in conductors shall enter buildings through: (1) an approved insulating tube, or (2) a hole giving 2 inches of clearance all around, or (3) a drilled window pane.

NEC 810-58 (b) Protective ground for transmitting stations as large as lead-in but not less than No. 10 copper, bronze, or copper-clad steel.

NEC 810-58 (c) Operating ground not less than No. 14 copper or equal.

Amateur Transmitting or Receiving Station

CAPACITORS

C = capacitance in farads   k = 1000
V = voltage   Xc = capacitive reactance
A = current   F = Hertz (cycles per second)

Capacitors in parallel
$$C_t = C_1 + C_2 + C_3 \text{ etc.}$$

Capacitors in series
$$\frac{1}{C_t} = \frac{1}{C_1} + \frac{1}{C_2} + \frac{1}{C_3} \quad \text{etc.}$$

Capacitive Reactance (Xc)
$$X_c = \frac{1}{2\pi FC}$$

Capacitance (C)
$$C = \frac{kVAR \; 10^3}{(2\pi F)(kV)^2} \quad = \quad \frac{10^6}{(2\pi F) \; Xc}$$

Capacitive kVAR's
$$kVAR = \frac{(2\pi F)C \; (kV)^2}{10^3} \quad \text{or} \quad \frac{10^3 \; (kV)^2}{Xc}$$

To determine capacitor rating when capacitor is operated at voltage or frequency below the rated value.
kVAR = Actual burden in kVA
E = Working voltage   C = Rated capacity in micro-farads   F = Working Hertz

$$kVAR = \frac{CE^2 \; 2\pi F}{10^3}$$

Capacitor rating decreases with the frequency and as the square of voltage if the frequency or voltage used is lower than that listed on the nameplate.

$$kVAR = \text{Rated } kVAR \times \left(\frac{\text{applied volts}}{\text{rated volts}}\right)^2$$

$$kVAR = \text{Rated } kVAR \times \left(\frac{\text{applied frequency}}{\text{rated frequency}}\right)$$

TRANSFORMER CONNECTIONS AND RATINGS

$$\text{Efficiency} = \frac{kVA \text{ (secondary)}}{kVA \text{ (primary)}} = \frac{\text{Output}}{\text{Input}}$$

Neglecting primary losses:
kVA (primary) = kVA (secondary)

Three single-phase transformers connected and operating as a three-phase unit:
Total kVA capacity = kVA of each x 3 or the total of the three units.

Two single-phase transformers connected as a three-phase open-delta bank:
Total kVA capacity = Total kVA of both units x .866

Three single-phase transformers operating as three-phase open-delta and one single-phase transformer is lost:
Total kVA = Total kVA of all units x .577

TRANSFORMER FORMULAS

Tp = Turns in primary   Ts = Turns in secondary
Ep = Primary voltage   Es = Secondary voltage
Ip = Primary current   Is = Secondary current
Neglecting transformer losses:
kVA primary (input) = kVA secondary (output) and
Ip Tp (amp turns of primary) = Is Ts (amp turns of secondary).

To find voltage, current and turns:

$$Ep \times Ip = Es \times Is \qquad Ep = \frac{Es \times Is}{Ip} \qquad Ip = \frac{Es \times Is}{Ep}$$

$$Es = \frac{Ep \times Ip}{Is} \qquad Is = \frac{Ep \times Ip}{Es}$$

$$Ip \times Tp = Is \times Ts \qquad Tp = \frac{Is \times Ts}{Ip} \qquad Ip = \frac{Is \times Ts}{Tp}$$

$$Ts = \frac{Ip \times Tp}{Is} \qquad Is = \frac{Ip \times Tp}{Ts}$$

$$Ep \times Ts = Es \times Tp \qquad Ep = \frac{Es \times Tp}{Ts} \qquad Tp = \frac{Ep \times Ts}{Es}$$

$$Es = \frac{Ep \times Ts}{Tp} \qquad Ts = \frac{Es \times Tp}{Ep}$$

AUTOTRANSFORMERS

Buck and Boost autotransformers (used as per manufacturer's specifications and installed as per NEC)
CHECK ALL LOCAL AND NATIONAL CODES RELATING TO THE USE OF AUTOTRANSFORMERS.
The basic transformer formula applies:
kVA input = kVA output (load)
Single-phase buck and boost transformer connected as an autotransformer to boost 208 volt to 240 volt or to buck 240 volt to 208 volt.
Total kVA capacity = 7.5 x kVA nameplate rating.
Two single-phase transformers connected as a buck or boost autotransformer to boost three-phase 208 volt primary to 230 volt secondary or to buck three-phase 230 volt primary to 208 volt secondary.
Total three-phase kVA capacity = (Total kVA nameplate rating x 10) x .866

See page 124 this book which shows a detailed drawing of a 1 kVA volt, 208/32 volt, 2-winding single-phase transformer connected as an autotransformer to boost the voltage from 208 V to 240 volt. This 1 kVA transformer feeds a 7.5 kVA load and conforms to NEC 210-9.

Student
Number_____

Instructor's
Name_____

INSTRUCTIONS: The following statements are either True or False. Draw a circle around the **T** if the statement is True, or around the **F** if it is False. The questions are to be answered, as nearly as possible, in accordance with the specific provisions of the NEC.

• True or False •

1. Hazardous concentrations of gas or vapor are not normally present in Class I, Division 1, locations. T F

2. Locations where concentrations of explosive dust are likely to be encountered are designated as Class II, Division 1. T F

3. A location where cloth is woven shall be designated as Class III, Division 1. T F

4. Seals must be installed in conduit runs passing between two adjacent Class I, Division 2, rooms. T F

5. General-use circuit breaker type lighting panels may be used in Class I, Division 2, locations. T F

6. Open squirrel-cage induction motors which have no sliding contacts may be used in Class I, Division 1, locations. T F

7. Askarel-filled transformers may be used in Class II, Division 2, locations. T F

8. Dust-tight receptacles for attachment plugs are acceptable in Class II, Division 1, locations. T F

9. In a commercial garage, a stockroom whose floor level is 18 inches higher than that of the garage would normally be considered a non-hazardous area. T F

10. Portable appliances used inside an aircraft hangar need be approved only for Class I, Division 2, locations. T F

11. Open outside areas within 20 feet horizontally from an underground gasoline tank fill pipe shall be considered as a Class I, Division 2, location upward to a height of 18 inches. T F

12. Open spaces within 20 feet horizontally of a gas-dispensing pump are designated as Class I, Division 1, locations. T F

Continued on Next Page

13. Spray booths may be illuminated through suitable glass panels. T F

14. In general, explosion-proof portable electric lamps may be used inside a paint spray booth during operation. T F

15. Storage batteries, motor-generator sets, or isolating transformers may be used to provide ungrounded circuits in anesthetizing locations. T F

16. Surgical fixtures more than 5 feet above the floor of an operating room may be installed on a grounded circuit. T F

17. The case of a low-voltage isolating-type transformer supplying circuits within an anesthetizing location shall not be grounded. T F

18. Resistance and reactor type dimmers for stage lighting may be placed in the grounded conductor of a circuit. T F

19. Lamps in cellulose nitrate film vaults may be installed in general-use fixtures provided they are equipped with suitable guards. T F

20. Branch circuits, which supply only electrical-discharge sign transformers shall not be rated in excess of 20 amps. T F

21. With outline lighting, stranded conductors need not be soldered to terminals when pin-type sockets are used. T F

22. A connection on the supply side of the service disconnect is never permitted as an emergency service. T F

23. A plug receptacle exclusively for the janitor's use may not be tapped from the emergency circuit wires. T F

24. Power in a Class I remote-control and signal circuit is limited. T F

25. Telephone wires are permitted on the same crossarm with light or power wires if separated therefrom by a distance not less than 2 feet. T F

TRADE COMPETENCY TEST NO. 5B

Student
Number _____

Instructor's
Name _____

INSTRUCTIONS: In the space provided at the right of each of the following questions, write **a, b, c,** or **d** to indicate which of the alternatives given makes it most nearly correct. The questions are to be answered, as nearly as possible, in accordance with the specific provisions of the NEC.

• Multiple-Choice •

1. Locations where combustible dust is normally in the air are designated as:

   (a) Class I, Division 2,   (b) Class II, Division 1,   (c) Class II, Division 2,   (d) Class III, Division 1   1. _____

2. Locations where combustible fibers are stored are designated as:

   (a) Class II, Division 2,   (b) Class III, Division 1,   (c) Class III, Division 2,   (d) non-hazardous   2. _____

3. For limited flexibility for motor connections in a Class I, Division 2, location, flexible conduit:

   (a) must be explosion-proof,   (b) shall not be used,   (c) may be standard flexible metal conduit,   (d) must be liquid-tight flexible conduit or equal   3. _____

4. Sealing compound is employed with mineral-insulated cable, in a Class I location for the purpose of:

   (a) preventing passage of gas or vapor,   (b) excluding moisture,   (c) limiting extent of a possible explosion,   (d) preventing escape of insulating powder   4. _____

5. The NEC provides that in Class I, Division 1, locations conduit seals shall be placed not farther from spark-producing devices than:

   (a) 18 inches,   (b) 24 inches,   (c) 12 inches,   (d) 30 inches   5. _____

6. In Class I, Division 1, locations the NEC requires conduit seals adjacent to boxes containing splices if the conduit size is equal to or larger than:

   (a) ¾ inch,   (b) 1½ inches,   (c) 1 inch,   (d) 2 inches   6. _____

Continued on Next Page

Blueprint Reading   163

TEAR OUT HERE

7. Fused isolating switches for transformers not intended to interrupt current in a Class 1, Division 2 location:

(a) may be of general-use type, (b) must be explosion-proof type, (c) must have interlocking devices, (d) may not have doors or openings in the hazardous area

7. _____

8. Where magnesium, aluminum, or aluminum-bronze powders may be present, transformers:

(a) must be dust-tight, (b) may be pipe-ventilated, (c) must be approved for Class II, Division I, locations, (d) are not allowed

8. _____

9. For general wiring in Class I, Division 1, locations, it is permissible to use:

(a) rigid metal conduit, (b) electrical metallic tubing, (c) flexible metal conduit, (d) all three

9. _____

10. In a Class II location, where electrically-conducting dust is present, flexible connections at motors could be made with:

(a) flexible metal conduit, (b) Type AC armored cable, (c) hard usage cord, (d) liquid-tight flexible metal conduit with approved fittings.

10. _____

11. In commercial garages generally the floor area to a height of 18 inches above grade is designated as:

(a) Class I, Division I, (b) Class I, Division 2, (c) Class II, Division 2, (d) Class II, Division 1

11. _____

12. Storerooms and similar areas adjacent to aircraft hangars but effectively isolated shall be designated as:

(a) Class I, Division 2, (b) Class II, Division 1, (c) Class II, Division 2, (d) non-hazardous

12. _____

13. Areas within 18" horizontally of gas-dispensing pumps are considered Class I, Division 1, locations to a height of:

(a) 12 inches, (b) 30 inches, (c) 18 inches, (d) 48 inches

13. _____

14. Locations where flammable paints are dried but in which the ventilating equipment is interlocked with the electrical equipment may be designated as:

(a) Class I, Division 2, (b) non-hazardous, (c) Class II, Division 2, (d) Class II, Division 1

14. _____

TEAR OUT HERE

15. A well-ventilated room where flammable anesthetics are stored in approved containers shall be considered to be a:

    (a) Class I, Division 1, area, (b) Class I, Division 2, (c) Class II, Division 2, area, (d) nonhazardous area

    15. _____

16. The Code requires that the primary voltage of an isolating transformer used in an anesthetizing location shall not exceed:

    (a) 300 volts, (b) 150 volts, (c) 250 volts, (d) 600 volts

    16. _____

17. In motion-picture studios, feeder conductors to the stage may be protected, with respect to ampacity, at a maximum value of:

    (a) 200%, (b) 250%, (c) 400%, (d) 500%

    17. _____

18. A branch circuit feeding a sign which has a combination of lamps and transformers shall not exceed the rating of:

    (a) 15 amps, (b) 20 amps, (c) 30 amps, (d) 50 amps

    18. _____

19. Each commercial building and occupancy with ground floor footage shall have at least one outside sign outlet rated at:

    (a) 15 amps, (b) 20 amps, (c) 25 amps, (d) 30 amps

    19. _____

20. The maximum fill in wireways or gutters used for sound-recording installations is:

    (a) 20%, (b) 58%, (c) 50%, (d) 75%

    20. _____

21. A storage battery supplying emergency lighting and power shall maintain not less than 87½ percent of full voltage at total load for a period of at least:

    (a) 2 hours, (b) 1½ hours, (c) 1 hour, (d) ¼ hour

    21. _____

22. The rating of a lampholder on a circuit which operates at a voltage less than 50 volts shall be at least:

    (a) 220 watts, (b) 660 watts, (c) 330 watts, (d) 550 watts

    22. _____

23. On circuits of 600 volts or less, overhead feeder spans up to 50 feet in length shall have copper conductors not smaller than:

    (a) No. 8, (b) No. 12, (c) No. 6, (d) No. 10

    23. _____

24. According to the National Electrical Code, conductors on poles, where not placed on racks or brackets, shall be separated not less than:

    (a) 6 inches, (b) 12 inches, (c) 18 inches, (d) 24 inches

    24. _____

TEAR OUT HERE

25. The Code provides that unshielded lead-in conductors of amateur transmitting stations shall clear the building surface which is wired over by a distance not less than:

   (a) 1 inch,   (b) 2 inches,   (c) 3 inches,   (d) 4 inches

   25. _____

TRADE COMPETENCY TEST NO. 5C

Student
Number_____

Instructor's
Name_____

INSTRUCTIONS: Complete each of the following statements, by writing the missing word or words in the space provided at the right of each question, to make a true statement. The questions are to be answered, as nearly as possible, in accordance with the specific provisions of the NEC.

• Completion •

1. In Class I, Division 2, locations, concentrations of explosive gases are normally _____ in closed containers or systems.

1. _____

2. Class III locations are those where _____ _____ fibers are present.

2. _____

3. Only _____ _____ flexible conduit is permitted in Class I, Division 1, areas.

3. _____

4. In a conduit run leaving a Class I, Division 1, location, the seal may be placed on either side of the _____ .

4. _____

5. A flexible cord may be used in a Class I location between a portable appliance and the _____ portion of its supply circuit.

5. _____

6. In Class I, Division 1, locations pendant fixtures with rigid conduit stems more than 12 inches long must be _____ .

6. _____

7. A horizontal run of conduit not less than _____ feet long may be employed as the sealing device in a Class II location.

7. _____

8. In a parking garage where only limited maintenance work is done, an area up to 18 inches is considered to be a _____ area.

8. _____

9. Where pendants are used in a commercial garage, flexible cords suitable for the type of service and approved for _____ usage shall be used.

9. _____

10. Mobile aircraft energizers shall carry a sign reading: "WARNING — KEEP _____ FEET CLEAR OF AIRCRAFT ENGINES AND FUEL TANK AREAS".

10. _____

11. Underground conduit runs within 20 feet of a gasoline dispensing location shall be considered to be in a Class I, Division _____, location.

11. _____

Continued on Next Page

12. The circuit to a single-phase motor on a gasoline dispensing pump shall have a _____ switch or disconnecting means.

12. _____

13. Flexible cords feeding 120-volt equipment within a hazardous anesthetizing location shall be of the _____ usage type.

13. _____

14. The secondary voltage of an isolating transformer used in an anesthetizing location may not exceed _____ volts.

14. _____

15. Surgical and other lighting fixtures above hazardous anesthetizing locations shall conform to NEC Section _____.

15. _____

16. Disappearing stage footlights shall have provisions for _____ disconnecting the current supply when the footlights are replaced in the recess designed for them.

16. _____

17. A night club lighting dimmer installed in an ungrounded conductor shall have overcurrent protection rated at no more than _____ percent of its own rating.

17. _____

18. Incandescent lamps less than 8 feet from the floor of a theatrical dressing room shall be equipped with _____ guards.

18. _____

19. A switch controlling a plug receptacle in a theatrical dressing room must be equipped with a _____ .

19. _____

20. Metal raceway for a single conductor from a sign transformer secondary shall not exceed a length of _____ feet.

20. _____

21. Electric discharge tubing operating at more than 7,500 volts shall clear the adjacent surface by not less than _____ inch(es).

21. _____

22. Transformers for outline lighting installations shall normally have secondary current rating not exceeding _____ milliamperes.

22. _____

23. A permanently installed electric sign shall be controlled by an _____ operable switch or circuit breaker.

23. _____

24. Audible and visual signal devices shall be employed, where practicable, to give warning of _____ of emergency power.

24. _____

25. Connections ahead of, but not _____ the normal power service disconnecting means, is permitted as a source of power for an emergency service.

25. _____

FINAL EXAMINATION — TEST NO. 6A

Student
Number _____

Instructor's
Name _____

INSTRUCTIONS: The following statements are either True or False. Draw a circle around the **T** if the statement is True, or around the **F** if it is False. The questions are to be answered, as nearly as possible, in accordance with the specific provisions of the NEC.

• True or False •

1. The smallest permissible hard-drawn copper service-drop conductor is No. 10.   T   F

2. The service disconnecting means may consist of not more than six switches or circuit breakers.   T   F

3. The grounding electrode conductor for a small single-family residence shall not be spliced.   T   F

4. The NEC states that any available wall space in a residence must be within 7 feet of a plug receptacle.   T   F

5. It is permissible to install 10 No. 12 conductors in a 4-inch by 4-inch by 1½-inch deep box that also contains one switch.   T   F

6. Switch enclosures, under certain conditions, may be used as junction boxes.   T   F

7. Under certain conditions, a 20-foot length of metal raceway need not be grounded.   T   F

8. An AC-DC general-use snap switch may be used to control tungsten-filament lamps in an apartment living room.   T   F

9. Embedded electric space heating cables shall not be spliced.   T   F

10. A separable connector may be used for ease in servicing a wall-mounted oven.   T   F

11. The screw shell of a fuseholder shall not be connected to the line side of the circuit.   T   F

12. A tap, for an individual lampholder or fixture more than 12 inches long may not be smaller than the branch circuit conductors.   T   F

13. Under certain conditions the neutral feeder conductor is permitted to be smaller than the ungrounded conductors.   T   F

14. Service-entrance conductors, in general, shall not be spliced.   T   F

Continued on Next Page

15. The maximum rating of a 250-volt cartridge fuse is 400 amps. T F

16. No. 18 copper wire may be used to ground a portable device connected to a 30-amp circuit. T F

17. Conductors in underfloor raceway systems shall not be tapped at individual outlets. T F

18. A header used with precast cellular concrete floor raceways shall not be run at right angles to the cells. T F

19. Cross-sectional area of all conductors in any part of an auxiliary gutter shall not exceed 15 percent of the gutter's cross-sectional area. T F

20. An AC general-use snap switch controlling an inductive load shall have an ampere rating equal to 150 percent of the load. T F

21. Secondary circuits of wound-rotor induction motors require overload protection. T F

22. A ½-hp pump motor controlled by a float switch requires running overcurrent protection. T F

23. Transformers are never required to have overcurrent devices in both primary and secondary circuits. T F

24. The primary windings of potential transformers installed indoors shall always be fused. T F

25. An 8-amp, 115-volt, single-phase sealed (hermetic-type) motor compressor must show the motor locked-rotor current on its nameplate. T F

26. A motor branch circuit short-circuit device may sometimes be rated in excess of 300 percent of full-load motor current. T F

27. The minimum size of Type THW copper conductors supplying a 70-amp, 480-volt transformer arc welder used on a 30 percent duty application is No. 8. T F

28. A rotating frequency changer is not classed as a generator with respect to its application in a dielectric heating device. T F

29. A 1/30-hp split-phase motor which drives a temperature recording machine requires running overcurrent protection. T F

30. The maximum size of flexible metallic tubing shall be 2 inches. T F

TEAR OUT HERE

National Electrical Code

FINAL EXAMINATION — TEST NO. 6B

Student
Number _____

Instructor's
Name _____

INSTRUCTIONS: In the space provided at the right of each of the following questions, write **a, b, c,** or **d** to indicate which of the alternatives given makes it most nearly correct. The questions are to be answered, as nearly as possible, in accordance with the specific provisions of the NEC.

• Multiple-Choice •

1. When the initial computed load for a one-family dwelling amounts to 10 kW, the rating of the service disconnecting means shall not be less than:

   (a) 70 amps, (b) 100 amps, (c) 50 amps, (d) 60 amps   1. _____

2. The minimum clearance from ground of service-drop conductors operating at 600 volts or less, which pass over public streets, alleys, roads, and driveways, on other than residential property, shall be:

   (a) 10 feet, (b) 12 feet, (c) 15 feet, (d) 18 feet   2. _____

3. Where the residential lighting load is computed on the 3 watts-per-square-foot basis the Code requires one 15-amp, 120-volt branch circuit for each:

   (a) 375 sq. ft. (b) 500 sq. ft. (c) 575 sq. ft. (d) 600 sq. ft.   3. _____

4. The service or feeder capacity necessary for four 10 kW electric ranges is:

   (a) 15 kW, (b) 17 kW, (c) 19 kW, (d) 21 kW   4. _____

5. Floor mounted flat conductor cable (Type FCC) shall be covered with carpet squares no larger than:

   (a) 20 inches square   (b) 24 inches square
   (c) 30 inches square   (d) 36 inches square   5. _____

6. The smallest permissible driven ground rod electrode made of nonferrous materials is:

   (a) ⅜ inch, (b) ½ inch, (c) ⅝ inch, (d) ¾ inch   6. _____

7. Heating cables installed on dry-board ceilings shall be secured at intervals not exceeding:

   (a) 10 inches, (b) 14 inches, (c) 16 inches, (d) 12 inches   7. _____

Continued on Next Page

8. The feeder load for three electric ranges, one 13 kW, one 18 kW, and one 20 kW, should be assessed at:

    (a) 17.5 kW, (b) 21 kW, (c) 19 kW, (d) 15 kW

    8. _____

9. A feeder supplying two 4-kW wall-mounted ovens and a 5-kW counter-mounted cooking unit shall have capacity for:

    (a) 9.35 kW, (b) 7.15 kW, (c) 10.4 kW, (d) 8.4 kW

    9. _____

10. The ampacity of a No. 3/0 Type THW copper conductor at a room temperature of 50 degrees C is:

    (a) 125 amps, (b) 150 amps, (c) 175 amps, (d) 200 amps

    10. _____

11. To supply a feeder for night lights in an apartment house, the required size of Type THW copper conductors for a continuous 100-amp load shall not be smaller than:

    (a) No. 0, (b) No. 1, (c) No. 2, (d) No. 3

    11. _____

12. The maximum rating of a cord-and-plug connected appliance used on a 20-amp branch circuit shall be:

    (a) 16 amps, (b) 10 amps, (c) 20 amps, (d) 12 amps

    12. _____

13. The Code lists the unit load for an office building as:

    (a) 2 watts per sq ft, (b) 2.5 watts per sq ft, (c) 3.5 watts per sq ft, (d) 5 watts per sq ft

    13. _____

14. The demand-factor which may be applied to the neutral feeder load for electric ranges is:

    (a) 40%, (b) 80%, (c) 70%, (d) 60%

    14. _____

15. Two or more services are permitted when the rating of the service is over:

    (a) 1500 amps, (b) 2000 amps, (c) 2500 amps, (d) 3000 amps

    15. _____

16. A lighting and power service has two 3/0 ungrounded copper conductors and a 1/0 neutral. A copper grounding electrode conductor attached to a water-pipe electrode shall not be smaller than:

    (a) No. 4, (b) No. 6, (c) No. 8, (d) No. 2

    16. _____

TEAR OUT HERE

National Electrical Code

17. A wood building, without water, is built on the earth without footings. If a 3000-amp, 460-volt service is installed and grounded by ground rods, the minimum size copper grounding electrode conductor would be:

    (a) No. 6, (b) No. 4, (c) No. 3/0, (d) 400 MCM

    17. _____

18. Where the automatic overcurrent device is set at 200 amps, the Code requires an aluminum equipment grounding conductor to be at least:

    (a) No. 4, (b) No. 6, (c) No. 8, (d) No. 2

    18. _____

19. The conductor fill of an underfloor raceway shall not exceed _____ percent of the interior area of the raceway.

    (a) 20, (b) 100, (c) 40, (d) 75

    19. _____

20. The largest size conductor permitted in underfloor raceway is:

    (a) 500 MCM, (b) No. 0, (c) 250 MCM, (d) determined by design

    20. _____

21. Branch-circuit conductors supplying a 50-hp, 3-phase, 460-volt induction motor shall be rated for at least:

    (a) 65 amps, (b) 130 amps, (c) 82 amps, (d) 91 amps

    21. _____

22. A circuit-breaker disconnecting means for this 50-hp motor shall have an ampere rating of at least:

    (a) 126 amps, (b) 75 amps, (c) 91 amps, (d) 63 amps

    22. _____

23. The absolute maximum size of nontime-delay fuses for branch-circuit overcurrent protection for this 50-hp motor would be:

    (a) 250 amps, (b) 200 amps, (c) 90 amps, (d) 300 amps

    23. _____

24. A 460-to-230-volt, 1.5 percent impedance, 100-amp rated primary current, askarel-insulated transformer is equipped with coordinated thermal overload protection and has proper secondary overcurrent protection. The primary circuit breaker shall not exceed:

    (a) 400 amps, (b) 250 amps, (c) 125 amps, (d) 600 amps

    24. _____

25. Where a 460/230-volt transformer (without coordinated thermal overload) has a 200-amp rated primary protected by a 500-amp fuse, the 230-volt secondary overcurrent device shall not exceed:

    (a) 400 amps, (b) 500 amps, (c) 600 amps, (d) 800 amps

    25. _____

TEAR OUT HERE

Blueprint Reading

173

26. A 5-hp, 3-phase, 230-volt squirrel-cage motor with no code letter is started at full voltage. The minimum ampacity of the branch-circuit conductor shall be:

    (a) 15.2 amps, (b) 18 amps, (c) 19 amps, (d) 20 amps

    26. _____

27. The maximum initial rating of branch-circuit nontime-delay fuses for this motor circuit shall be:

    (a) 40 amps, (b) 45 amps, (c) 50 amps, (d) 60 amps

    27. _____

28. If this motor is marked 40 degrees C and has a nameplate rating of 15 amps, the setting of an adjustable overload device shall not exceed:

    (a) 19 amps, (b) 20 amps, (c) 21 amps, (d) 25 amps

    28. _____

29. The rating of a fuse used as overload protection for the above motor would be:

    (a) 45 amps, (b) 40 amps, (c) 25 amps, (d) 20 amps

    29. _____

30. With no overcurrent protection in the secondary, the primary overcurrent device for a 600-volt dry-type transformer with a rated primary current of 100 amps shall not exceed:

    (a) 125 amps, (b) 250 amps, (c) 600 amps, (d) 150 amps

    30. _____

TEAR OUT HERE

FINAL EXAMINATION — TEST NO. 6C

Student
Number _____

Instructor's
Name _____

INSTRUCTIONS: Complete each of the following statements, by writing the missing word or words in the space provided at the right of each question, to make a true statement. The questions are to be answered, as nearly as possible, in accordance with the specific provisions of the NEC.

• Completion •

1. The clearance of service-drop conductors over residential driveways, that are not subject to truck traffic, shall be not less than _____ feet. This service is 120/240 volts.

   1. _____

2. The copper grounding electrode conductor for a 350 MCM, 120/240 volt service installation, if run to a water pipe electrode, shall not be smaller than No. _____.

   2. _____

3. If the copper grounding electrode conductor for this service is run to a made electrode it shall not be smaller than No. _____.

   3. _____

4. Under the optional calculation for a one-family residence, central space-heating load is assessed at _____ percent.

   4. _____

5. Type AC cable, except where fished or to fixtures within an accessible ceiling shall be secured every _____ feet.

   5. _____

6. The first 3000 watts of computed load in a multifamily dwelling shall be assessed at _____ percent.

   6. _____

7. The first 20,000 watts of computed load in a hotel shall be assessed at _____ percent.

   7. _____

8. When used on 20-amp circuits, receptacles shall be rated not less than _____ amps.

   8. _____

9. Leads of a 120-volt heating cable are marked with _____ .

   9. _____

10. When initially installed, flexible cords shall be used only in _____ lengths.

    10. _____

11. Except in dwellings each _____ feet or fraction of multioutlet assembly strip is usually considered a load of 180 volt-amps.

    11. _____

12. Voltage drop of power feeders should be limited to _____ percent.

    12. _____

13. A 120/240-volt feeder must have a rating of at least _____ amps to supply a 9 kW noncontinuous and a 12 kW continuous load.

    13. _____

14. If the allowable ampacity of a conductor is 55 amps, it will be considered as protected by a _____ amp fuse.

    14. _____

Continued on Next Page

TEAR OUT HERE

15. The ampacity of No. 2 Type THWN-THHN dual rated copper conductors, in rigid conduit, installed in a wet location is _____ amps.

15. _____

16. Liquidtight flexible nonmetallic conduit shall be limited in length to _____ feet.

16. _____

17. Metal surface raceways that are used for both signaling and lighting circuits shall have two separate compartments, each identified by contrasting colors of the _____ finish.

17. _____

18. Splices and taps in underfloor raceway installations shall be made only in _____ _____.

18. _____

19. Auxiliary gutters used in wet locations shall be of approved _____ construction.

19. _____

20. The motor _____ shall be in sight of the motor and the driven machinery location or shall be capable of being locked in the open position.

20. _____

21. Where thermal cutouts are used for motor overload protection for a three-phase motor, a total of _____ thermal cutouts shall be used.

21. _____

22. Receptacles in marinas and boatyards that provide shore power to boats, shall be rated not less than 20 Amps and shall be of the locking and _____ types.

22. _____

23. Nonmedical X-ray, high-voltage _____ cables shall be grounded.

23. _____

24. The rating of a single receptacle, installed on a 20-amp branch circuit, shall have a rating of _____ amps.

24. _____

25. The ampacity of conductors supplying an oscillator-type, induction-heating unit need not exceed its _____ rating.

25. _____

26. The full load current of a 5-hp, 3-phase, 208-volt induction motor is _____.

26. _____

27. Radius of bends in mineral-insulated cable shall be not less than _____ times cable diameter.

27. _____

28. All exposed noncurrent-carrying metal parts of a data processing system shall be _____.

28. _____

29. Horsepower rated switches are not required as a disconnecting means for 300-volt stationary motors of _____ or less.

29. _____

30. The controller disconnect shall be in sight of the _____.

30. _____

National Electrical Code

ENCLOSURES – ELECTRICAL EQUIPMENT (1000 VOLTS MAXIMUM)
NATIONAL ELECTRICAL MANUFACTURERS ASSOCIATION (NEMA)

INTRODUCTION

An enclosure is a surrounding case constructed to provide a degree of protection to personnel against incidental contact with the enclosed equipment and to provide a degree of protection to the enclosed equipment against specified environmental conditons.

A brief description of the more common types of enclosures used by the electrical industry relating to their environmental capabilities follows. Refer to the appropriate sections of this standards publication for more information regarding applications, features, and design tests.

NEMA Standard 1-10-1979.

NEC Art. 100 Basic definition of an enclosure is a case or housing of apparatus to prevent personnel from accidentally contacting energized parts, or to protect the equipment from physical damage.

DEFINITIONS PERTAINING TO NONHAZARDOUS LOCATIONS
Type 1 Enclosure

Type 1 enclosures are intended for indoor use primarily to provide a degree of protection against contact with the enclosed equipment.

NEMA Standard 1-10-1979.

NEMA TYPE I
ENCLOSURE

(Courtesy of Square D Company)

Type 2 Enclosure

Type 2 enclosures are intended for indoor use primarily to provide a degree of protection against limited amounts of falling water and dirt.

NEMA Standard 1-10-1979.

Type 3 Enclosure

Type 3 enclosures are intended for outdoor use primarily to provide a degree of protection against windblown dust, rain, sleet, and external ice formation.

NEMA Standard 1-10-1979.

Type 3R Enclosure

Type 3R enclosures are intended for outdoor use primarily to provide a degree of protection against falling rain, sleet, and external ice formation.

NEMA Standard 1-10-1979.

NEMA TYPE 3R
ENCLOSURE

(Courtesy of Square D Company)

Type 3S Enclosure

Type 3S enclosures are intended for outdoor use primarily to provide a degree of protection against windblown dust, rain, sleet, and provide for operation of external mechanisms when ice laden.

NEMA Standard 1-10-1979.

Type 4 Enclosure

Type 4 enclosures are intended for indoor or outdoor use primarily to provide a degree of protection against windblown dust and rain, splashing water, and hose-directed water.

NEMA Standard 1-10-1979.

Type 4X Enclosure

Type 4X enclosures are intended for indoor or outdoor use primarily to provide a degree of protection against corrosion, windblown dust and rain, splashing water, and hose-directed water.

NEMA Standard 1-10-1979.

ENCLOSURES – ELECTRICAL EQUIPMENT (1000 VOLTS MAXIMUM)
NATIONAL ELECTRICAL MANUFACTURERS ASSOCIATION (NEMA) – Continued

Type 5 Enclosure
Type 5 enclosures are intended for indoor use primarily to provide a degree of protection against dust and falling dirt.

NEMA Standard 1-10-1979.

Type 6 Enclosure
Type 6 enclosures are intended for indoor or outdoor use primarily to provide a degree of protection against the entry of water during occasional temporary submersion at a limited depth.

NEMA Standard 1-10-1979.

Type 6P Enclosure
Type 6P enclosures are intended for indoor or outdoor use primarily to provide a degree of protection against the entry of water during prolonged submersion at a limited depth.

NEMA Standard 1-10-1979.

Type 11 Enclosure
Type 11 enclosures are intended for indoor use primarily to provide, by oil immersion, a degree of protection to enclosed equipment against the corrisive effects of liquids and gases.

NEMA Standard 1-10-1979.

Type 12 Enclosure
Type 12 enclosures are intended for indoor use primarily to provide a degree of protection against dust, falling dirt, and dripping non-corrosive liquids.

NEMA Standard 1-10-1979.

NEMA TYPE 12 ENCLOSURE

(Courtesy of Square D Company)

Type 12K Enclosure
Type 12K enclosures with knockouts are intended for indoor use primarily to provide a degree of protection against dust, falling dirt, and dripping noncorrosive liquids other than at knockouts.

NEMA Standard 1-10-1979.

Type 13 Enclosure
Type 13 enclosures are intended for indoor use primarily to provide a degree of protection against dust, spraying of water, oil, and non-corrosive coolant.

NEMA Standard 1-10-1979.

DEFINITIONS PERTAINING TO HAZARDOUS (CLASSIFIED) LOCATIONS
Type 7 Enclosure
Type 7 enclosures are for use indoors in locations classified as Class I, Groups A, B, C, or D, as defined in the *National Electrical Code*.

NEMA Standard 1-10-1979.

NEMA TYPE 7 & 9 ENCLOSURE

(Courtesy of Square D Company)

Type 8 Enclosure
Type 8 enclosures are for indoor or outdoor use in locations classified as Class I, Groups A, B, C, or D, as defined in the *National Electrical Code*.

NEMA Standard 1-10-1979.

Type 9 Enclosure
Type 9 enclosures are for use in indoor locations classified as Class II, Groups E, F, or G, as defined in the *National Electrical Code*.

NEMA Standard 1-10-1979.

Type 10 Enclosure
Type 10 enclosures are constructed to meet the applicable requirements of the Mine Safety and Health Administration.

NEMA Standard 1-10-1979.

APPENDIX

Electrical Symbols Commonly Used on Blueprints

GENERAL OUTLETS

Ceiling   Wall

- ◯   ⊸◯   Outlet.
- Ⓑ   ⊸Ⓑ   Blanked Outlet.
- Ⓓ   Drop Cord.
- Ⓔ   ⊸Ⓔ   Electrical Outlet; for use only when circle used alone might be confused with columns, plumbing symbols, etc.
- Ⓕ   ⊸Ⓕ   Fan Outlet.
- Ⓙ   ⊸Ⓙ   Junction Box.
- Ⓛ   ⊸Ⓛ   Lamp Holder.
- Ⓛ$_{PS}$ ⊸Ⓛ$_{PS}$ Lamp Holder with Pull Switch.
- Ⓢ   ⊸Ⓢ   Pull Switch.
- Ⓥ   ⊸Ⓥ   Outlet for Vapor Discharge Lamp.
- Ⓧ   ⊸Ⓧ   Exit Light Outlet.
- Ⓒ   ⊸Ⓒ   Clock Outlet. (Specify Voltage.)

CONVENIENCE OUTLETS

- Duplex Convenience Outlet.
- Convenience Outlet other than Duplex. 1=Single, 3=Triplex, etc.
- Weatherproof Convenience Outlet.
- Range Outlet.
- Switch and Convenience Outlet.
- Radio and Convenience Outlet.
- Special Purpose Outlet. (Des. in Spec.)
- Floor Outlet.

SWITCH OUTLETS

- S   Single Pole Switch.
- S$_2$ Double Pole Switch.
- S$_3$ Three-Way Switch.
- S$_4$ Four-Way Switch.
- S$_D$ Automatic Door Switch.
- S$_E$ Electrolier Switch.
- S$_K$ Key Operated Switch.
- S$_P$ Switch and Pilot Lamp.
- S$_{CB}$ Circuit Breaker.
- S$_{WCB}$ Weatherproof Circuit Breaker.
- S$_{MC}$ Momentary Contact Switch.
- S$_{RC}$ Remote Control Switch.
- S$_{WP}$ Weatherproof Switch.
- S$_F$ Fused Switch.
- S$_{WF}$ Weatherproof Fused Switch.

SPECIAL OUTLETS

- ◯$_{a,b,c,etc}$
- ⊸$_{a,b,c,etc}$
- S$_{a,b,c,etc}$

Any Standard Symbol as given above with the addition of a lower case subscript letter may be used to designate some special variation of Standard Equipment of particular interest in a specific set of Architectural Plans.

When used, they must be listed in the Key of Symbols on each drawing and if necessary further described in the specifications.

Courtesy of American National Standards Institute New York, N.Y.

PANELS, CIRCUITS, AND MISCELLANEOUS

- ■   Lighting Panel.
- ▨   Power Panel.
- ——   Branch Circuit; Concealed in Ceiling or Wall.
- - - -   Branch Circuit; Concealed in Floor.
- -----   Branch Circuit; Exposed.
- →→   Home Run to Panel Board. Indicate number of Circuits by number of arrows.
  Note: Any circuit without further designation indicates a two-wire circuit. For a greater number of wires indicate as follows ─⧸⧸⧸─ (3 wires) ⧸⧸ ⧸⧸ (4 wires), etc.
- ——   Feeders. Note: Use heavy lines and designate by number corresponding to listing in Feeder Schedule.
- ⧮□⧮   Underfloor Duct and Junction Box. Triple System. For double or single systems, eliminate 1 or 2 lines. This symbol equally adaptable to auxiliary system layouts.
- Ⓖ   Generator.
- Ⓜ   Motor.
- Ⓘ   Instrument.
- Ⓣ   Power Transformer. (Or draw to scale.)
- ⊠   Controller.
- ⎚   Isolating Switch.

AUXILIARY SYSTEMS

- ⊡   Push Button.
- ⊏⧸   Buzzer.
- ⊏◻   Bell.
- ◇   Annunciator.
- ◀   Outside Telephone.
- ⋈   Interconnecting Telephone.
- ⋈   Telephone Switchboard.
- Ⓣ   Bell Ringing Transformer.
- Ⓓ   Electric Door Opener.
- F▷   Fire Alarm Bell.
- F   Fire Alarm Station.
- ⊠   City Fire Alarm Station.
- FA   Fire Alarm Central Station.
- FS   Automatic Fire Alarm Device.
- W   Watchman's Station.
- [W]   Watchman's Central Station.
- H   Horn.
- N   Nurse's Signal Plug.
- M   Maid's Signal Plug.
- R   Radio Outlet.
- SC   Signal Central Station.
- ◻   Interconnection Box.
- �|ᵗ   Battery.
- - - -   Auxiliary System Circuits.
  Note: Any line without further designation indicates a 2-Wire System. For a greater number of wires designate with numerals in manner similar to - - 12-No. 18W-¾" C., or designate by number corresponding to listing in Schedule.
- ◻$_{a,b,c}$ Special Auxiliary Outlets. Subscript letters refer to notes on plans or detailed description in specs.

Blueprint Reading   179

ELECTRICAL SYMBOLS

ELECTRICAL SYMBOLS: Normally a list of Electrical Symbols is submitted with each set of electrical blueprints. Most of the material and equipment on a blueprint can be readily identified by using the standard ANSI Electrical Symbols, but in some cases where special material or equipment is used, such as Automatic Fire Alarm Systems, Smoke Detectors or receptacles for special equipment, it becomes necessary for the engineer to concoct a symbol which will serve to identify these special items. The following is a typical application of Electrical Symbols which would be submitted with a blueprint.

INCANDESCENT OR H.I.D. FIXTURE

INCANDESCENT OR H.I.D. WALL BRACKET

FLUORESCENT FIXTURE

EXIT LIGHT

FIXTURE TYPE
CIRCUIT
SWITCHING
} TYPICAL OUTLET DESIGNATION

JUNCTION BOX

CLOCK OUTLET ....7'-6"

DELTA

PHASE

DUPLEX RECEPTACLE ....1'-0"

DUPLEX RECEPTACLE SPLIT WIRED ....1'-0"

120/208, 1φ COMBINATION RECEPTACLE ..1'-0"

FLOOR OUTLET (ATTACH ADDITIONAL SYMBOL FOR RECEPT. TEL., ETC.)

SPECIAL OUTLET—AS NOTED

240V, 3W, 4P RECEPTACLE—AMPS AS NOTED

240V, 2W, 3P RECEPTACLE—AMPS AS NOTED

120V, 2W 3P, 30-AMP RECEPTACLE

GROUND

FUSE

CIRCUIT BREAKER

S SINGLE POLE SWITCH ....4'-0"

S₂ DOUBLE POLE SWITCH ....4'-0"

S₃ THREE-WAY SWITCH ....4'-0"

S₄ FOUR-WAY SWITCH ....4'-0"

S_D DIMMER SWITCH ....4'-0"

S_K KEY OPERATED SWITCH ....4'-0"

S_P SWITCH AND PILOT LIGHT ....4'-0"

S_MC MOMENTARY CONTACT SWITCH ....4'-0"

BRANCH CIRCUIT PANEL

DISTRIBUTION PANEL

EXPOSED CONDUIT

CONDUIT CONCEALED IN CEILING OR WALLS

CONDUIT CONCEALED IN OR UNDER FLOOR

SURFACE METAL RACEWAY

CONDUIT UP

CONDUIT DOWN
CROSSLINES INDICATE NUMBER OF WIRES—SEE SPECIFICATIONS

UNDERFLOOR DUCT & JUNCTION BOX, NUMBER OF LINES INDICATES NUMBER OF DUCTS

MOTOR—NUMBER, IF ANY, REFERS TO MOTOR SCHEDULE

S_T THERMAL SWITCH

MOTOR STARTER

SAFETY SWITCH

AQUASTAT ....5'-0"

THERMOSTAT ....5'-0"

HUMIDISTAT

CONTROL DEVICE—LETTERS INDICATE SPECIAL FUNCTION SUCH AS:
HOA—HAND OFF AUTO
PBPL—PUSH BUTTON, PILOT LIGHT

REMOTE CONTROL SWITCH OR CONTACTOR

ALL MOUNTING HEIGHTS ARE AS LISTED ABOVE UNLESS NOTED OTHERWISE ON PLANS.

T	TRANSFORMER	◀	TELEPHONE OUTLET ....1'-0''
⊡	PUSH BUTTON	◁	INTERCOM OUTLET
□/	BUZZER	Ⓢ	SPEAKER
□כ	BELL	Ⓣ	TELEVISION OUTLET
□⊢	CHIME	Ⓜ	MICROPHONE OUTLET

FIRE ALARM SYMBOLS

F	FIRE ALARM SENDING STATION ....5'-0''	FA	FIRE ALARM PANEL
F◁	FIRE ALARM HORN ....7'-6''	F⊢	FIRE ALARM PRE-SIGNAL CHIME
Fכ	FIRE ALARM BELL ....7'-6''	A_S	AUTOMATIC FIRE ALARM STATION—IONIZATION SMOKE DETECTOR
A	AUTOMATIC FIRE ALARM STATION	F_D	MAGNETIC DOOR HOLDER
A_H	AUTOMATIC FIRE ALARM STATION—HIGH TEMP.		
ELR	FIRE ALARM END OF LINE RESISTOR		

ALL MOUNTING HEIGHTS ARE AS LISTED ABOVE UNLESS NOTED OTHERWISE ON PLANS.

ABBREVIATIONS

A or AMP	AMPERE	MAN or MNL	MANUAL
AC	ALTERNATING CURRENT	MCC	MOTOR CONTROL CENTER
APPROX	APPROXIMATE	MECH	MECHANICAL
AWG	AMERICAN WIRE GAGE	MIN	MINIMUM
		MTR or MOT	MOTOR
BLDG	BUILDING	MS	MOTOR SWITCH
		MTD	MOUNTED
C or CND	CONDUIT		
CB	CIRCUIT BREAKER	NEUT	NEUTRAL
CAB	CABINET	NO	NUMBER
CIR or CKT	CIRCUIT		
CLG	CEILING	OC	OVERCURRENT
CO	COMPANY	OUT	OUTLET
COMPR or CPRSR	COMPRESSOR		
CONN	CONNECTED OR CONNECTION	PB	PUSH BUTTON
CONTR	CONTRACTOR	PF	POWER FACTOR
		PL	PILOT LIGHT
DC	DIRECT CURRENT	PNL	PANEL
DISTR	DISTRIBUTION	PR	PAIR
DT	DUST TIGHT	PVC	POLYVINYL CHLORIDE
EMT	ELECTRICAL METALLIC TUBING	RECEPT or RCPT	RECEPTACLE
EP	EXPLOSION PROOF	RM	ROOM
EXH	EXHAUST	RT	RAINTIGHT
EXIST	EXISTING		
		SE	SERVICE ENTRANCE
FNSH	FINISH	SHT or SH	SHEET
FL	FLOOR	SP	SPARE
FLA	FULL LOAD AMPS	STR or START	STARTER
FPN*	FINE PRINT NOTE	SW	SWITCH
FUT	FUTURE		
		TEL	TELEPHONE
GFCI	GROUND FAULT CIRCUIT INTERRUPTER	TRANS or XFMR	TRANSFORMER
GRND or GND	GROUND	TERM	TERMINAL
HOA	HAND-OFF-AUTO	V	VOLTAGE
HTG	HEATING	VD	VOLTAGE DROP
		VT	VAPOR TIGHT
IMC	INTERMEDIATE METAL CONDUIT		
		WP or WTRPRF	WEATHERPROOF
JB	JUNCTION BOX	W	WATT
LOC	LOCATION		
LT	LIGHT		
LTG	LIGHTING		
LTS	LIGHTS		

Abbreviations are often submitted with a blueprint so persons using the blueprint can readily identify abbreviations used.

*This abbreviation is not considered a standard, but has been used in this text-workbook.

BASES FOR INCANDESCENT LAMPS

Listed below are the dimensions of six common bases for incandescent lamps.

	Depth	Diameter
1. Mogul	1 5/8	1 1/2
2. Admedium	1 1/8	1 1/8
3. Medium	1 1/16	1
4. Intermediate	25/32	21/32
5. Candelabra	5/8	5/8
6. Miniature	15/32	3/8

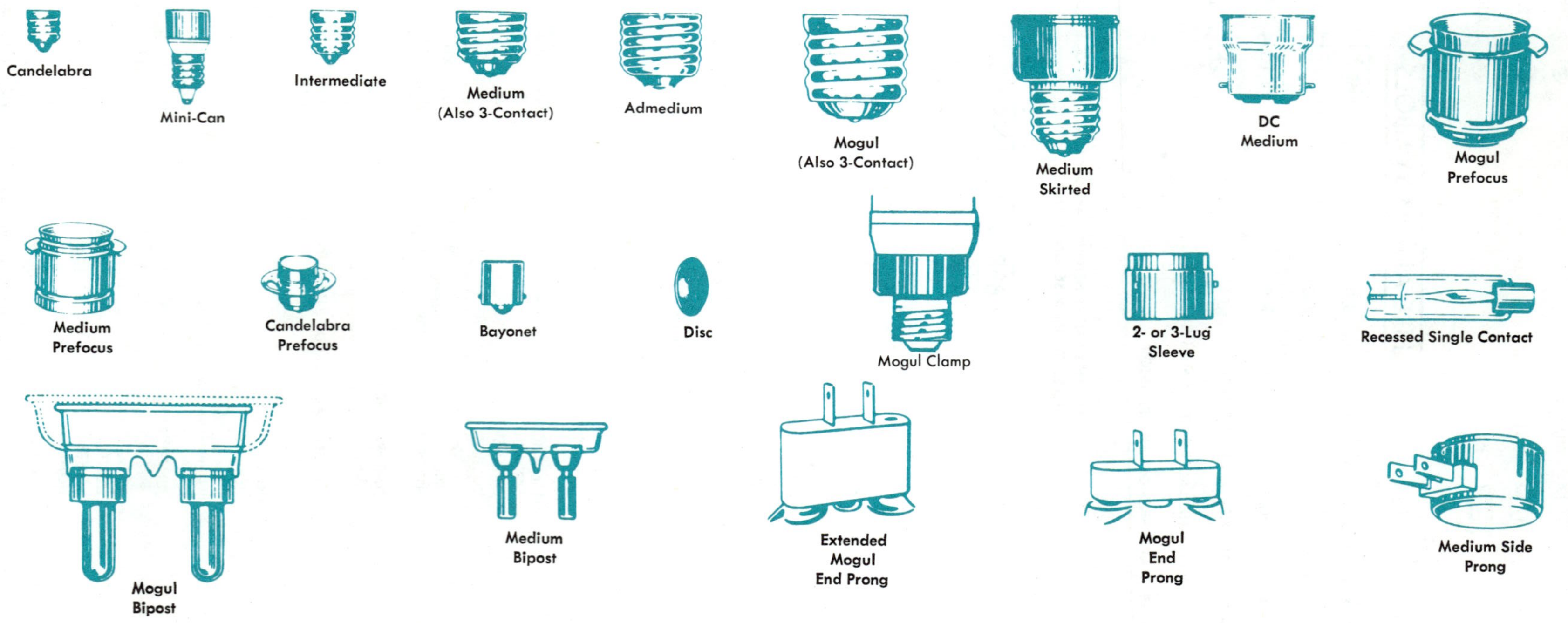

Shown below are different types of bases for incandescent lamps. The bayonet, candelabra, and intermediate base are used on the small-sized (miniature) lamps. The medium base, used on general-service lamps of 300 watts or less, is the most common type. The mogul base is used on sizes of 300 watts and up. The admedium is slightly larger in diameter than the medium and is used on some of the mercury Mazda lamps. The three-contact base is used with a three-lite type of lamp. The disc base is used on lumiline lamps.

The medium and mogul prefocused bases are used on certain types of concentrated-filament lamps, such as those for picture projection and aviation service where it is desirable to have the light source accurately located. The medium bipin base is for fluorescent lamps. The medium bipost base is made in 500, 750, and 1,000 watt lamp sizes for use principally with indirect fixtures, where it allows a better radiation of heat than is obtainable with the mogul-base type. For the large size lamps of 1,500 watts and for floodlights, the mogul bipost is the standard.

NEC 410-32 The identified conductor, where used, shall be connected to the screw-shell of a lampholder.

NEC 410-53 An incandescent lamp for general use on lighting branch circuits shall not be equipped with a medium base when rated over 300 watts, nor with a mogul base when rated over 1,500 watts. Above 1,500 watts, special approved bases or other devices shall be used.

NEC 210-24 Branch circuits rated 30, 40 and 50 amps having two or more outlets may supply fixed lighting units with heavy-duty (mogul) lampholders in other than dwelling occupancies.

NEC 210-6 The voltage on branch circuits supplying lampholders of 15-amp or less rating shall not exceed 150 volts. (See Exceptions.)

National Electrical Code

APPENDIX

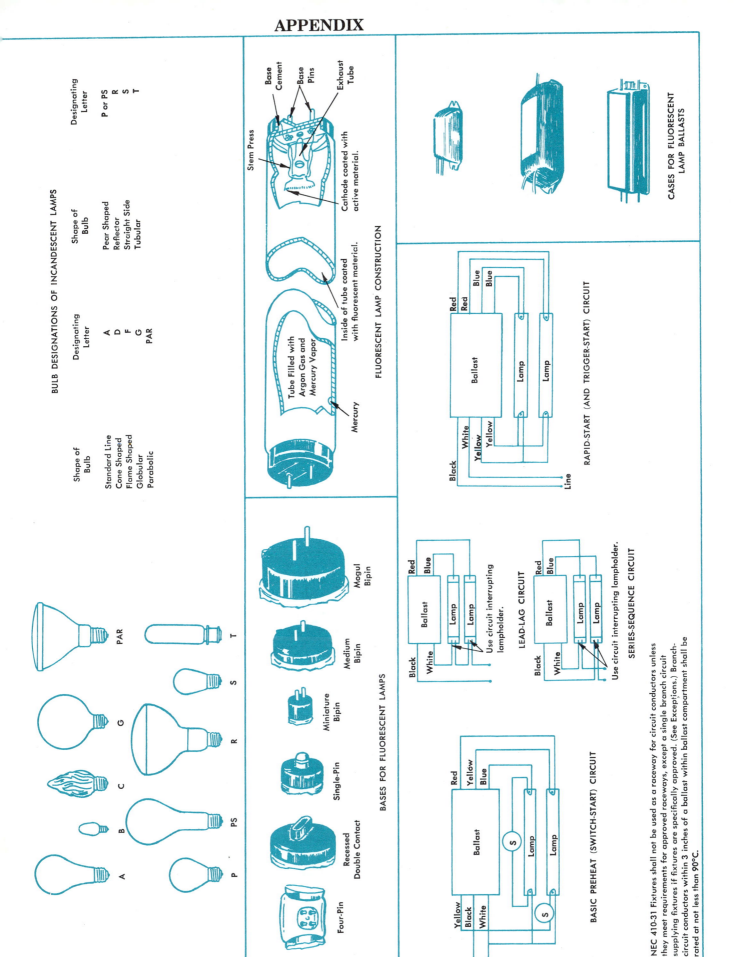

BULB DESIGNATIONS OF INCANDESCENT LAMPS

Shape of Bulb	Designating Letter
Standard Line	A
Cone Shaped	C
Flame Shaped	F
Globular	G
Parabolic	PAR

Shape of Bulb	Designating Letter
Pear Shaped	P or PS
Reflector	R
Straight Side	S
Tubular	T

FLUORESCENT LAMP CONSTRUCTION

Base Cement
Base Pins
Exhaust Tube
Stem Press
Cathode coated with active material.
Inside of tube coated with fluorescent material.
Tube Filled with Argon Gas and Mercury Vapor.
Mercury

CASES FOR FLUORESCENT LAMP BALLASTS

BASES FOR FLUORESCENT LAMPS

Mogul Bipin
Medium Bipin
Miniature Bipin
Single-Pin
Recessed Double Contact
Four-Pin

RAPID-START (AND TRIGGER-START) CIRCUIT

Red
Red
Blue
Blue
Ballast
Lamp
Lamp
Black
White
Yellow
Yellow
Line

LEAD-LAG CIRCUIT

Red
Blue
Ballast
Lamp
Lamp
Black
White
Use circuit interrupting lampholder.

SERIES-SEQUENCE CIRCUIT

Red
Blue
Ballast
Lamp
Lamp
Black
White
Use circuit interrupting lampholder.

BASIC PREHEAT (SWITCH-START) CIRCUIT

Red
Yellow
Blue
Ballast
S
S
Lamp
Lamp
Yellow
Black
White
Line

NEC 410-31 Fixtures shall not be used as a raceway for circuit conductors unless they meet requirements for approved raceways, except a single branch circuit supplying fixtures if fixtures are specifically approved. (See Exceptions.) Branch-circuit conductors within 3 inches of a ballast within ballast compartment shall be rated at not less than 90°C.

CONFIGURATIONS FOR GENERAL-PURPOSE NONLOCKING PLUGS AND RECEPTACLES

		15 AMPERE		20 AMPERE		30 AMPERE		50 AMPERE		60 AMPERE	
		RECEPTACLE	PLUG	RECEPTACLE	PLUG	RECEPTACLE	PLUG	RECEPTACLE	PLUG	RECEPTACLE	PLUG
2-POLE 2-WIRE	1 / 125 V	1-15R	1-15P								
	2 / 250 V		2-15P	2-20R	2-20P	2-30R	2-30P				
	3 / 277 V	(RESERVED FOR FUTURE CONFIGURATIONS)									
	4 / 600 V	(RESERVED FOR FUTURE CONFIGURATIONS)									
2-POLE 3-WIRE GROUNDING	5 / 125 V	5-15R	5-15P	5-20R	5-20P	5-30R	5-30P	5-50R	5-50P		
	6 / 250 V	6-15R	6-15P	6-20R	6-20P	6-30R	6-30P	6-50R	6-50P		
	7 / 277 V AC	7-15R	7-15P	7-20R	7-20P	7-30R	7-30P	7-50R	7-50P		
	24 / 347 V AC	24-15R	24-15P	24-20R	24-20P	24-30R	24-30P	24-50R	24-50P		
	8 / 480 V AC	(RESERVED FOR FUTURE CONFIGURATIONS)									
	9 / 600 V AC	(RESERVED FOR FUTURE CONFIGURATIONS)									
3-POLE 3-WIRE	10 / 125/250 V			10-20R	10-20P	10-30R	10-30P	10-50R	10-50P		
	11 / 3ø 250 V	11-15R	11-15P	11-20R	11-20P	11-30R	11-30P	11-50R	11-50P		
	12 / 3ø 480 V	(RESERVED FOR FUTURE CONFIGURATIONS)									
	13 / 3ø 600 V	(RESERVED FOR FUTURE CONFIGURATIONS)									
3-POLE 4-WIRE GROUNDING	14 / 125/250 V	14-15R	14-15P	14-20R	14-20P	14-30R	14-30P	14-50R	14-50P	14-60R	14-60P
	15 / 3ø 250 V	15-15R	15-15P	15-20R	15-20P	15-30R	15-30P	15-50R	15-50P	15-60R	15-60P
	16 / 3ø 480 V	(RESERVED FOR FUTURE CONFIGURATIONS)									
	17 / 3ø 600 V	(RESERVED FOR FUTURE CONFIGURATIONS)									
4-POLE 4-WIRE	18 / 3ø 208Y/120 V	18-15R	18-15P	18-20R	18-20P	18-30R	18-30P	18-50R	18-50P	18-60R	18-60P
	19 / 3ø 480Y/277 V	(RESERVED FOR FUTURE CONFIGURATIONS)									
	20 / 3ø 600Y/347 V	(RESERVED FOR FUTURE CONFIGURATIONS)									
4-POLE 5-WIRE GROUNDING	21 / 3ø 208Y/120 V	(RESERVED FOR FUTURE CONFIGURATIONS)									
	22 / 3ø 480Y/277 V	(RESERVED FOR FUTURE CONFIGURATIONS)									
	23 / 3ø 600Y/347 V	(RESERVED FOR FUTURE CONFIGURATIONS)									

Courtesy National Electrical Manufacturers Association (NEMA)

CONFIGURATIONS FOR LOCKING TYPE PLUGS AND RECEPTACLES

Courtesy National Electrical Manufacturers Association (NEMA)

50 AMPERE / 60 AMPERE Locking Configurations

		50 AMPERE RECEPTACLE	50 AMPERE PLUG	60 AMPERE RECEPTACLE	60 AMPERE PLUG
2 POLE 3 WIRE GROUNDING	125 V — L5	L5-50R	L5-50P	L5-60R	L5-60P
	250 V — L6	L6-50R	L6-50P	L6-60R	L6-60P
	277 V AC — L7	L7-50R	L7-50P	L7-60R	L7-60P
	480 V AC — L8	L8-50R	L8-50P	L8-60R	L8-60P
	600 V AC — L9	L9-50R	L9-50P	L9-60R	L9-60P
3 POLE 4 WIRE GROUNDING	125/250 V — L14	L14-50R	L14-50P	L14-60R	L14-60P
	3Ø 250 V — L15	L15-50R	L15-50P	L15-60R	L15-60P
	3Ø 480 V — L16	L16-50R	L16-50P	L16-60R	L16-60P
	3Ø 600 V — L17	L17-50R	L17-50P	L17-60R	L17-60P
4 POLE 5 WIRE GROUNDING	3Ø 208Y/120V — L21	L21-50R	L21-50P	L21-60R	L21-60P
	3Ø 480Y/277V — L22	L22-50R	L22-50P	L22-60R	L22-60P
	3Ø 600Y/347V — L23	L23-50R	L23-50P	L23-60R	L23-60P

CONFIGURATIONS FOR LOCKING TYPE PLUGS AND RECEPTACLES

15 / 20 / 30 AMPERE Locking Configurations

		15 AMPERE RECEPTACLE	15 AMPERE PLUG	20 AMPERE RECEPTACLE	20 AMPERE PLUG	30 AMPERE RECEPTACLE	30 AMPERE PLUG
2-POLE 2-WIRE	125 V — L1	L1-15R	L1-15P				
	250 V — L2			L2-20R	L2-20P		
	277 V AC — L3	FUTURE	FUTURE	FUTURE	FUTURE	FUTURE	FUTURE
	600 V — L4	FUTURE	FUTURE	FUTURE	FUTURE	FUTURE	FUTURE
2-POLE 3-WIRE GROUNDING	125 V — L5	L5-15R	L5-15P	L5-20R	L5-20P	L5-30R	L5-30P
	250 V — L6	L6-15R	L6-15P	L6-20R	L6-20P	L6-30R	L6-30P
	277 V AC — L7	L7-15R	L7-15P	L7-20R	L7-20P	L7-30R	L7-30P
	480 V AC — L8			L8-20R	L8-20P	L8-30R	L8-30P
	600 V AC — L9			L9-20R	L9-20P	L9-30R	L9-30P
3-POLE 3-WIRE	125/250 V — L10			L10-20R	L10-20P	L10-30R	L10-30P
	3Ø 250 V — L11	L11-15R	L11-15P	L11-20R	L11-20P	L11-30R	L11-30P
	3Ø 480 V — L12			L12-20R	L12-20P	L12-30R	L12-30P
	3Ø 600 V — L13					L13-30R	L13-30P
3-POLE 4-WIRE GROUNDING	125/250 V — L14			L14-20R	L14-20P	L14-30R	L14-30P
	3Ø 250 V — L15			L15-20R	L15-20P	L15-30R	L15-30P
	3Ø 480 V — L16			L16-20R	L16-20P	L16-30R	L16-30P
	3Ø 600 V — L17					L17-30R	L17-30P
4-POLE 4-WIRE	3Ø 208Y/120 V — L18			L18-20R	L18-20P	L18-30R	L18-30P
	3Ø 480Y/277 V — L19			L19-20R	L19-20P	L19-30R	L19-30P
	3Ø 600Y/347 V — L20			L20-20R	L20-20P	L20-30R	L20-30P
4-POLE 5-WIRE GROUNDING	3Ø 208Y/120 V — L21			L21-20R	L21-20P	L21-30R	L21-30P
	3Ø 480Y/277 V — L22			L22-20R	L22-20P	L22-30R	L22-30P
	3Ø 600Y/347 V — L23			L23-20R	L23-20P	L23-30R	L23-30P

ALTERNATING CURRENT (AC) CIRCUITS

ALTERNATING CURRENT OR VOLTAGE CHANGES WITH TIME, RISING FROM 0 TO MAXIMUM VALUE IN ONE DIRECTION, DECREASING TO 0, THEN RISING TO THE SAME MAXIMUM VALUE IN THE OPPOSITE DIRECTION, AGAIN DECREASING TO 0, THEN REPEATING THESE VALUES AT EQUAL INTERVALS OF TIME. IN THE FORE-GOING, THE NUMBER OF CYCLES PER SECOND IS CALLED FREQUENCY (F) OR HERTZ (HZ). IN AC CIRCUITS THE EFFECTIVE VALUES OF CURRENT OR VOLTAGE ARE ALWAYS USED UNLESS OTHERWISE SPECIFIED.

EFFECTIVE VALUE = .707 x MAXIMUM VALUE

DC CIRCUITS--BASIC OHM'S LAW

E = VOLTS   R = RESISTANCE   I = AMPS OR CURRENT

$E = IR$ or $I = \dfrac{E}{R}$ or $R = \dfrac{E}{I}$

Resistance in Series Circuit
R_t = Total circuit resistance (ohms)
R_1, R_2, etc. = Each resistance in circuit
$R_t = R_1 + R_2 + R_3$ etc.

Resistance in Parallel Circuit
$\dfrac{1}{R_t} = \dfrac{1}{R_1} + \dfrac{1}{R_2} + \dfrac{1}{R_3}$ etc.

Amps in a Series Circuit
Same amps flow through all parts of circuit.

Amps in a Parallel Circuit
$I_t = I_1 + I_2 + I_3$ etc.

Volts in a Parallel Circuit
Applied volts are the same in all parts of circuit.

Volts in a Series Circuit
Applied volts equal to total volt drop for each load.
$E_t = E_1 + E_2 + E_3$ etc.

Electrical Power (Watts = W)
$W = EI = I^2 R = \dfrac{E^2}{R}$

To find I, E and R: $I = \dfrac{W}{E} = \sqrt{\dfrac{W}{R}}$; $E = \dfrac{W}{I} = \sqrt{RW}$; $R = \dfrac{E^2}{W} = \dfrac{W}{I^2}$

Convert Mechanical Power to Electrical Power
HP = Horsepower = 746 watts

$HP = \dfrac{Watts}{746}$, or Watts x .00134

$HP = \dfrac{kilowatts}{.746}$ or kilowatts x 1.34

Efficiency (Ratio of energy output to energy input)
$Eff = \dfrac{output}{input} \times 100\%$

$output = \dfrac{input \times eff}{100\%}$

Torque (Expressed in pound-feet or pounds of force at a certain radius.)
Torque (T) and horsepower (HP)
RPM = Speed (revolutions per minute)
$HP = \dfrac{6.28 \times RPM \times T}{33,000}$; $HP = \dfrac{T \times RPM}{5250}$

$T = \dfrac{33,000 \times HP}{6.28 \times RPM}$

Motors --To find speed, frequency and poles of AC motors:
RPM = Revolutions per minute
F = Frequency in Hertz (cycles per second)
P = Number of field poles

$RPM = \dfrac{F \times 120}{P}$; $F = \dfrac{RPM \times P}{120}$; $P = \dfrac{F \times 120}{RPM}$

Impedance:
Impedance in an AC circuit is similar to resistance in a DC circuit in that both are the total opposition to the flow or current.   In AC circuits, Ohm's law could read: "The current is directly proportional to the applied voltage and inversely proportional to the impedance."
Impedance could be defined as the total resultant ohmic combination of resistance and reactance.
Impedance is the vector sum of resistance, capacitive reactance and inductive reactance.
Z = Impedance   F = Hertz (cycles per second)
Xc = Capacitive Reactance   C = Capacitance (farads)
X_L = Inductive reactance   L = Inductance (henrys)
R = Resistance π = 3.14 (Greek letter pi)

$Z = \sqrt{R^2 + (X_L - X_C)^2} = 2 FL$; $X_C = \dfrac{1}{2\pi FC}$

MISCELLANEOUS FORMULAS

To find CM (circular mil area) of a conductor, square the diameter.
CM = Diameter in Inches x 1,000   (Also see Chapter 9, Table 8--NEC)

To find CM of a bus bar:
$CM = \dfrac{(\text{width in inches} \times 1000) \times (\text{thickness in inches} \times 1000)}{.7854}$

CM = (width in inches x 1000) x (thickness in inches x 1000) x 1.2732
To find the sq mil area when the CM is known: sq mil = CM x .7854

To find the CM area when the sq mil is known:
$CM = \dfrac{\text{sq mil}}{.7854}$

Temperature Conversions
C = Centigrade degrees   F = Fahrenheit degrees
(Also see Note 13 to Ampacity Tables in NEC.)
$F = \left(\dfrac{9}{5} \times C\right) + 32$ $C = \dfrac{5}{9} \times (F - 32)$

ELECTRICAL FORMULAS

AC CIRCUITS--OHM'S LAW

$E = IZ$ $I = \dfrac{E}{Z}$ $Z = \dfrac{E}{I}$

Power Factor:
The power factor of an AC circuit is the ratio between the true power in watts and the apparent power in volt-amperes. Power factor is normally expressed in percent.

PF is actually the cosine of the phase angle that the current leads or lags the voltage in an AC circuit.

PF = Power Factor
$PF = \dfrac{\text{True Power}}{\text{Apparent Power}} = \dfrac{\text{Watts}}{\text{Volt-Amps}} = \dfrac{kW}{KVA} = \dfrac{R}{Z}$

Single-Phase AC Circuits
$kVA = \dfrac{E_I}{1,000} = \dfrac{kW}{PF}$ $kW = kVA \times PF$

$I = \dfrac{W}{E \times PF}$ $E = \dfrac{W}{I \times PF}$ $PF = \dfrac{W}{E_I}$

Three-Phase Circuits (Balanced Star or Wye)
In (neutral current) = 0
I_l (line current) = I_p (phase current)
E_l (line voltage = 1.73 x E_p (phase voltage)
E_p (phase voltage) = $\dfrac{E_I \text{ (line voltage)}}{1.73}$ = $E_I \times 0.577$

Formula for current in the neutral of a three-phase, unbalanced Star or Wye at unity power factor:

In = Neutral current   Ib = Phase B current
Ia = Phase A current   Ic = Phase C current

$In = \sqrt{Ia^2 + Ib^2 + Ic^2 - IaIb - IbIc - IcIa}$

Three-Phase Circuits (Balanced Delta)
$I_l = 1.73 \times I_p$
$I_p = \dfrac{I_l}{1.73} = I_l \times 0.577$
$E_p = E_l$

Watts (Power) in Three-Phase Circuits
W = watts (true power)
VA = volt-amps (apparent power)
E = line volts
I = line amps
W = 1.73 x E x I x PF
VA = 1.73 x E x I
$E = \dfrac{W}{1.73 \times I \times PF} = \dfrac{0.577 \times W}{I \times PF}$

$I = \dfrac{W}{1.73 \times E \times PF} = \dfrac{0.577 \times W}{E \times PF}$

$PF = \dfrac{W}{1.73 \times E \times I} = \dfrac{0.577 \times W}{E \times I}$

Watt (Power) Loss in AC or DC circuits.
$W = I^2 R$ $I = \sqrt{\dfrac{W}{R}}$ $R = \dfrac{W}{I^2}$

APPENDIX

TYPICAL FAULT CURRENT FORM

DEPARTMENT OF INSPECTIONS
City of Minneapolis

To _____ Date _____
  (Electrical Contractor)

_____ Electrical Permit No. _____
  (Address)

The following information is requested to determine whether or not the electrical

equipment installed at _____
                   (street address)   (Name of occupant)

Electrical Permit No. _____, meets the requirements of the

National Electrical Code as it pertains to available short-circuit current.

Electrical Inspector _____ Phone _____

Trans. size _____ Imp _____ Sec Volt _____ Phase _____ 3-4 Wire _____

Service Conductors Size _____ Copper _____ Aluminum _____ Length _____

Available Fault current at **line side of service disconnect** _____

Type, size and fault current interrupting rating of overcurrent device used at main

panel or panels _____

_____

Available fault current* at power panel #1 _____

Available fault current* at power panel #2 _____

Available fault current* at lighting panel #1 _____

Available fault current* at lighting panel #2 _____

Available fault current* at lighting panel #3 _____

Available fault current* at lighting panel #4 _____

Available fault current* at lighting panel #5 _____

Available fault current* at emergency panel _____

Use back of form for data on additional panels. Show one-line diagram of service
and all related equipment on back of form also.

*If panels are protected by fuses, indicate the fault current interrupting rating and
the let-thru current of fuses and also include fuse type and size. For circuit
breakers indicate the fault current interrupting rating and the let-thru current,
voltage, type and size.

(See NEC-110-10 "Circuit Impedence and Other Characteristics" for specific code requirements.) (Also see NEC 110-9, 230-65 and 240-1)

Electrical Contractor: _____ Date: _____

Address: _____ Phone _____

Form No. 32483

(Courtesy of Department of Inspections – Minneapolis, Minnesota)

Index

A

Abbreviations, blueprint, 181
AC system, grounded conductor for, 13
Adapters, fuse, 20
Additions to buildings, 57
Aerial cable, 54
Agricultural building, 34
Air conditioning equipment, 59
Air plenums, wiring, 88
Aircraft hangar, 148
Aluminum conductors, 43
Aluminum sheath, 26
Aluminum wire, 80-81
Amateur transmitting and receiving stations, 159
American Wire Gage (AWG), 43
Ammeter, 114
Ampacities of wires, 43, 45
Amplifiers, 154
Analysis, circuit, 109, 111-112
Anesthetizing locations, 151
Antenna systems, 159
Apartment, drawing of, 44, 47
Appliance,
    branch circuit protection, 57
    cord-connected, 58
    disconnecting means, 32
    motor, 57
Appliance loads, fixed, 58
Appliance panelboard 50, 54, 86
Arc welder,
    motor-generator, 108
    transformer, 108, 129
Architectural blueprints, 71
Armored cable, 26
Arresters, surge, 125
Askarel transformers, 124
Attic wiring, 27
Autotransformer, 124
Autotransformer-type dimmer, 153
Auxiliary gutters, 89, 90
    for sound recording, 154

B

Basement joists, 27
Basement plan, 77-81
Battery, storage, 155
Blueprint, architectural, 71
Blueprint symbols, electrical, 179-181
Boatyards, ground fault interrupters, 19
Bonding jumper, 8-9, 10, 11, 87
Bonding services, 8-9
Bonding swimming pools, 132
Booth, spray painting, 150
Box supports, threaded, 85
Boxes,
    angle pulls, 89
    floor, 86

grounding, 87, 88
junction, 17, 89
metal, 17
outlet, 16, 17
pull, 89
straight pulls, 89
switch, 17
Bracket fixture wiring, stage, 153
Braid-covered insulated conductors, 128
Branch circuit, 54
    outside, 158
    multiwire, 94
Branch circuit cable, 114
    Type UF, 27
Branch circuit load calculations, 94
Branch circuit protection of appliances, 57
Branch circuit service calculations, 2
Branch circuit tap rules, 87
Branch circuit with more than two motors, 119
Building,
    agricultural, 34
    manufactured, 34
Building openings,
    service drop clearance, 5
Bulk-storage plant, 149
Bushing, grounding, 8-9, 10, 14
Busways, 115
BX, 26, 27

C

Cable,
    aerial, 54
    armored, type AC, 26
    branch circuit, 27, 114
    exposed, 28
    FCC, 84
    fishing of, 27, 28
    flat conductor, 84
    heating, 55, 56
    metal-clad, type MC, 26
    metal-sheathed, 114
    mineral-insulated, 114
    nonmetallic-sheathed, 27, 28
    service entrance, 5
    traveling, 121
    Type SE, 5
    underground service, 83
Cable clamp, 16
Cable connectors for stage wiring, 153
Cable strap, 5
Cable trays, 152
Cablebus system, 115
Calculations,
    branch circuit load, 94
    building addition loads, 57
    farm loads, 116
    feeder Loads, 94, 108

motor feeder, 119
multi-family dwelling load, 47-48
multi-family dwelling service load, 48-49
service, 79-80, 108-109, 112-113
short-circuit currents, 23
Capacitors, 128
Cartridge fuses, 7, 21, 22
Cellular floor raceways, 91
Chain supported fixtures, 58
Circuit,
    branch, 158
    combining classes of, 156
    communication, 158
    instrument, 116
    motor, 103-109
    motor control, 118
    output, 130
    power-limited, 156
    remote control, 156
    signaling, 156
Circuit analysis, industrial, 109, 111-112
Circuit breakers, 18, 84
Circuit conductor, overcurrent protection, 128
Class I locations, 144, 145, 146
Class III locations, 147
Class II locations, 146
Classes of dangerous operations, 141-147
Clearances
    service drops, 4, 5
    service laterals, 4
Closets, lighting fixtures in, 31
Combustible dust, 146
Combustible flyings, 147
Commercial locations, 69-94
Commercial repair garage, 148
Common neutral for feeders, 94
Communication circuits, 158
Concealed conductors, 157
Concealed knob-and-tube wiring, 25
Concrete floor, heated, 56
Concrete floor raceways, 91
Conductor, 30
    AC system, 13
    aluminum, 43
    application of, 45
    braid-covered insulated, 128
    circuit, 128
    commercial, 69
    concealed sign, 157
    construction of, 45
    copper, 43
    deflection of, 85
    general wiring, 30
    grounded, 11
    grounding electrode, 8-9, 10, 13, 83
    grounding equipment, 14
    identified, 57

National Electrical Code

insulated, 26, 28
neutral, 8-9, 50
number in conduit, 89
overcurrent protection, 84
protection, pole, 83
secondary, 127
service-over 600 V, 127
service-entrance, 5, 6, 14, 82
sizes, 43
vertical supports, 88
Conductor clearances, overhead, 132
Conduit, 80-81
    conductors in, 89
    electrical plastic, 29
    flexible metal, 28, 85
    intermediate metal, 28
    liquidtight flexible, 28, 114
    nonmetallic, 14
    rigid metal, 28
    rigid nonmetallic, 28, 29
Conduit in concrete, 114
Conduit grounding bushing, 14
Conduit size, service, 80
Conduit in wet locations, 114
Configurations for plugs and
    receptacles, 184-185
Connectors,
    cable, 153
    pressure, 51
Construction sites, ground fault
    interrupters, 19
Control, flue damper, 153
Control circuits, motor, 118
Control hoistway, 121
Controllers, motor, 119
Cooking equipment, 33
Copper conductors, 43
Cord- and plug-connected equipment,
    grounding, 52
Cord connected appliances, 58
Cords,
    flexible, 45-46, 86
    overcurrent protection, 85
Countertop ranges, 33
Cove lighting, 92
Covers,
    fusetron box, 20
    outlet box, 16
Crane, 120
Crane motors, 106-107
Current-carrying values, 43
Current-limiting type fuses, 21, 23

D

Data processing systems, 88
Deflection of conductors, 85
De-icing equipment, 56
Details, 46, 49
    electrical, 70
Dielectric heating, 130
Differential transformer, 19
Dimmer,
    autotransformer type, 153
    stage lighting, 153
Direct current motor, 118

Disconnect,
    maximum number of, 6
    rating of, 6
Disconnecting means, 155
    appliances, 32
    dryers, 33
    motor, 117
    motor appliance, 57
    motor generator, 117
    ranges, 33
    service, 6, 7, 82, 127
    thermostatic devices, 56
Distribution system, overhead, 53
Distribution panelboards, 50
Drawings, 81
Dressing rooms, 153
Drip loops, 5
Dryer, grounding, 59
Dryer, paint, 150
Dry-type transformer, 124
Dual element fuse, 20, 21
Duct accessories, 29
Duct fittings, 29
Dumbwaiters, 121
Duplex convenience outlet, 91
Dust, combustible, 146
Dust accumulation, 146
Dwellings,
    multi-family, 43-59
    single-family, 1-34
Dwelling unit, ground fault
    interrupters, 19
    optional calculation, 3

E

Edison-base fuses, 20
Electric-discharge lighting, 53, 93
Electric-discharge signs, 93
Electric-discharge tubing, 157
Electric dryers, disconnect means for,
    33
Electric heaters, installing, 57
Electrical metallic tubing, 28
Electrical nonmetallic tubing, 28
Electrical plans, 71
Electrodes, 51
Elevations, 46, 49
Elevator, 121
Elevator motor, 105-106
Elevator wiring, 121
Emergency system, 155
Enclosures, electrical equipment, 177
Enclosures, switch, 51
Equipment, commercial, 69
Equipment bonding conductor, 8
Equipment bonding jumpers, 10
Equipment in concrete, 114
Equipment grounding, fixed, 87
Equipment grounding conductors, 14
Equipment in wet locations, 114
Escalators, 121
Explosion-proof area, 144
Exposed cable, 28
Extension rings, 16
Extensions, nonmetallic, 54

F

Face plates, 32
Farm loads, 116
Fast-acting fuse, 21, 22
Fault current, available, 23
Fault current form, 187
FCC cable, 84
Feeder, 94
    electric ranges, 58
    multi-family dwelling, 50
    outside, 158
    underground, 27, 114
Feeder calculations, 2
    industrial, 108
    motor, 119
Feeder in place of ground fault
    interrupters, 19
Feeder load calculations, 94
Feeder riser diagram, 70
Feeder supplying more than one
    motor, 106
Feeder tap, 87
Feeder tap rule, transformer, 126
Feeder to foot tap rule, 87
Festoons, 153
Fibers, ignitible, 147
Field bending rigid nonmetallic
    conduit, 29
Film storage vaults, 154
Finishing, spray, 150
Fire alarm symbols, blueprint, 181
First floor plan, 74, 76-77
Fished cable, 27, 28
Fixed appliance loads, 58
Fixture,
    fluorescent, 92
    hanging, 31, 58
    incandescent, 32
    lighting, 31, 114
    prewired, 32
    recessed, 32
    type IC, 32
    show window, 92
Fixture mounting, 92
Fixture as raceway, 92
Fixture supports, 92
Fixture symbols, 70
Fixture wire, 45
    overcurrent protection, 85
Flammable gases or liquids, 144, 145
Flat conductor cable, 84
Flexible conduit, liquidtight, 28
Flexible cords, 45, 46, 86
Flexible metal conduit, 28, 85
Flexible Metallic Tubing, 28
Floor boxes, 86
Flue damper control, 153
Fluorescent fixtures,
    connecting, 92
    wiring, 88
Fluorescent lamps, 183
Formulas, 60, 160, 186
Fuses, 7
    ampere ferrule, 22
    ampere knife-blade type, 22

cartridge, 21, 22
cable limiters, 24
Class G, 22
Class H, 22
Class J, 21
Class K, 21
Class L, 22
Class R, 22
Class T, 24
current-limiting type, 21, 23
dual element, 20, 21
fast-acting, 21, 22
limitron, 21, 22
low-leak, 21
plug, 20
time-delay, 21, 22
Type S, 20
Type SC, 22
Type W, 20
Fused isolating switch, 127
Fuseholder, 20, 22
Fusetron box cover units, 20

G

Garages, 148
Gasoline dispensing station, 149
Generator, 120
Generator set, 155
GFCI, 19
Greenfield, 27
Ground clamp, 30
Ground fault detection system, 15
Ground fault interrupters, 19
Ground fault protection, 15
Ground rod, 10
Ground rod clamp, 10
Grounded conductors, 11
    for AC system, 13
Grounded systems, 12
Grounding,
    boxes, 87, 88
    bushing, 8-9, 10, 14
    conductors, 8, 9, 10, 14
    cord- and plug-connected
        equipment, 52
    dryers, 59
    electrodes, 51
    electrode conductor, 8-9, 10, 11, 13,
        83
    electrode system, 30
    fixed equipment, 87
    metal enclosures, 52
    metallic shield, 128
    multiphase system, 51
    ranges, 59
    receptacles, 52
    separately derived system, 11
    service, 7, 8-9, 14
    swimming pools, 131
    two or more buildings from single
        service equipment, 12

H

Hangar, aircraft, 148
Hanging fixtures, 58
Hazardous locations, 141-159

Hazards, reducing, 141-142
Heaters, installing, 57
Heating, industrial, 58
Heating appliances, infrared, 115
Heating cables, 55, 56
Heating equipment, overcurrent
    protection for, 58
Heating panels, 55, 56
Heavy-duty lampholders, 84
Hermetic motor, 118
High-leg, identification of, 83
High-voltage installation, 125
Hoists, 120
Hoistway, control, 121

I

Identified conductor, 57
Impedance, 23, 122
Incandescent fixtures, 32
Incandescent lamps, 182-183
Indoor transformers, 124
Induction heating, 130
Industrial heating, 58
Industrial locations, 103-132
Infrared lamps, 58, 115
Instrument circuits, 116
Instrument transformers, 116
Insulated conductors, 26, 28
Insulation,
    conductors, 43, 45
Intermediate metal conduit, 28
Irons, 93
Isolating switches, 127

J-K

Jumper,
    bonding, 8, 9, 10, 87
    equipment bonding, 10
    main bonding, 10
Junction boxes, 17, 89
Knob-and-tube wiring, 25

L

Lampholders, 31
    heavy-duty, 84
    screw-shell, 92
    weatherproof, 31
Lamps,
    fluorescent, 183
    incandescent, 182-183
    infrared, 58
Lighting,
    cove, 92
    electric-discharge, 53, 93
    emergency, 155
    outline, 157
    stage, 152
Lighting fixtures, 31, 114
Lighting panelboards, 50, 54, 86
Lighting system, 480-volt, 80
Limitron fuse, 21, 22
Liquidtight flexible metal conduit,
    114

Liquidtight flexible nonmetallic
    conduit, 28
Load calculations,
    branch circuits, 94
    feeders, 94
    multi-family dwelling, 47-48
    panels, 75-77, 79
    single-family dwelling, 2
Load side of service, 14
Loads,
    branch circuits, 54
    building additions, 57
    farm, 116
    fixed appliance, 58
    noncoincident, 58
    service, 108-109, 112-113
Loop, switch, 57
Loop wiring, 91
Loud speaker systems, 69, 71
Low-leak fuse, 21
Lugs, 51

M

Machine room, 121
Machine tools, 107, 120
"Made" electrodes, 51
Main bonding jumpers, 10
Main bonding service, 8
Main service, commercial, 69
Manufactured building, 34
Manufactured wiring systems, 53
Marinas, ground fault interrupters,
    19
Masonry floors, heated, 56
Masonry walls and fished cable, 27
Melting equipment, 56
Metal clad cable, type MC, 26
Metal conduit,
    flexible, 28, 85
    intermediate, 28
    rigid, 28
Metal enclosures, grounding, 52
Metal floor raceways, 91
Metal outlet boxes, 17
Metal-sheathed cable, 114
Metal-working tools, 107, 120
Metallic shielding, grounding, 128
Mineral-insulated metal-sheathed
    cable, 114
Mobile homes, 131
    ground fault interrupters, 19
Motion picture projector, 154
Motion picture studio, 153
Motor,
    crane, 106-107
    direct current, 118
    elevator, 105-106
    general, 117
    hermetic, 118
    over 600 volts, 120
    part-winding, 118
    running protection, 117
    synchronous, 107
    wound-rotor, 118
Motor appliance, disconnect means
    for, 57

Motor circuit, 103-109
Motor circuit capacitor, 128
Motor control circuits, 118
Motor controllers, 119
Motor disconnect, 117
Motor feeder calculations, 119
Motor-generator arc welder, 108
Motor-generator disconnect, 117
Motor-generator equipment, 130
Motor-generator welder, 129
Motor rules, 119
Motor-running overload protection, 120
Mounting fixtures, 92
Moving walk, 121
Multi-family dwellings, 43-59
    drawing, 44, 47
Multiphase system grounding, 51
Multiwire branch circuits, 94

N

Nameplate information, air conditioner, 59
Neon signs, 93
Neon tubing, 157
Neutral, common for feeders, 94
Neutral conductor, 8-9, 50
Noncoincident loads, 58
Nonflammable transformers, 124
Nonmetallic conduit, 14, 28
Nonmetallic extensions, 54
Nonmetallic outlet box, 16
Nonmetallic sheathed cable, 26, 27, 28

O

Office furnishings, 133
Ohmmeter, 114
Oil-insulated transformers, 124
Open runs, 128
Optical Fiber Cables, 134
Organs, electric, 154
Outdoor de-icing and melting equipment, 56
Outdoor transformers, 124
Outlet boxes, 16, 17
Outline lighting, 157
Output circuits, 130
Oven, wall-mounted, 33
Overcurrent devices, 23, 128
Overcurrent protection, 11, 21, 22
    cartridge fuses, 21, 22
    circuit conductors, 128
    conductors, 84
    cords, 85
    fixture wire, 85
    heating equipment, 58
    panelboards, 86
    services over 600 volts, 125
    sound equipment, 154
    transformers, 123
Overhead conductor clearances, swimming pool, 132
Overhead distribution system, 53
Overhead services, 6
Overload, motor, 117, 120

P

Pad-mount transformer, 23
Paint drying, 150
Panel, heating, 55, 56
Panel load calculations, 75-77, 79
Panelboard,
    lighting and appliance, 20, 54, 86
    distribution, 50
    overcurrent protection of, 86
    service, 50
Parallel transformers, 124
Parking garage, 148
Parks, ground fault interrupters, 19
Part-winding motor, 118
Plan,
    basement, 77-81
    electrical, 71
    first floor, 74, 76-77
    second floor, 71-73, 74-75
Plant, bulk-storage, 149
Plaster rings, 16
Plastic conduit, 29
Plastic tubing, 29
Plenums, wiring, 88
Plug configurations, 184, 185
Plug fuses, 20
Pole, 83
Potential transformers, 122
Power installation, restaurant, 109-113
Power-limited circuits, 156
Prewired fixtures, 32
Projector, motion picture, 154
Protection,
    conductor, 83, 84, 128
    cords, 85
    fixture wire, 85
    ground fault, 15
    ground fault interrupters, 19
    motor, 117, 120
    overcurrent, 21, 22, 58
    overload, 117, 120
    panelboards, 86
    running motors, 117, 120
    services, 125
    sound equipment, 154
    transformers, overcurrent, 123
Pull boxes, 89
PVC, 29

R

Raceway,
    cellular concrete, 91
    cellular metal, 91
    continuity of, 88
    in fixtures, 92
    floor, 91
    surface, 90
    underfloor, 90, 91
Raceways and temperature changes, 88
Radio equipment, 159
Ranges,
    countertop, 33
    disconnect means for, 33

electric, 58
    grounding, 59
    installation of, 93
Ratings,
    lampholders, 84
    receptacle, 84
    snap switches, 85
Reactors, 129
Reactor-type dimmer, stage, 153
Receiving equipment, 159
Receptacle, 52
    weatherproof, 31
Receptacle configurations, 184, 185
Receptacle grounding, 52
Receptacle plates, 32
Receptacle ratings, 84
Recessed auxiliary gutters, 154
Recessed fixtures, 32
Recording studio, 154
Recreational vehicles, 19, 131
Remote control circuits, 156
Resistance-type dimmer, 153
Resistance welder, 107-108, 129
Resistors, 129
Rigid metal conduit, 28
Rigid nonmetallic conduit, 28, 29
Ring,
    extension, 16
    plaster, 16
Romex, 26, 28
Roof spaces, wiring, 27
Runs, open, 128

S

Schedules, 81
Screw-shell lampholders, 92
Second floor plan, 71-73, 74-75
Secondary conductors as service conductors, 127
Secondary ties, 122
Sensor, ground fault, 15
Service, 69, 82, 155
    multi-family dwelling, 50
    number of, 6, 83
    overhead, 6
    underground, 6, 83
    wiring, 127
Service bonding, 8-9
Service cable, underground, 83
Service calculation, 79-80
    industrial, 108-109, 112-113
    multi-family dwelling, 48-49
    single-family dwelling, 3
Service conductors, 127
Service conduit size, 80
Service details, 7
Service disconnecting means, 6, 7, 82, 127
Service drop, 13
    clearances, 4, 5
Service-entrance conductors, 5, 6, 14, 82
Service-entrance conduit, 8, 9
Service equipment, 11, 82
Service grounding, 7, 8-9

Service head, 4, 5
Service lateral clearances, 4
Service load calculation, 108-109, 112-113
Service load side, 14
Service mast fittings, 4
Service overcurrent protection, 125
Service station, 149
Service switch, 6, 7
Shielding tape, 128
Show window fixtures, 92
Show window signs, 158
Show window wiring, 86
Showcases, cord-connected, 92
Sign, 157
    electric-discharge, 93
    neon, 93
    show window, 158
Signaling circuits, 156
Single-family dwellings, 1-34
Single service equipment, 12
Snap switches, 85, 86
Snow melting equipment, 56
Solar photovoltaic systems, 134
Solder lug, 51
Solderless lugs, 51
Sound equipment, 154
Sound recording, 154
Space heating, 55, 56, 58
Specialized locations, 141-159
Specifications, store building, 69, 71
Spray booth, 150
Storage battery, 155
Storage garage, 148
Studio, TV or motion picture, 153
Supports,
    box, 85
    conductor, 88
    fixture, 92
Surface extensions, nonmetallic, 54
Surface raceway, 90
Surge arresters, 125
Swimming pools, 19, 132
Switch boxes, 17
Switch enclosures, 51
Switch loop, 57
Switchboard, 85, 116
Switches, 53
    accessibility, 90
    fused isolating, 127
    isolating, 127
    service, 7
    snap, 85, 86
    three-way & four-way, 52
    transfer, 155
Symbols,
    electrical blueprint, 179-181
    fixtures, 70
    wiring diagram, 2
Synchronous motor, 107

T

Tap feeder, 87
Telephone systems, 69, 71
Television equipment, 159
Temperature and raceways, 88

Temporary wiring, 85
Terminal bar, 20
Terminal connections, 51
Terminals for neon tubing, 157
Theaters, 152, 153
Thermostatic devices, 56
Three-pole circuit breaker, 18
Ties, secondary, 122
Time-delay fuse, 21, 22
Tools,
    machine, 120
    metal-working, 110
Transfer switches, emergency, 155
Transformer, 10, 11
    askarel, 124
    differential, 19
    dry-type, 124
    formulas, 160
    impedance, 122
    indoor, 124
    instrument, 116
    nonflammable, 124
    oil-insulated, 124
    outdoor, 124
    overcurrent protection, 123
    pad-mount, 23
    parallel, 124
    potential, 122
    sign, 53
Transformer arc welder, 108, 129
Transformer Feeder Tap Rule, 126
Transformer vault, 125
Transformers in parallel, 124
Transmitting station, 159
Trays, cable, 152
Tubing,
    electrical metallic, 28
    electrical plastic, 29
    electric-discharge, 157
    neon, 157
TV studio, 153
Two-pole circuit breaker, 18

U

Underfloor raceway, 90, 91
Underground feeder, 27, 114
Underground service, 6, 83
Unit equipment for emergency lighting, 155
Utility supply system, 13

V

Vaults,
    film storage, 154
    transformer, 125
Voltage between adjacent switches, 53
Voltage drop,
    branch circuits, 30

    calculating, 75
    feeders, 30
    formulas, 60
Voltage to lighting fixture, 114
Voltage rating, circuit breakers, 18
Voltmeter, 114

W

Walks, moving, 121
Weatherproof enclosure for switchboard, 85
Weatherproof lampholders, 31
Weatherproof receptacles, 31
Welder,
    arc, 108
    motor-generator, 108, 129
    resistance, 107-108, 129
    transformer, 108, 129
Wet locations, 114
Window fixtures, 92
Windows, show, 86
Wire,
    aluminum, 80-81
    fixture, 45, 85
Wire ampacities, 43, 45
Wireways, 115
    for sound recording, 154
Wiring,
    attic, 27
    bracket fixture for theater, 153
    fluorescent fixtures, 88
    knob-and-tube, 25
    plenums, 88
    roof spaces, 27
    show windows, 86
    stages in theaters, 153
    temporary, 85
Wiring diagram symbols, 2
Wiring methods, commercial, 69
Wiring services, 127
Wiring systems, manufactured, 53
Wound-rotor motor, 118

X

X-ray unit, 107, 151